Plate and Shell Models

Robert Nzengwa

Plate and Shell Models

Variational Methods in Plate and Shell Theory

Springer

Robert Nzengwa
University of Douala
Douala, Cameroon

ISBN 978-981-97-2779-7 ISBN 978-981-97-2780-3 (eBook)
https://doi.org/10.1007/978-981-97-2780-3

This Springer imprint is published by the registered company Springer Nature Singapore Pte Ltd.
The registered company address is: 152 Beach Road, #21-01/04 Gateway East, Singapore 189721, Singapore

To my Mandjibo/Moungo
To my Family
To my Friends

Acknowledgements

A shell is a three-dimensional (3D) structure in which one dimension, referred herein to as the thickness, is relatively smaller than the two others. A flat or curved metal sheet is an excellent example. It can be observed that the curved metal sheet (shell) offers a better resistance than the flat one (plate). Therefore, a shell structure certainly resists by the mechanical properties of its constitutive material, but it also resists by its geometric shape. This last property justifies the growing interest in using shell structures for design, in very many fields in modern engineering. Roofing very large spaces without columns, arch dams, cooling towers, storage tanks, water towers, bridge structures or even pavements are non-exhaustive examples in civil engineering. Aircraft fuselage and marine vessels in aircraft and marine construction, industrial hoses, engine parts, pressure pipes, car hulls, etc. are examples of shell structures. We cannot herein mention all the examples of structures defined as shells or plates. The challenge for designers (architects and structural engineers) is how to predict the mechanical behaviour of these structures under various loadings (static, dynamic, thermal, etc.) during their service period. The relative smallness of the third dimension, the thickness, has naturally inspired scientists to orient their research of a solution by describing the 3D behaviour with parameters defined on the shell's mid-surface. In other words, it is about defining a two-dimensional (2D) model to predict the 3D behaviour of the shell. One should be able to solve the governing equations thus obtained. The first attempts at 2D modelling of a shell date back to the nineteenth century with the works of Kirchhoff (1850) (see, for example, [61]) on plate theory and Aron [9] in 1874, on shells. Between 1888 and 1963, Love [68] published many papers. More historical details on this topic are found in Naghdi [75] or Love [68]. The hypotheses gathered under the name "Kirchhoff-Love hypotheses" have led to the Kirchhoff-Love (K-L) model which is widely used to calculate thin shells. Let h be the thickness of a shell and R the least radius of curvature in absolute value. We define $\chi = h/2R$ as the characteristic ratio of the shell which should be strictly less than 1. It is well known by practitioners that the K-L model becomes inappropriate when χ is greater than a critical value χ_0 situated between 0.15 and 0.2. Transverse stresses through the thickness become significant and difficult to predict. Very many numerical methods turn out inefficient. Models developed by Reissner and Mindlin

(R-M) (cf. [70, 93, 94]) have proven better efficiency than (K-L) on "*moderately thin*" plates by calculating transverse shear stresses through the thickness. Nevertheless, they do not predict pinch stress, and numerical calculations become inefficient when the thickness tends to zero. Many other models were developed during the second half of the twentieth century to overcome these difficulties, some more or less rigorous or simply heuristic. Despite continuous improvements in these models or their mathematical justification (see Naghdi [74], Koiter [62], Ciarlet et al. [30] and [31]), many problems still remain open. In any way, the different mathematical approaches in shell theory (double series, integration of 3D equations, multi-scale dilatation and limit analysis, etc.) are continuously more elaborated and less familiar to structural engineers. Because of the improvements in high-performance numerical methods and the tremendous computing capacity of computers available nowadays, architects design structures with very sophisticated shapes and require a very high level of reliability that structural engineers should respect. How to transmit these calculating methods based on new knowledge, to structural engineers who are very chilly to highly elaborated mathematical theories, is a permanent challenge to researchers and professors in applied mechanics. The aim of this textbook is to present to structural engineers models in shell theory, some of them deduced from 3D elasticity and validated on benchmarks, in their variational form suitable for implementing Finite Element Methods and numerical calculations. In order to avoid getting lost in unnecessary digressions, we refer the reader to specialized articles or books, concerning demonstrations which are purely mathematical. On the other hand, we have taken care to expose, if only succinctly, any demonstration likely to equip the reader on the resolution of practical problems. Master's students in physics, mechanics and mathematics will discover with wonder, we hope, the applications of certain concepts which have been taught to them in a spirit of finesse and rigour. Practitioners in structural design analysis, as for them, will find herein a panoply of results which they will be satisfied to apply directly to analyse elastic shell structures.

This book is organized into six chapters. Chapter 1 is a brief introduction to 3D and 2D curvilinear media. Essential notions on tensor algebra and differential geometry of surfaces are treated. This chapter presents basic elements for the next chapters. Chapter 2 is devoted to establishing the general equilibrium equations of shells. Kinematics deduced by Nzengwa et al. [87] from an asymptotic analysis of a multi-scale dilatation (*multiple scaling*) of a 3D elastic shell problem was applied. In addition to terms found in the classical K-L and R-M models, the equations contain also terms related to the change of the third fundamental form of the shell's mid-surface. The energy impact of this tensor, called the Gauss deformation tensor, is remarkable in numerical calculations presented in Chap. 5. Governing equations of transverse shear and pinch stresses through-the-thickness are established. These equations are ordinary differential equations defined in specific functional spaces with initial and final conditions. Elastic dynamic analysis is studied in Chap. 3. Results obtained by Nzengwa [87] and [88], deduced from the 3D equivalent problem, show the existence of very many inertia terms, generally neglected in classical thin shell theory. Moreover, these additional terms contain and justify the corrective transverse inertia term suggested in 1967 by the Russian engineer Morozov [72], to be considered in aircraft wing vibration analysis.

During the deformation of a thick shell, the variation in thickness creates significant transverse stresses. The model studied in 2005 by Nzengwa [90] is presented. Governing equations of the distribution of the transverse stresses are established. All the equations obtained, thanks to the theory of compact operators, are presented in variational forms suitable for numerical calculations. Thin shells constitute a large class of structures to which numerous works and books have been devoted. Although the equations of shells and plates are always quite complex, requiring appropriate numerical methods, with some simplifying assumptions, a certain number of problems in thin plates and shells are solved analytically. In Chap. 4, membrane, bending or mixed (membrane and flectional) theories of this class will be presented with applications in current engineering. Orthotropic plates or membrane prestressed plates are studied. Huber and Von Karman equations obtained are expressed in a variational form. As said earlier, shell equations require appropriate numerical methods. Even in the variational form, the equations contain second order derivatives which need at least C^1 [33] finite elements. Moreover, continuity of deformation and stress at the common edge of elements should be ensured. These requirements have led to the creation of very sophisticated finite elements highly memory-greedy. Recent numerical methods are presented in Chap. 5. The Gradient Recovery method ((GR), Naga et al. [7]) is applied to these equations, requiring only finite elements of class C^0. Results are compared to those of different finite elements such as Discrete Kirchhoff Triangles (DKT) on benchmarks. The GR is not only memoryless-greedy, but also converges more rapidly to reference results. Another recent method, the Strain Deformation approach (SD), Zeighampour et al. [106], has been applied with the same conclusions. These two methods have shown that the characteristic parameter $2\chi = h/R$, which delimits the boundary between thin and thick shells, is around $\sqrt{0.1}$ and also beyond this critical value, strain energy or additional rigidity contributed by the change in Gauss curvature becomes significant. Chapter 6 is devoted to stiffened shells, thermoelastic shells under thermal load in steady state and also to anisotropic shells. This last category is largely addressed because of the growing interest in composite structured or unstructured materials. This is particularly the case of periodic 3D structures such as elevation towers, 2D such as embankments reinforced, for example, by a regular solid bentonite intrusion grid or 1D such as periodic stratified media. Their homogenized moduli are easily determined by applying the two-scale convergence theorem established in 1989 by Nguetseng [82], which give them a homogeneous anisotropic structure. Then the 2D shell model is derived as in the above paragraph. Some authors have proposed particular kinematics resulting from some expansion series according to the thickness parameter. These heuristic models are efficient in solving some specific problems but cannot be applied elsewhere without precautions. Finally, some shells, for example conical, can only be described correctly by a thickening in a direction different from the normal to the mid-surface. Their equilibrium equations are better analysed by using the theory of oriented surfaces developed by the Cosserat brothers [37]. A brief presentation of these heuristic models and Cosserat shells ends this chapter.

Two appendices are added at the end of the text. The first appendix is devoted to essential knowledge in 3D linear elasticity that has been used throughout the text. The second appendix is a summary of the essential points of each chapter of the textbook for fast learning of master's research students or fast use of practitioners in shell structures.

Throughout this document, we shall make use of the repeated index convention in both curvilinear and Cartesian coordinate systems. Derivations will sometimes be noted with the symbol "," and will be understood in the classical sense or in the sense of distribution of functional analysis, assuming that all the conditions for their validity are met. The domains in $\mathbb{R}^n$, $n = 1, 2, 3$ are deemed bounded and sufficiently smooth. We shall make use of the spaces of square-integrable functions $L^2(\Omega)$ or $L^2(S)$ and sometimes with their first and second derivatives also square-integrable. They will be equipped with their natural norms. Further notations will be indicated throughout the document.

I want to thank my former students Nkongho A. Joseph, Feumo Achille Germain, Ngatcha Ndengna Arno Roland, Djopkop Kouanang Landry and others, who, for many years, have encouraged me to write this document. I thank particularly Feumo Achille Germain and Ngatcha Ndengna Arno Roland who have spent a lot of time formatting the text. I am indebted to Prof. Gabriel Nguetseng whose peer review and suggestions led to the rearrangement of certain paragraphs. Finally, I especially thank Prof. Philippe G. Ciarlet, who for the first time offered me the opportunity to face the theory of shells.

Contents

List of Figures

Chapter 1
Curvilinear Media

A curvilinear media is a domain in $\mathbb{R}^n$ $n = 1,\ 2,\ 3$ better described by a non-cartesian coordinate system. We can cite for example a curve, a circle, a disc, a sphere, a cylinder, a curved surface, a surface of revolution, etc. In general, real structures are often composed of sub-structures of different shapes which cannot be easily described point by point with cartesian coordinates. Without trying to develop the theory on manifolds appropriate in this case, we shall limit ourselves to essential notions such as tensors, local bases and tangent spaces which are necessary to describe the evolution of these structures.

1.1 Geometry of a 3D Curvilinear Media

1.1.1 Parametrization and Covariant Base

Let $X = \left(X^1,\ X^2,\ X^3\right)$ or $X = (X_1,\ X_2,\ X_3)$ be the coordinates of a material point M defined in a cartesian base by

$$\begin{aligned} \overrightarrow{OM} &= Z_1\,\overrightarrow{i_1} + Z_2\,\overrightarrow{i_2} + Z_3\,\overrightarrow{i_3} \\ &= \quad Z_i\,(X)\,\overrightarrow{i_i} \end{aligned} \tag{1.1}$$

where $Z = (Z_i)$ is a C^1 bijective application. Then $det\nabla Z \neq 0$

Definition of the covariant base Let $G_i = \overrightarrow{OM}_{,i} = \frac{\partial \overrightarrow{OM}}{\partial X^i} = Z_{k,i}\,\overrightarrow{i_k}$; G_i is the vector described by column i of the matrix

R. Nzengwa, *Plate and Shell Models*,
https://doi.org/10.1007/978-981-97-2780-3_1

$$\nabla Z = \begin{bmatrix} Z_{1,1} & Z_{1,2} & Z_{1,3} \\ Z_{2,1} & Z_{2,2} & Z_{2,3} \\ Z_{3,1} & Z_{3,2} & Z_{3,3} \end{bmatrix} \tag{1.2}$$

and the system $\{G_1, G_2, G_3\}$ constitutes a base because $(G_1, G_2, G_3) = det\nabla Z \neq 0$. Therefore $\{G_1, G_2, G_3\}$ constitutes a base in $\mathbb{R}^3$. It is the covariant base and from (1.2)

$$G_1 = \begin{bmatrix} Z_{1,1} \\ Z_{2,1} \\ Z_{3,1} \end{bmatrix}; \quad G_2 = \begin{bmatrix} Z_{1,2} \\ Z_{2,2} \\ Z_{3,2} \end{bmatrix}; \quad G_3 = \begin{bmatrix} Z_{1,3} \\ Z_{2,3} \\ Z_{3,3} \end{bmatrix} \tag{1.3}$$

1.1.2 Metric Tensor, Line Element and Contravariant Base

Definition of the metric tensor Let dM denote a tangent vector. Then $dM = dX^1G_1 + dX^2G_2 + dX^3G_3 = dX^iG_i$. A line element is defined by $dM \cdot dM = dX^iG_i \cdot dX^jG_j = G_i \cdot G_j dX^i dX^j = G_{ij} dX^i dX^j$ with $G_{ij} = G_i \cdot G_j = G_{ji}$. Let us calculate the determinant of the tensor (G_{ij}). We have

$$\left(G_{ij}\right) = {}^t\nabla Z \cdot \nabla Z \tag{1.4}$$

therefore

$$det\left(G_{ij}\right) = det^t\nabla Z det\nabla Z = (det\nabla Z)^2 > 0 \tag{1.5}$$

The matrix (G_{ij}) is symmetric and positive defined at each point. It therefore defines a metric tensor (Fig. 1.1).

Definition of the contravariant base In a cartesian orthonormal base the scalar product of two vectors $V = V_ie_i$ and $W = W_je_j$ is

$$\begin{aligned} V.W &= (V_ie_i) \cdot \left(W_je_j\right) \\ &= V_iW_je_i \cdot e_j \\ &= V_iW_i \end{aligned} \tag{1.6}$$

because $e_i.e_j = \delta_{ij}$. Let $V = dX^idG_i$, $W = dX^jG_j$. Then $V.W = dX^idX^jG_{ij}$. In order to obtain the same simple form as above we define the vectors G^i such that $G^i \cdot G_j = \delta^i_j$. We deduce from

$$\nabla Z \left(\nabla Z^{-1}\right) = \left(\nabla Z^{-1}\right) \nabla Z = I \tag{1.7}$$

that

$$G^i = \frac{dX^i}{dZ_k}\overrightarrow{i_k} = (\nabla Z)^{-1}_{ik}\overrightarrow{i_k} \tag{1.8}$$

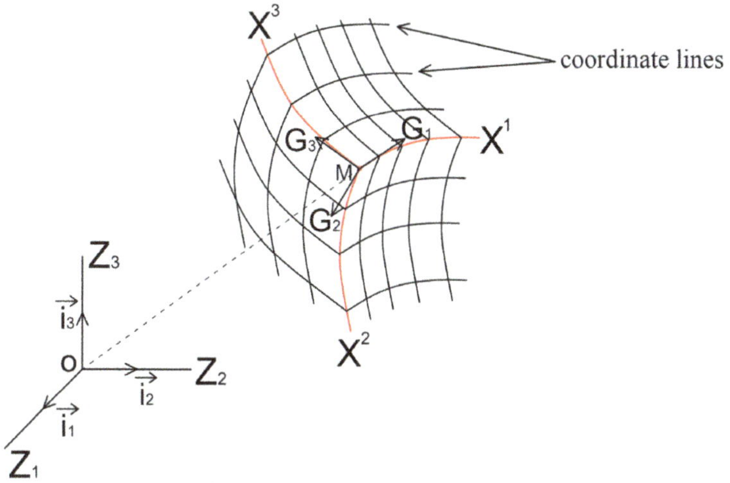

Fig. 1.1 3D curvilinear system

Therefore $\{G^1, G^2, G^3\}$ also constitutes a base in $\mathbb{R}^3$. It is the contravariant base which also defines the contravariant metric tensor

$$G^{ij} = G^i.G^j \tag{1.9}$$

A vector can be expressed in any of the two dual bases, i.e. $dM = dX^i G_i = dX_j G^j$. Therefore

$$dM.dM = dX_j G^j.dX_i G^i = G^{ij} dX_i dX_j = dX^i G_i.dX_j G^j = dX^i dX_i \tag{1.10}$$

A vector in one base can be expressed in the other by using the metric tensors as follows. Let $G^i = \alpha^j G_j$. From $G^i.G^j = \alpha^j = G^{ij}$ we deduce that

$$G^i = G^{ij} G_j \text{ and } G_i = G_{ij} G^j \tag{1.11}$$

1.1.3 Area Element, Volume Element and Change of Variance

Let $G = det\left(G_{ij}\right)$, then

$$det\left(G^{ij}\right) = \frac{1}{G}, \quad G = (det\nabla Z)^2, \quad \frac{1}{G} = \left(det\nabla Z^{-1}\right)^2 \tag{1.12}$$

Let an area vector element be defined by a vector product $dX^i G_i \times dX^j G_j = \alpha_k G^k dX^i dX^j$. From

$$\left(G_i \times G_j\right) \cdot G_k = \alpha_k = \sqrt{G} e_{ijk} \tag{1.13}$$

we obtain

$$dX^i G_i \times dX^j G_j = \sqrt{G} e_{ijk} G^k dX^i dX^j \tag{1.14}$$

where e_{ijk} is the covariant permutation symbol. Similarly

$$dX_i G^i \times dX_j G^j = \frac{1}{\sqrt{G}} e^{ijk} G_k dX_i dX_j \tag{1.15}$$

and e^{ijk} is the contravariant permutation symbol. A volume element dV defined from three vectors is

$$\begin{aligned} \left(dX^i G_i,\, dX^j G_j,\, dX^k G_k\right) &= \left(G_i,\, G_j,\, G_k\right) dX^i dX^j dX^k \\ &= e_{ijk} \sqrt{G} dX^i dX^j dX^k \end{aligned} \tag{1.16}$$

which yields

$$dV = \sqrt{G} dX^1 dX^2 dX^3 \tag{1.17}$$

Definition of the elementary area vector transport general formula Let (x) denote the coordinate of a position vector of a point in the deformed configuration, the elementary area vector is

$$\vec{\nu}\, ds = \left(\nu_k ds g^k\right) = \left(ds_k g^k\right) \tag{1.18}$$

where $\vec{\nu} = \nu_k g^k$ is the unitary normal vector to the surface. In the reference configuration with (X) as coordinate, the elementary area vector is

$$\vec{\nu^o}\, dS = \left(\nu_K^o dS G^K\right) = \left(dS_K G^K\right) \tag{1.19}$$

We have

$$\vec{\nu}\, ds = g_i \times g_j dx^i dx^j = \sqrt{g} e_{ijk} dx^i dx^j g^k \tag{1.20}$$

In the same way, we have

$$\vec{\nu^o}\, dS = G_I \times G_J dX^I dX^J \tag{1.21}$$

$$\nu_K^0 dS = \sqrt{G} e_{IJK} dX^I dX^J = dS_K \tag{1.22}$$

It follows that

$$
\begin{aligned}
ds_k &= \sqrt{g}e_{ijk}\frac{\partial x^i}{\partial X^I}\frac{\partial x^j}{\partial X^J}dX^I dX^J \\
&= \frac{\sqrt{g}}{\sqrt{G}}\left(\frac{e_{ijk}}{e_{IJK}}\frac{\partial x^i}{\partial X^I}\frac{\partial x^j}{\partial X^J}\frac{\partial x^k}{\partial X^K}\right)\frac{\partial X^K}{\partial x^k}dS_K
\end{aligned}
\tag{1.23}
$$

and

$$
ds_k = \frac{\sqrt{g}}{\sqrt{G}}JX,_k^K dS_K,\ J = det(x,_K^k) \tag{1.24}
$$

We deduce from this expression that if the initial surface is the border of the domain of the parameters (which is a cartesian coordinate system), then the transformed surface is the curvilinear surface and

$$
J = det(x,_K^k) = det(\delta_K^k) = 1 \text{ and } ds_k = \sqrt{G}dS_K \tag{1.25}
$$

Relations between tensor components in different bases Let a tensor be defined by its covariant components $T = \left(T_{ij}\right) = T_{ij}G^i \otimes G^j$, we have

$$
T_{ij}G^i \otimes G^j = T_{ij}G^{ik}G_k \otimes G^{jl}G_l = T_{ij}G^{ik}G^{jl}G_k \otimes G_l \tag{1.26}
$$

It follows that

$$
\begin{aligned}
T^{kl} &= G^{ki}G^{lj}T_{ij}, \quad T_{ij} = G_{ki}G_{lj}T^{kl} \\
T_j^k &= G^{ki}T_{ij}, \quad T_j^i = G^{ik}T_{kj} \\
T^{ij}T_{ij} &= G^{ik}G^{lj}T_{kl}T_{ij} = T_k^iT_i^k = T_j^iT_i^j
\end{aligned}
\tag{1.27}
$$

1.1.4 Covariant Derivation and Christoffel Symbols

We have $G_{i,j} = \Gamma_{ijk}G^k$, where $\Gamma_{ijk} = G_{i,j} \cdot G_k$ is the Christofel symbol. We also have

$$
\begin{aligned}
\Gamma_{ijk} &= \tfrac{1}{2}\left(G_{ik,j} + G_{jk,i} - G_{ij,k}\right) \\
G_{i,j} &= \Gamma_{ijk}G^{kl}G_l = \Gamma_{ij}^lG_l
\end{aligned}
\tag{1.28}
$$

and Γ_{ij}^l is also a Christoffel symbol, the most commonly used. From

$$
\left(G^i \cdot G_l\right),_j = \left(\delta_l^i\right),_j = 0 \tag{1.29}
$$

we deduce that

$$
G,_j^i \cdot G_l + G^iG_{l,j} = 0; \quad G,_j^i \cdot G_l = -G_{l,j} \cdot G^i = -\Gamma_{lj}^i \tag{1.30}
$$

and

$$G,^i_j = \alpha_l G^l = -\Gamma^i_{jl} G^l$$

Let $V = V^i G_i = V_i G^i$. Then

$$\begin{aligned}V,_j = & \left(V^i G_i\right),_j = V,^i_j G_i + V^i G_{i,j} \\ & = V,^i_j G_i + V^i \Gamma^l_{ij} G_l = V^l_{,j} G_l + V^i \Gamma^l_{ij} G_l \\ = & \left(V^l_{,j} + V^i \Gamma^l_{ij}\right) G_l = V^l_{/j} G_l\end{aligned} \tag{1.31}$$

In the same way, we obtain

$$V,_j = \left(V_i G^i\right),_j = V_{l/j} G^l$$

with

$$V^l_{/j} = V^l_{,j} + V^i \Gamma^l_{ij} \quad V_{l/j} = V_{l,j} - \Gamma^k_{lj} V_k \tag{1.32}$$

Exercise 1: Proof that $V^i_{/i} = \frac{1}{\sqrt{G}}\left(\sqrt{G} V^i\right),_i$ and $\int_\Omega V^i_{/i} d\Omega = \int_{\partial\Omega} V^i n_i dS$

Covariant derivation of a tensor Let us consider the tensor $T = \left(T_{ij}\right) = T_{ij} G^i \otimes G^j$ we have

$$\begin{aligned}\left(T_{ij} G^i \otimes G^j\right),_k = & T_{ij,k} G^i \otimes G^j + T_{ij} G^i_{,k} \otimes G^j + T_{ij} G^i \otimes G^j_{,k} \\ = & T_{ij,k} G^i \otimes G^j - \Gamma^i_{lk} T_{ij} G^l \otimes G^j - T_{ij} \Gamma^j_{kl} G^i \otimes G^l \\ = & T_{ij,k} G^i \otimes G^j - \Gamma^m_{ik} T_{mj} G^i \otimes G^j - T_{im} \Gamma^m_{kj} G^i \otimes G^j \\ = & \left(T_{ij,k} - \Gamma^m_{ik} T_{mj} - \Gamma^m_{kj} T_{im}\right) G^i \otimes G^j\end{aligned} \tag{1.33}$$

We obtain

$$T_{ij/k} = T_{ij,k} - \Gamma^m_{ik} T_{mj} - \Gamma^m_{kj} T_{im} \tag{1.34}$$

In the same way

$$T^{ij}_{/k} = T^{ij}_{,k} + \Gamma^i_{km} T^{mj} + \Gamma^j_{km} T^{im} \text{and } T^i_{j/k} = T^i_{j,k} + \Gamma^i_{km} T^m_j - \Gamma^m_{kj} T^i_m \tag{1.35}$$

In order to calculate $V^k_{/ji}$ we apply the same formula on the tensor $U^k_j = V^k_{/j}$ and obtain

$$\begin{aligned}V^k_{/ji} = & U^k_{j/i} = U^k_{j,i} + \Gamma^k_{im} U^m_j - \Gamma^m_{ij} U^k_m \\ & = (V,^k_j + V^l \Gamma^k_{lj}),_i + \Gamma^k_{im}(V,^m_j + V^l \Gamma^m_{lj}) - \Gamma^m_{ij}(V,^k_m + V^l \Gamma^k_{lm})\end{aligned} \tag{1.36}$$

$$\begin{aligned}V^k_{/ji} - V^k_{/ij} = & (\Gamma^k_{lj,i} - \Gamma^k_{li,j}) V^l + (\Gamma^k_{im} \Gamma^m_{lj} - \Gamma^k_{jm} \Gamma^m_{li}) V^l \\ = & R^k_{lij} V^l\end{aligned}$$

$$R^k_{lij} = \Gamma^k_{lj,i} - \Gamma^k_{li;j} + \Gamma^k_{im}\Gamma^m_{lj} - \Gamma^k_{jm}\Gamma^m_{li} \tag{1.37}$$

R^k_{lij} is the Riemann-Christoffel tensor. In the same way, we also obtain

$$V_{k/ji} - V_{k/ij} = R^l_{kij} V_l, \quad A_{ij/kl} - A_{ij/lk} = R^m_{ikl} A_{mj} + R^m_{jkl} A_{im} \tag{1.38}$$

Exercise 2: Proof that $G_{ij/k} = G^{ij}_{/k} = 0$

Correction of exercise 1) Proof of $V^i_{/i} = \frac{1}{\sqrt{G}}\left(\sqrt{G}V^i\right)_{,i}$. We have

$$\sqrt{G} = \left(G_i,\ G_j,\ G_k\right) e^{ijk}; \quad \frac{1}{G} = \left(G^1,\ G^2,\ G^3\right)$$

$$\begin{aligned}\left(\sqrt{G}V^i\right)_{,i} = & \quad \left(\left(G_1,\ G_2,\ G_3\right) V^i\right)_{,i} \\ = & \left(G_{1,i},\ G_2,\ G_3\right) V^i + \left(G_1,\ G_{2,i},\ G_3\right) V^i \\ & + \left(G_1,\ G_2,\ G_{3,i}\right) V^i + \left(G_1,\ G_2,\ G_3\right) V^i_{,i}\end{aligned}$$

$$G_{1,i} = \Gamma^l_{i1} G_l$$

Therefore

$$\left(\Gamma^l_{i1} G_l,\ G_2,\ G_3\right) = \quad \Gamma^1_{i1}\sqrt{G}$$

$$\begin{aligned}\left(\sqrt{G}V^i\right)_{,i}\frac{1}{\sqrt{G}} &= \Gamma^1_{1i} V^i + \Gamma^2_{2i} V^i + \Gamma^3_{3i} V^i + V^i_{,i} \\ &= \quad \Gamma^k_{ki} V^i + V^i_{,i}\end{aligned} \tag{1.39}$$

$$V^i_{/j} = V^i_{,j} + \Gamma^i_{jk} V^k, \quad V^i_{/i} = V^i_{,i} + \Gamma^i_{ki} V^k = V^i_{,i} + \Gamma^k_{ki} V^i \tag{1.40}$$

(2) $\int_\Omega V^i_{/i} d\Omega = \int_\Omega \frac{1}{\sqrt{G}}\left(\sqrt{G}V^i\right)_{,i} d\Omega, \quad d\Omega = \left(G_1,\ G_2,\ G_3\right) dX = \sqrt{G}dX$. Let $\overline{\Omega}$ be the domain occupied by $X = \left(X_1,\ X_2,\ X_3\right)$ So

$$\begin{aligned}\int_\Omega V^i_{/i} d\Omega &= \int_{\overline{\Omega}} \frac{1}{\sqrt{G}}\left(\sqrt{G}V^i\right)_{,i}\sqrt{G}dX \\ &= \int_{\overline{\Omega}}\left(\sqrt{G}V^i\right)_{,i} dX \\ &= \int_{\overline{\Omega}} div\left(\sqrt{G}V\right) dX \\ &= \int_{\partial\overline{\Omega}} \sqrt{G}V^i \overline{n}_i d\overline{S}\end{aligned}$$

$$= \int_{\partial\overline{\Omega}} V^i \overline{n}_i \sqrt{G} d\overline{S}$$
$$= \int_{\partial\Omega} V^i n_i dS \tag{1.41}$$

(3) Proof of $G^{ij}_{/k} = G_{ij/k} = 0$

$$G^{ij}_{/k} = G^{ij}_{,k} + \Gamma^i_{km} G^{mj} + \Gamma^j_{km} G^{im} \quad ,$$

$$G^{ij}_{,k} = \left(G^i \cdot G^j\right)_{,k} = G^i_{,k} \cdot G^j + G^i \cdot G^j_{,k} = -\Gamma^i_{km} G^{mj} - \Gamma^j_{km} G^{im}$$

We obtain the result by substitution.

Example of a curvilinear media $\overrightarrow{OM_1} = r cos\theta_1 \overrightarrow{i_1} + r sin\theta_1 \overrightarrow{i_2} + \theta_2 \overrightarrow{i_3}$ θ_i are angles or $\overrightarrow{OM_2} = r cos\theta_1 \overrightarrow{i_1} + r sin\theta_1 \overrightarrow{i_2} + z \overrightarrow{i_3}$. We have $X = (r, \theta_1, \theta_2)$ or $X = (r, \theta_1, z)$

$$G_1 = \begin{pmatrix} cos\theta_1 \\ sin\theta_1 \\ 0 \end{pmatrix}; \quad G_2 = \begin{pmatrix} -r sin\theta_1 \\ r cos\theta_1 \\ 0 \end{pmatrix}; \quad G_3 = \begin{pmatrix} 0 \\ 0 \\ 1 \end{pmatrix}$$

$$\left(G_{ij}\right) = \begin{bmatrix} 1 & 0 & 0 \\ 0 & r^2 & 0 \\ 0 & 0 & 1 \end{bmatrix} \quad \left(G^{ij}\right) = \begin{bmatrix} 1 & 0 & 0 \\ 0 & \frac{1}{r^2} & 0 \\ 0 & 0 & 1 \end{bmatrix}$$

$$G = r^2; \; \sqrt{G} = r; \; G^1 = G_1; \; G^2 = \frac{1}{r} G_2; \; G^3 = G_3$$

Exercise: calculate: $\Gamma^1_{jk}, \Gamma^2_{jk}, \Gamma^3_{jk}$

1.1.5 Strain Tensor

Different parametric systems

This is appropriate in the case of large deformation such as forming by rolling (Fig. 1.2).

In the reference configuration a line element dS is

$$dS^2 = dM.dM = G_{IJ} dX^I dX^J \tag{1.42}$$

In the deformed configuration where the coordinates of a point are $x = (x_1, x_2, x_3)$ or $x = (x^1, x^2, x^3)$

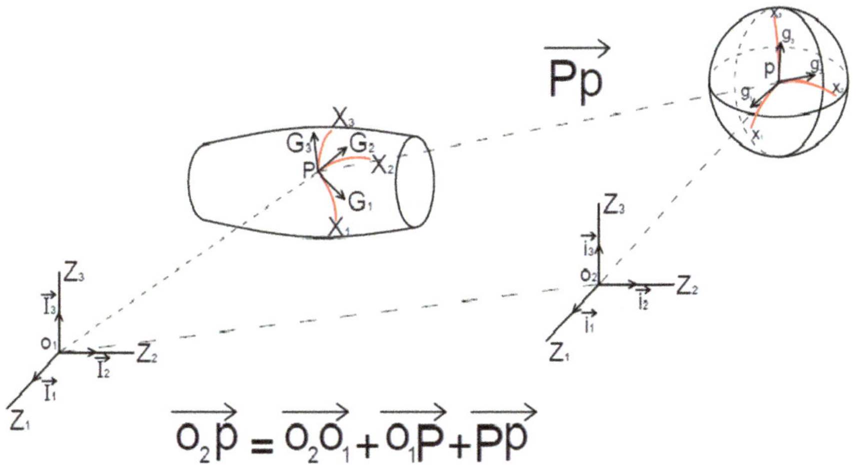

Fig. 1.2 Different coordinate systems

$$ds^2 = g_{ij} dx^i dx^j \tag{1.43}$$

We have

$$\begin{aligned} ds^2 - dS^2 = & \quad g_{ij} dx^i dx^j - G_{IJ} dX^I dX^J \\ = & \; g_{ij} \tfrac{\partial x^i}{\partial X^I} \cdot \tfrac{\partial x^j}{\partial X^J} dX^I dX^J - G_{IJ} dX^I dX^J \\ = & \quad \left(g_{ij} \tfrac{\partial x^i}{\partial X^I} \cdot \tfrac{\partial x^j}{\partial X^J} - G_{IJ} \right) dX^I dX^J \\ = & \quad \left(g_{ij} - G_{IJ} \tfrac{\partial X^I}{\partial x^i} \tfrac{\partial X^J}{\partial x^j} \right) dx^i dx^j \end{aligned} \tag{1.44}$$

We define the Lagrange strain tensor

$$E_{IJ}(X) = \frac{1}{2} \left(g_{ij} \frac{\partial x^i}{\partial X^I} \frac{\partial x^j}{\partial X^J} - G_{IJ} \right)(X) \tag{1.45}$$

and the Euler strain tensor

$$E_{ij}(x) = \frac{1}{2} \left(g_{ij} - G_{IJ} \frac{\partial X^I}{\partial x^i} \frac{\partial X^J}{\partial x^j} \right)(x) \tag{1.46}$$

Convected parametric system

This system is appropriate in small strain (or infinitesimal deformation) description (Fig. 1.3).
Coordinates are the same in the reference and deformed configurations, i.e. $x = X$

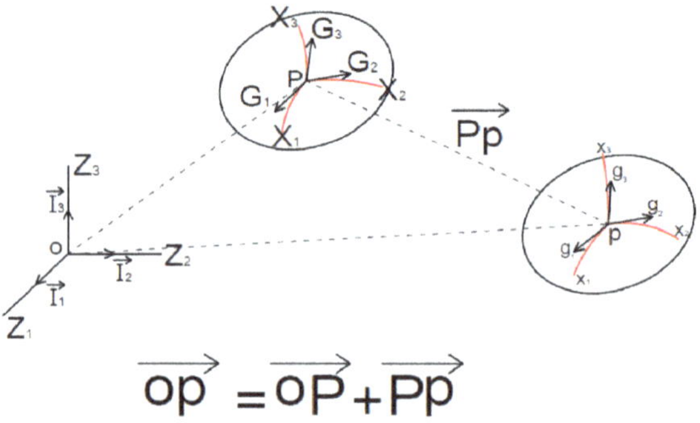

Fig. 1.3 Convected coordinate system

$$E_{ij}(x) = \frac{1}{2}\left(g_{ij} - G_{ij}\right) = \frac{1}{2}\left(g_i \cdot g_j - G_i \cdot G_j\right) \tag{1.47}$$

$$E_{IJ}(X) = \frac{1}{2}\left(g_{IJ} - G_{IJ}\right) = \frac{1}{2}\left(g_I \cdot g_J - G_I \cdot G_J\right) \tag{1.48}$$

Let us consider P and p, position vectors of a material point in the reference and deformed configurations, respectively. Let O be the origin of the coordinate system. Then $u = \overrightarrow{P_P}$ is the displacement vector and

$$\overrightarrow{O_P} = \overrightarrow{OP} + u \tag{1.49}$$

where

$$u = u^k G_k = u_k G^k = u^k g_k = u_k g^k \tag{1.50}$$

$$\overrightarrow{Op}_{,K} = \frac{\partial \overrightarrow{Op}}{\partial X^K} = g_k = \overrightarrow{OP}_{,K} + u_{,K} = G_K + u_{,K} \tag{1.51}$$

$$u_{,K} = u_{M/K} G^M = u^M_{/K} G_M \tag{1.52}$$

(upper case index refers to the reference configuration), the Lagrange strain tensor reads

$$E_{KL} = \frac{1}{2}\left(u_{L/K} + u_{K/L} + u_{M/K} u^M_{/L}\right)(X) \tag{1.53}$$

From

$$g_k - u_{,k} = G_k, \; u_{,k} = u_{m/k} g^m = u^m_{/k} g_m \tag{1.54}$$

Euler's strain tensor reads

$$E_{kl} = \frac{1}{2}\left(u_{l/k} + u_{k/l} - u_{m/k}u^{m}_{/l}\right)(x) \tag{1.55}$$

Example in a cartesian coordinate system
$G_K = G^K = \vec{i_k}$, $G_{IJ} = G_{JI} = \delta_{IJ}$, $\Phi = id + u = x_0 + u$, $g_k = \Phi,_k$ $g_{ij} = \Phi,_i \cdot \Phi,_j$. It follows that

$$\begin{aligned} g_{ij} = & \left(x_{0,i} + u,_i\right) \cdot \left(x_{0,j} + u,_j\right) \\ = & \, x_{0,i} \cdot x_{0,j} + x_{0,i} \cdot u,_j + x_{0,j} \cdot u,_i + u,_i \cdot u,_j \\ = & \, \delta_{ij} + u_{i,j} + u_{j,i} + u_{k,i}u_{k,j} \end{aligned} \tag{1.56}$$

We thus obtain the cartesian formula

$$E_{ij} = \frac{1}{2}\left(g_{ij} - G_{ij}\right) = \frac{1}{2}\left(u_{i,j} + u_{j,i} + u_{k,i}u_{k,j}\right) \tag{1.57}$$

1.1.6 Equilibrium Equation

The equilibrium equation of a 3D curvilinear media, occupied by a linearly elastic material, under mixed boundary conditions, is defined by

$$\begin{aligned} \sigma^{ij}_{/j} + f^i & = 0 \text{ in } \Omega \\ \sigma^{ij} n_j & = p^i \text{ on } \Gamma_1 \text{ and } \partial\Omega = \Gamma_1 \cup \Gamma_0 \\ u = & \, 0 \quad on \ \Gamma_0 \\ \sigma^{ij}(u) & = \bar{\lambda}\epsilon^k_k(u)G^{ij} + 2\bar{\mu}\epsilon^{ij}(u), \ \bar{\lambda}, \ \ \bar{\mu} > 0 \\ \epsilon_{ij}(w) & = \tfrac{1}{2}\left(w_{i/j} + w_{j/i}\right) \end{aligned} \tag{1.58}$$

($n = n_j G^j$ is the outer unit normal vector, $\bar{\lambda}$ and $\bar{\mu}$ are elastic moduli (Lamé's constants)). Its variational formulation reads
Find

$$u \in U_{ad} = \left\{w_i : \Omega \to \mathbb{R}, \quad w_i \in L^2(\Omega), \ w_{i/j} \in L^2(\Omega), \ w_i = 0 \text{ on } \Gamma_0\right\}$$

such that

$$\int_\Omega \sigma^{ij}(u)\epsilon_{ij}(v)d\Omega = \int_\Omega f^i(x)v_i(x)d\Omega + \int_{\Gamma_1} p^i(x)v_i(x)d\Gamma \tag{1.59}$$

for any tangent vector (or virtual displacement) $v \in U_{ad}$. This variational equation has a unique solution in the space of admissible displacements U_{ad}.

Exercise: Proof that the variational equation has a unique solution by establishing Korn's inequality and applying Lax-Milgram's lemma.

1.2 Surface Geometry

1.2.1 Parametrization, Covariant and Contravariant Bases, Fundamental Forms

Covariant base

Let us consider a surface parameterized by $X = (X^1, X^2) = (X^\alpha)$ or $X = (X_1, X_2) = (X_\alpha)$. Let m be a generic point on the surface (Fig. 1.4), we denote

$$A_\alpha = \overrightarrow{Om}_{,\alpha} \text{ and } A_3 = \frac{A_1 \times A_2}{\mid A_1 \times A_2 \mid} \tag{1.60}$$

The system $\{A_1, A_2, A_3\}$ constitutes a base in $\mathbb{R}^3$. Let $V = A_\alpha dX^\alpha$,

$$\begin{aligned} dS^2 = \quad & V \cdot V = A_\alpha dX^\alpha \cdot A_\beta dX^\beta \\ & = A_\alpha . A_\beta dX^\alpha dX^\beta = A_{\alpha\beta} dX^\alpha dX^\beta \end{aligned} \tag{1.61}$$

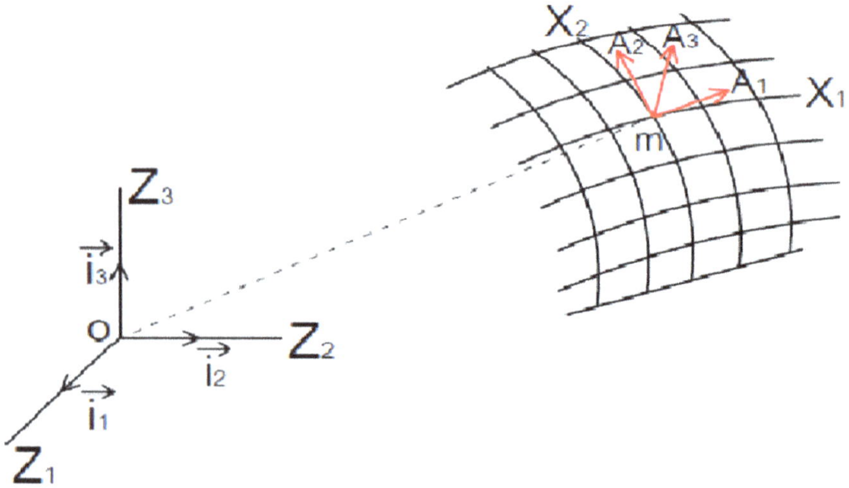

Fig. 1.4 Surface coordinate

$(A_{\alpha\beta})$ is a metric tensor because $A_{..}$ is symmetric and $A = det(A_{\alpha\beta}) > 0$. In fact

$$\begin{pmatrix} A_1 \\ A_2 \\ A_3 \end{pmatrix} \begin{pmatrix} A_1 & A_2 & A_3 \end{pmatrix} = \begin{bmatrix} A_{11} & A_{12} & 0 \\ A_{21} & A_{22} & 0 \\ 0 & 0 & 1 \end{bmatrix} \tag{1.62}$$

We have

$$\left(A_1, A_2, A_3 \right) = A_3 \cdot A_1 \times A_2 = \frac{A_1 \times A_2}{\mid A_1 \times A_2 \mid} \cdot A_1 \times A_2 = \mid A_1 \times A_2 \mid \tag{1.63}$$

We deduce that $A = det(A_{\alpha\beta}) = (A_3 \cdot A_1 \times A_2)^2 = \mid A_1 \times A_2 \mid^2$

First fundamental form and contravariant base

The tensor $(A_{\alpha\beta})$ is the first fundamental form of the surface S. The contravariant metric tensor is defined by

$$\left(A^{\alpha\beta}\right) = \left(A_{\alpha\beta}\right)^{-1} \tag{1.64}$$

We have

$$A^{\alpha\beta} A_{\beta\gamma} = \delta^{\alpha}_{\gamma}, \quad \frac{1}{A} = det\left(A^{\alpha\beta}\right) \tag{1.65}$$

Let us define

$$A^{\alpha} = A^{\alpha\beta} A_{\beta}, \; A^3 = A_3 \tag{1.66}$$

Then $A^{\alpha}.A_{\beta} = \delta^{\alpha}_{\beta}$. Therefore $\left\{A^1, A^2, A^3\right\}$ also constitutes a base, the contravariant base.

Area element

An area element is defined by

$$dS = \mid A_1 \times A_2 \mid dX^1 dX^2 = \sqrt{A} dX^1 dX^2 = \left(A_1, A_2, A_3\right) dX^1 dX^2 \tag{1.67}$$

We have

$$\begin{aligned} A_{\alpha} \times A_{\beta} &= \sqrt{A} e_{\alpha\beta} A_3, \quad A^{\alpha} \times A^{\beta} = \tfrac{1}{\sqrt{A}} e^{\alpha\beta} A^3 \\ A^3 \times A_{\beta} &= \sqrt{A} e_{\rho\beta} A^{\rho}, \quad A_3 \times A^{\alpha} = \tfrac{1}{\sqrt{A}} e^{\alpha\beta} A_{\beta} \end{aligned} \tag{1.68}$$

where $e_{\alpha\beta}$, $e^{\alpha\beta}$ are the 2D permutation symbols. As developed in 3D, tensor components satisfy the same relations, i.e. $T^{\alpha} = A^{\alpha\rho} T_{\rho}$; $T_{\rho} = A_{\rho\gamma} T^{\gamma}$;

$$T_{\alpha\beta} = A_{\alpha\rho} A_{\gamma\beta} T^{\rho\gamma}, \; T^{\alpha\beta} = A^{\alpha\rho} A^{\gamma\beta} T_{\rho\gamma}, \; T^{\alpha}_{\beta} = A^{\alpha\rho} T_{\rho\beta};$$

$$T^{\alpha\beta} T_{\alpha\beta} = A^{\alpha\rho} A^{\gamma\beta} T_{\rho\gamma} T_{\alpha\beta} = T^{\alpha}_{\gamma} T^{\gamma}_{\alpha} = T^{\alpha}_{\beta} T^{\beta}_{\alpha} \tag{1.69}$$

Second fundamental form and curvature tensor

We define the symmetric tensor $B_{..}$ by its components

$$B_{\alpha\beta} = A_3 \cdot A_{\alpha,\beta} = \tfrac{1}{2}\left(A_3 \cdot A_{\alpha,\beta} + A_3 \cdot A_{\beta,\alpha}\right) = A_3 \cdot \partial^2_{\alpha\beta}\overrightarrow{Om}$$
$$= \quad -\tfrac{1}{2}\left(A_\alpha \cdot A_{3,\beta} + A_\beta \cdot A_{3,\alpha}\right) = -A_\alpha \cdot A_{3,\beta} \tag{1.70}$$

The tensor $B_{..} = (B_{\alpha\beta})$ is the second fundamental form of the surface S. We also consider the tensor

$$B^\alpha_\beta = A^{\alpha\rho} B_{\rho\beta} \tag{1.71}$$

$B_{..}$ gives the different radii of curvature if the base vectors A_α are not unit vectors, while $B^._.$ gives the different curvatures.

Interpretation:

$$dS^2 = dm \cdot dm = A_\alpha dX^\alpha \cdot A_\beta dX^\beta$$
$$= A_{\alpha\beta} dX^\alpha dX^\beta$$

We have

$$1 = A_\alpha \frac{dX^\alpha}{dS} \cdot A_\beta \frac{dX^\beta}{dS} \tag{1.72}$$

We deduce that the vector $\vec{t} = \frac{dX^\alpha}{dS} A_\alpha$ is a unit vector. From $\vec{t} \cdot \vec{t} = 1$ it follows that $d\vec{t}/dS$ is perpendicular to $\vec{t}$. Therefore $d\vec{t}/dS = \mu A_3$. We obtain by derivation

$$\frac{d\vec{t}}{dS} = \frac{d^2X^\alpha}{dS^2} A_\alpha + \frac{dX^\alpha}{dS} A_{\alpha,\beta} \frac{dX^\beta}{dS} \tag{1.73}$$

and

$$A_3 \cdot \frac{d\vec{t}}{dS} = A_{\alpha,\beta} \cdot A_3 \frac{dX^\alpha}{dS} \cdot \frac{dX^\beta}{dS} = B_{\alpha\beta} \frac{dX^\alpha}{dS} \cdot \frac{dX^\beta}{dS} = B_{\alpha\beta} t^\alpha t^\beta = \mu \tag{1.74}$$

In order to determine the curvature at a point m with respect to the unit tangent vector of an arbitrary curve $\vec{t} = \frac{dX^\alpha}{dS} A_\alpha$, we should consider the curve traced on the surface S, by the plane generated by $\vec{t}$ and A_3. Let C be the point such that $\overrightarrow{mC} = \mu A_3$. C is the centre of normal curvature. It is the centre of the auscultating circle, i.e. the largest tangent circle to the curve, at the point m and $R = \pm Cm$ is its radius. The curvature $\mu = 1/R$ is positive if C is on the side pointed by A_3. It is otherwise negative. Therefore, the sign of the curvature depends on the orientation of the unit normal vector A_3 (Fig. 1.5).

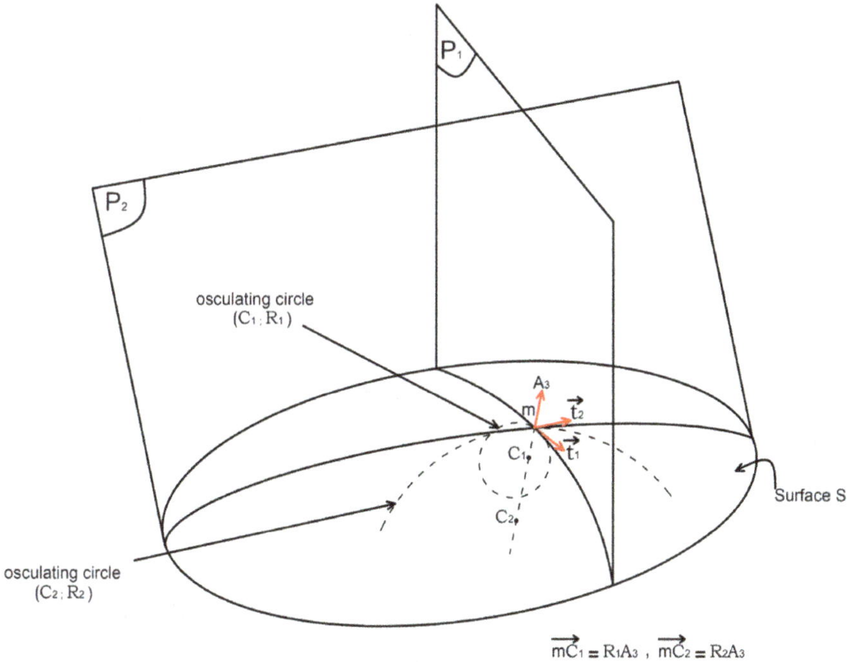

Fig. 1.5 Curvature

Diagonal base

Let $\vec{t} = t^{\alpha} A_{\alpha}$ be an eigenvector of the eigenvalue λ, then $B\vec{t} = \lambda\vec{t}$ which gives

$$\left(B_{\alpha\beta} t^{\beta}\right) = \lambda t_{\alpha} = \lambda A_{\alpha\rho} t^{\rho}$$

$$A^{\rho\alpha} B_{\alpha\beta} t^{\beta} = \lambda t^{\rho}$$

$$B^{\rho}_{\beta} t^{\beta} = \lambda t^{\rho}, \quad \left(B^{\cdot}_{\cdot} - \lambda I\right)\vec{t} = 0$$

then

$$det\left(B^{\cdot}_{\cdot} - \lambda I\right) = 0 \quad B^{\cdot}_{\cdot} = \begin{bmatrix} 1/r_1 & 0 \\ 0 & 1/r_2 \end{bmatrix} = \begin{bmatrix} \frac{1}{R_{min}} & 0 \\ 0 & \frac{1}{R_{max}} \end{bmatrix} \tag{1.75}$$

In an orthonormal diagonal base, the eigenvalues are arranged as follows $\frac{1}{R_{min}}, \frac{1}{R_{max}}$. We define the mean curvature

$$\bar{H} = \frac{1}{2}\left(\frac{1}{R_{min}} + \frac{1}{R_{max}}\right) = \frac{1}{2} B^{\alpha}_{\alpha} \tag{1.76}$$

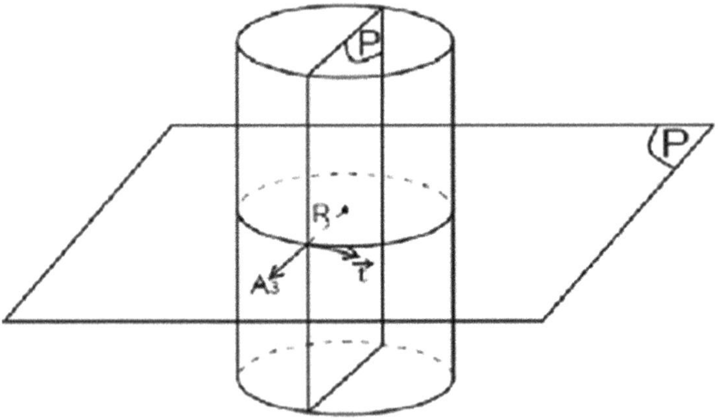

Fig. 1.6 Curvature of a cylindrical surface

and the Gaussian curvature

$$\bar{G} = \frac{1}{R_{min}} \times \frac{1}{R_{max}} = det B^{\cdot}_{\cdot} = B^1_1 B^2_2 - B^1_2 B^2_1 = det(A^{\alpha\beta}) \times det(B_{\beta\rho}) \quad (1.77)$$

Example 1: Cylinder with radius R

$$\overrightarrow{Om} = (Rcos\theta,\ Rsin\theta,\ z)\ and\ B_{cyl} = \begin{bmatrix} \frac{1}{R} & 0 \\ 0 & \frac{1}{\infty} \end{bmatrix} = \begin{bmatrix} \frac{1}{R} & 0 \\ 0 & 0 \end{bmatrix}$$

The curvature is positive if the unit normal vector points towards the centre. It is otherwise negative (Fig. 1.6).

Example 2: Sphere with radius R
$\overrightarrow{Om} = (Rsin\theta_1 cos\theta_2,\ Rsin\theta_1 sin\theta_2,\ Rcos\theta_1)$ and

$$B_{..sphère} = \begin{bmatrix} \frac{1}{R} & 0 \\ 0 & \frac{1}{R} \end{bmatrix}$$

The curvature is positive if the unit normal vector points towards the centre. It is otherwise negative (Fig. 1.7).

Example 3: Hyperbolic surface

$$R_s < 0,\ R_\varphi > 0\ and\ B_{..hyb} = \begin{bmatrix} \frac{1}{R_s} & 0 \\ 0 & \frac{1}{R_\varphi} \end{bmatrix}$$

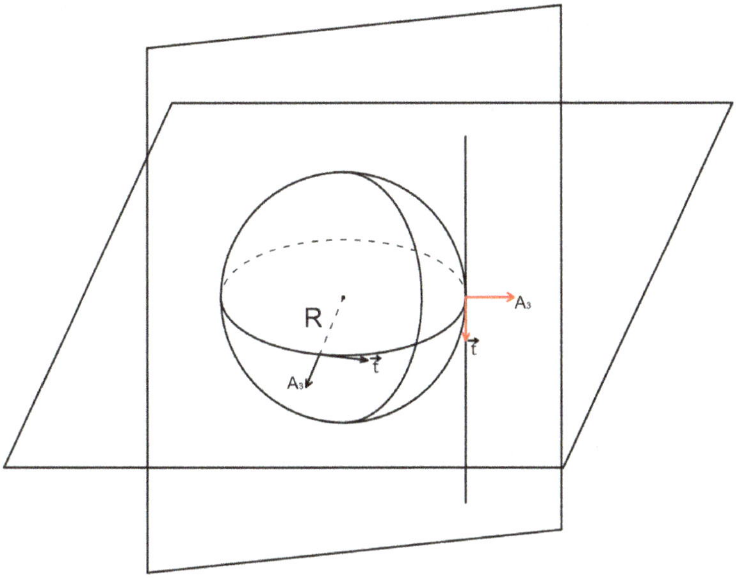

Fig. 1.7 Curvature of a spherical surface

Third fundamental form

The third fundamental form is defined by

$$C_{\alpha\beta} = B_{\alpha}^{\rho} B_{\rho\beta} = A^{\rho\gamma} B_{\alpha\gamma} B_{\rho\beta} = B_{\alpha\gamma} A^{\rho\gamma} B_{\rho\beta} = B_{\alpha\gamma} B_{\beta}^{\gamma} = B_{\beta}^{\gamma} B_{\alpha\gamma} = C_{\beta\alpha}$$

We can remark that in an orthonormal diagonal base (physical base), the determinant of the tensor $C_{..}$ is $\bar{G}^2$, the square of the Gaussian curvature (Fig. 1.8).

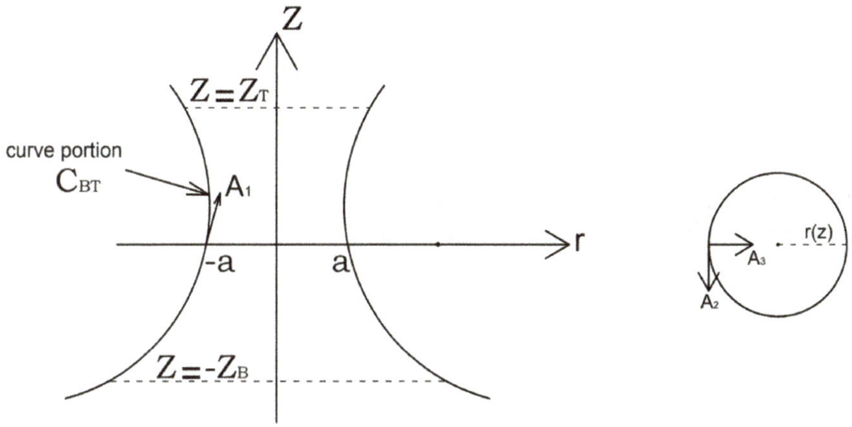

Fig. 1.8 Surface of revolution

1.2.2 2D Covariant Derivation

Let $A_{\alpha,\beta} = X^{\rho} A_{\rho} + X^3 A_3$. By multiplying both sides by A^{λ} and $A_3 = A^3$, respectively, we obtain

$$A_{\alpha,\beta} = \Gamma^{\lambda}_{\alpha\beta} A_{\lambda} + B_{\alpha\beta} A_3$$

where the Christofel symbol $\Gamma^{\lambda}_{\alpha\beta}$ on the surface is defined by

$$\Gamma^{\rho}_{\alpha\beta} = A_{\alpha,\beta}.A^{\rho} = A^{\rho\gamma}\Gamma_{\alpha\beta\gamma} = \Gamma^{\rho}_{\beta\alpha}$$

$$\Gamma_{\alpha\beta\rho} = \frac{1}{2}\left(A_{\rho\beta,\alpha} + A_{\alpha\rho,\beta} - A_{\alpha\beta,\rho}\right) \tag{1.78}$$

Its unit is $1/m$. We deduce as above that $A^{\alpha}_{,\beta} = -\Gamma^{\alpha}_{\beta\lambda} A^{\lambda} + B^{\alpha}_{\beta} A^3$. From $A_3.A_3 = 1$, it follows that

$$A_{3,\alpha} = X^{\lambda} A_{\lambda} = X_{\lambda} A^{\lambda} \ and\ A_{3,\alpha} \cdot A_{\lambda} = X_{\lambda} \tag{1.79}$$

From which

$$A_{3,\alpha} = -B^{\lambda}_{\alpha} A_{\lambda} = -B_{\alpha\lambda} A^{\lambda} \tag{1.80}$$

Let $v = v^{\lambda} A_{\lambda} + v^3 A_3 = v_{\lambda} A^{\lambda} + v_3 A^3$. We have

$$\begin{aligned} v,_{\alpha} = & \quad v,^{\lambda}_{\alpha} A_{\lambda} + v^{\lambda} A_{\lambda,\alpha} + v,^{3}_{\alpha} A_3 + v^3 A_{3,\alpha} \\ = & \left(v,^{\lambda}_{\alpha} + v^{\gamma}\Gamma^{\lambda}_{\gamma\alpha} - B^{\lambda}_{\alpha} v^3\right) A_{\lambda} + \left(v,^{3}_{\alpha} + B_{\lambda\alpha} v^{\lambda}\right) A_3 \\ = & \left(v_{\lambda},_{\alpha} - \Gamma^{\gamma}_{\lambda\alpha} v_{\gamma} - B_{\alpha\lambda} v_3\right) A^{\lambda} + \left(v_3,_{\alpha} + B^{\lambda}_{\alpha} v_{\lambda}\right) A^3 \end{aligned} \tag{1.81}$$

We denote the covariant derivative

$$\nabla_{\alpha} v^{\lambda} = v,^{\lambda}_{\alpha} + \Gamma^{\lambda}_{\gamma\alpha} v^{\gamma} \quad \nabla_{\alpha} v_{\lambda} = v_{\lambda,\alpha} - \Gamma^{\gamma}_{\lambda\alpha} v_{\gamma} \tag{1.82}$$

So

$$\begin{aligned} v,_{\alpha} &= \left(\nabla_{\alpha} v^{\lambda} - B^{\lambda}_{\alpha} v^3\right) A_{\lambda} + \left(v,^{3}_{\alpha} + B_{\lambda\alpha} v^{\lambda}\right) A_3 \\ &= \left(\nabla_{\alpha} v_{\lambda} - B_{\alpha\lambda} v_3\right) A^{\lambda} + \left(v_{3,\alpha} + B^{\lambda}_{\alpha} v_{\lambda}\right) A^3 \end{aligned} \tag{1.83}$$

We can also calculate the second-order covariant derivation $v,_{\alpha\beta}$ by proceeding in the same way. In order to calculate the second derivative $v_3,_{\alpha\beta}$, we consider the vector field

$$
\begin{aligned}
(v_{3,\alpha}\, A^{\alpha})_{,\beta} = &\quad (v_{3,\alpha\beta}\, A^{\alpha} + v_{3,\lambda}\, A^{\lambda}_{,\beta} \\
&= v_{3,\alpha\beta}\, A^{\alpha} - v_{3,\lambda}\, \Gamma^{\lambda}_{\alpha\beta} A^{\alpha} - v_{3,\alpha}\, B^{\alpha}_{,\beta} A^{3} \\
&= \ (v_{3,\alpha\beta} - v_{3,\lambda}\, \Gamma^{\lambda}_{\beta\alpha}) A^{\alpha} - v_{3,\alpha}\, B^{\alpha}_{,\beta} A^{3}
\end{aligned}
\tag{1.84}
$$

The projection of this vector on the tangent plane is the second derivative, i.e.

$$
(v_{3,\alpha\beta} - v_{3,\lambda}\, \Gamma^{\lambda}_{\beta\alpha}) = \nabla_{\alpha\beta} v_3 = \nabla_{\beta\alpha} v_3 \tag{1.85}
$$

The covariant derivation of a tensor, expressed in a tensor base, is defined as follows. Let

$$
T = (T_{\alpha\beta}) = T_{\alpha\beta} A^{\alpha} \otimes A^{\beta} = T^{\alpha\beta} A_{\alpha} \otimes A_{\beta} = (T^{\alpha\beta}) \tag{1.86}
$$

Derivating the tensor T as above we obtain

$$
\nabla_{\lambda} T_{\alpha\beta} = T_{\alpha\beta,\lambda} - \Gamma^{\gamma}_{\alpha\lambda} T_{\gamma\beta} - \Gamma^{\gamma}_{\beta\lambda} T_{\alpha\gamma}
$$

$$
\nabla_{\lambda} T^{\alpha\beta} = T^{\alpha\beta}_{,\lambda} + \Gamma^{\alpha}_{\lambda\gamma} T^{\gamma\beta} + \Gamma^{\beta}_{\lambda\gamma} T^{\alpha\gamma}
$$

Exercise 1 Calculate $\nabla_{\alpha} T^{\alpha\beta}$, $\nabla_{\alpha} T^{\beta}_{\rho}$, $\nabla_{\alpha} T^{\alpha\beta}_{\gamma}$,$\nabla_{\gamma} T^{\alpha\beta}_{\rho\delta}$

Exercise 2 Show that $\nabla_{\alpha} A_{..} = \nabla_{\alpha} A^{..} = 0$ where $A_{..}$ and $A^{..}$ are the metric tensors of a surface
Exercise 3 Show that $\nabla_{\alpha} v^{\alpha} = \frac{1}{\sqrt{A}} \left(\sqrt{A} v^{\alpha}\right)_{,\alpha}$
Exercise 4 Show that $\int_s \nabla_{\alpha} v^{\alpha} dS = \int_{\partial s} v^{\alpha} \nu_{\alpha} dc$. Use the vector field $v = \left(v^1, v^2, 0\right)$ defined in a one metre thick cylindrical domain centred on the surface.
Example: The sphere with radius R:
$\overrightarrow{om} = R\,(sin\theta_1 cos\theta_2,\ sin\theta_1 sin\theta_2,\ cos\theta_1)$. The coordinates are $X = (\theta_1, \theta_2)$. We have

$$
A_1 = R \begin{pmatrix} cos\theta_1\ cos\theta_2 \\ cos\theta_1\ sin\theta_2 \\ -\ \ sin\theta_1 \end{pmatrix}, \quad A_2 = R \begin{pmatrix} -sin\theta_1\ sin\theta_2 \\ sin\theta_1\ \ cos\theta_2 \\ 0 \end{pmatrix}
$$

$$
A_3 = \begin{pmatrix} sin\theta_1\ cos\theta_2 \\ sin\theta_1\ sin\theta_2 \\ cos\theta_1 \end{pmatrix}; \quad A_3 \cdot A_3 = 1 \tag{1.87}
$$

$$
\left(A_{\alpha\beta}\right) = R^2 \begin{bmatrix} 1 & 0 \\ 0 & sin^2\theta_1 \end{bmatrix}, \quad \left(A^{\alpha\beta}\right) = \frac{1}{R^2} \begin{bmatrix} 1 & 0 \\ 0 & \frac{1}{sin^2\theta^1} \end{bmatrix}, \tag{1.88}
$$

$$
A = R^4 sin^2\theta_1 = det(A_{\alpha\beta})
$$

$$\Gamma^1_{\alpha\beta} = \begin{bmatrix} 0 & 0 \\ 0 & -sin\theta_1 cos\theta_1 \end{bmatrix}, \quad \Gamma^2_{\alpha\beta} = \begin{bmatrix} 0 & cotg\theta_1 \\ cotg\theta_1 & 0 \end{bmatrix} \tag{1.89}$$

$$B_{..} = \begin{bmatrix} -R & 0 \\ 0 & -Rsin^2\theta^1 \end{bmatrix}, \quad B_{..} = B_{\alpha\beta} A^\alpha \otimes A^\beta \tag{1.90}$$

$$| A_\alpha | = r_\alpha \text{ and } \bar{A}_\alpha = \frac{A_\alpha}{r_\alpha} = \bar{A}^\alpha, \quad A^\alpha \cdot A_\alpha = \left(\frac{1}{r_\alpha}\bar{A}^\alpha\right) \cdot \left(r_\alpha \bar{A}_\alpha\right) \tag{1.91}$$

Therefore

$$\begin{gathered} B_{..} = B_{\alpha\beta} A^\alpha \otimes A^\beta \quad B_{..} = B_{\alpha\beta} \tfrac{1}{r_\alpha} \bar{A}^\alpha \otimes \tfrac{1}{r_\beta} \bar{A}^\beta = \bar{B}_{\alpha\beta} \bar{A}^\alpha \otimes \bar{A}^\beta \\ \bar{B}_{\alpha\beta} = \tfrac{1}{r_\alpha} \cdot \tfrac{1}{r_\beta} B_{\alpha\beta} \end{gathered} \tag{1.92}$$

On a sphere,

$$\bar{B}_{11} = \frac{1}{R} \cdot \frac{1}{R} B_{11} = \frac{1}{R^2} B_{11} \;\; for \;\; A_3 \text{ pointing outside the sphere } \; B_{11} = -R \tag{1.93}$$

and

$$\begin{gathered} \bar{B}_{11} = -\tfrac{1}{R}; \quad \bar{B}_{22} = \tfrac{1}{Rsin\theta_1} \cdot \tfrac{1}{Rsin\theta_1} B_{22} = \tfrac{1}{R^2 sin^2\theta_1} \left(-Rsin^2\theta_1\right) \\ \text{and } \; \bar{B}_{22} = -\tfrac{1}{R} \end{gathered} \tag{1.94}$$

Corrections of exercises

Exercise 1: Calculate $\nabla_\alpha A^{\alpha\beta}$
Indeed, $\nabla_\lambda A^{\alpha\beta} = A^{\alpha\beta}_{,\lambda} + \Gamma^\alpha_{\lambda\gamma} A^{\gamma\beta} + \Gamma^\beta_{\lambda\gamma} A^{\alpha\gamma}$; letting $\lambda = \alpha$, we obtain

$$\nabla_\alpha A^{\alpha\beta} = A^{\alpha\beta}_{,\alpha} + \Gamma^\alpha_{\alpha\gamma} A^{\gamma\beta} + \Gamma^\beta_{\alpha\gamma} A^{\alpha\gamma}$$

Exercise 2 We have

$$\nabla_\lambda A^{\alpha\beta} = A^{\alpha\beta}_{,\lambda} + \Gamma^\alpha_{\lambda\gamma} A^{\gamma\beta} + \Gamma^\beta_{\lambda\gamma} A^{\alpha\gamma} \tag{1.95}$$

But

$$\begin{aligned} A^{\alpha\beta}_{,\lambda} &= \left(A^\alpha \cdot A^\beta\right)_{,\lambda} = A^\alpha_{,\lambda} . A^\beta + A^\alpha \cdot A^\beta_{,\lambda} \\ &= A^\beta . \left(-\Gamma^\alpha_{\lambda\gamma} A^\gamma + B^\alpha_\lambda A^3\right) + A^\alpha . \left(-\Gamma^\beta_{\lambda\gamma} A^\gamma + B^\gamma_\lambda A^3\right) \\ &= -\Gamma^\alpha_{\lambda\gamma} A^{\gamma\beta} - \Gamma^\beta_{\lambda\gamma} A^{\gamma\alpha} \end{aligned} \tag{1.96}$$

(1.96) in (1.95) gives

$$\nabla_\lambda A^{\alpha\beta} = -\Gamma^\alpha_{\lambda\gamma} A^{\gamma\beta} - \Gamma^\beta_{\lambda\gamma} A^{\gamma\alpha} + \Gamma^\alpha_{\lambda\gamma} A^{\gamma\beta} + \Gamma^\beta_{\lambda\gamma} A^{\alpha\gamma} = 0 \tag{1.97}$$

$$\nabla_\lambda A_{\alpha\beta} = A_{\alpha\beta,\lambda} - \Gamma^\gamma_{\alpha\lambda} A_{\gamma\beta} - \Gamma^\gamma_{\beta\lambda} A_{\alpha\gamma} \tag{1.98}$$

But

$$\begin{aligned} A_{\alpha\beta,\lambda} &= A_{\alpha,\lambda}.A_\beta + A_{\beta,\lambda}.A_\alpha \\ &= \left(\Gamma^\gamma_{\alpha\lambda} A_\gamma + B_{\alpha\lambda} A_3\right).A_\beta + \left(\Gamma^\gamma_{\beta\lambda} A_\gamma + B_{\beta\lambda} A_3\right).A_\alpha \\ &= \Gamma^\gamma_{\alpha\lambda} A_{\gamma\beta} + \Gamma^\gamma_{\beta\lambda} A_{\gamma\alpha} \end{aligned} \tag{1.99}$$

(1.99) in (1.98) gives

$$\nabla_\lambda A_{\alpha\beta} = \Gamma^\gamma_{\alpha\lambda} A_{\gamma\beta} + \Gamma^\gamma_{\beta\lambda} A_{\gamma\alpha} - \Gamma^\gamma_{\alpha\lambda} A_{\gamma\beta} - \Gamma^\gamma_{\beta\lambda} A_{\alpha\gamma} = 0 \tag{1.100}$$

$$\nabla_\lambda A^{\alpha\beta} = \nabla_\lambda A_{\alpha\beta} = 0 \tag{1.101}$$

Exercice 3 Show that $\nabla_\alpha v^\alpha = \frac{1}{\sqrt{A}} \left(\sqrt{A} v^\alpha\right)_{,\alpha}$

$$\nabla_\alpha v^\alpha = v^\alpha_{,\alpha} + \Gamma^\alpha_{\beta\alpha} v^\beta \tag{1.102}$$

$$\begin{aligned} \sqrt{A} &= det\left\{A_1,\ A_2,\ A_3\right\} = \left(A_1,\ A_2,\ A_3\right) \\ \left(\sqrt{A}\right)_{,\alpha} &= \left(A_{1,\alpha},\ A_2,\ A_3\right) + \left(A_1,\ A_{2,\alpha},\ A_3\right) + \left(A_1,\ A_2,\ A_{3,\alpha}\right) \end{aligned} \tag{1.103}$$

with $A_{\beta,\alpha} = \Gamma^\lambda_{\beta\alpha} A_\lambda + B_{\beta\alpha} A_3$ and $A_{3,\alpha} = -B^\lambda_\alpha A_\lambda$. Therefore

$$\begin{aligned} \left(\sqrt{A}\right)_{,\alpha} &= \Gamma^1_{1\alpha} \left(A_1,\ A_2,\ A_3\right) + \Gamma^2_{2\alpha} \left(A_1,\ A_2,\ A_3\right) \\ &= \sqrt{A}\Gamma^\beta_{\beta\alpha} \\ \left(\sqrt{A} v^\alpha\right)_{,\alpha} &= \sqrt{A} v^\alpha_{,\alpha} + \sqrt{A}\Gamma^\beta_{\beta\alpha} v^\alpha = \sqrt{A}\left(v^\alpha_{,\alpha} + \Gamma^\beta_{\beta\alpha} v^\alpha\right) \end{aligned} \tag{1.104}$$

$$\left(\sqrt{A} v^\alpha\right)_{,\alpha} = \sqrt{A} \nabla_\alpha v^\alpha \tag{1.105}$$

Exercise 4 Show that $\int_s \nabla_\alpha v^\alpha dS = \int_{\partial s} v^\alpha \nu_\alpha dc$ from (1.105)
We consider the vector field $v = \left(v^1, v^2, 0\right)$ in the domain $\Omega = S \times [-\frac{1}{2}, \frac{1}{2}]$, $\partial\Omega = \{z = -1/2\} \cup \{z = 1/2\} \cup \Sigma$ and $\sqrt{A} v = \left(\sqrt{A} v^1, \sqrt{A} v^2, 0\right)$. Let us denote as above the domain occupied by the coordinates (x, z) by $\bar{\Omega}$, then

$$\int_{\frac{-1}{2}}^{\frac{1}{2}} \int_S \nabla_\alpha v^\alpha dS dz = \int_{\frac{-1}{2}}^{\frac{1}{2}} dz \int_S \frac{1}{\sqrt{A}} \left(\sqrt{A} v\right)_{,\alpha} dS$$

But

$$div\left(\sqrt{A} v\right) = \left(\sqrt{A} v^\alpha\right)_{,\alpha}$$

and

$$\begin{aligned}
\int_{\frac{-1}{2}}^{\frac{1}{2}} dz \int_S \tfrac{1}{\sqrt{A}} \left(\sqrt{A} v^\alpha\right)_{,\alpha} \sqrt{A} dx^1 dx^2 = &\ \int_{\frac{-1}{2}}^{\frac{1}{2}} dz \int_S \left(\sqrt{A} v^\alpha\right)_{,\alpha} dx^1 dx^2 \\
= &\ \int_{\bar{\Omega}} div\left(\sqrt{A} v\right) dx^1 dx^2 dz \\
= &\ \int_{\bar{\Sigma}} v^\alpha \sqrt{A} \bar{\nu}_\alpha d\bar{\Sigma} \\
= &\ \int_\Sigma v^\alpha \nu_\alpha d\Sigma \\
= &\ \int_{\frac{-1}{2}}^{\frac{1}{2}} dz \int_{\partial S} v^\alpha \nu_\alpha dc = \int_{\partial S} v^\alpha \nu_\alpha dc
\end{aligned} \tag{1.106}$$

1.2.3 Variation of Fundamental Forms

We shall use the convected parametric coordinates system throughout, i.e. $X = \left(X^1, X^2\right)$, $x = (x_1, x_2) = X$. The variation of the first fundamental form (or metric tensor) between the reference and deformed configurations is defined by the variation of line elements as follows:

$$\frac{1}{2}\left(dm \cdot dm - dm_0 \cdot dm_0\right) = \frac{1}{2}\left(a_{\alpha\beta} - A_{\alpha\beta}\right) dX^\alpha dX^\beta \tag{1.107}$$

The strain tensor (or the change in metric tensor) is

$$E_{\alpha\beta} = \frac{1}{2}\left(a_{\alpha\beta} - A_{\alpha\beta}\right) \tag{1.108}$$

Let us consider the displacement vector, expressed in the reference configuration base as follows:

$$U = U^\alpha A_\alpha + U^3 A_3 = U_\alpha A^\alpha + U_3 A^3 \tag{1.109}$$

or in the deformed configuration by

$$u = u^{\alpha} a_{\alpha} + u^3 a_3 = u_{\alpha} a^{\alpha} + u_3 a^3 \tag{1.110}$$

Then

$$\begin{aligned} E_{\alpha\beta} = & \tfrac{1}{2}\left(a_{\alpha}.a_{\beta} - A_{\alpha}\cdot A_{\beta}\right) \\ &= \tfrac{1}{2}\left[(om_0 + U)_{,\alpha}\cdot(om_0 + U)_{,\beta} - A_{\alpha}\cdot A_{\beta}\right] \\ &= \tfrac{1}{2}\left[a_{\alpha}\cdot a_{\beta} - (om - u)_{,\alpha}\cdot(om - u)_{,\beta}\right] \end{aligned} \tag{1.111}$$

$$E_{\alpha\beta} = \frac{1}{2}((A_{\alpha} + U_{,\alpha})\cdot(A_{\beta} + U_{,\beta}) - A_{\alpha}\cdot A_{\beta})$$

$$\begin{aligned} E_{\alpha\beta} &= \tfrac{1}{2}(A_{\alpha}\cdot U_{,\beta} + A_{\beta}\cdot U_{,\alpha} + U_{,\alpha}\cdot U_{,\beta}) \\ &= \tfrac{1}{2}(a_{\alpha}\cdot u_{,\beta} + a_{\beta}\cdot u_{,\alpha} - u_{,\alpha}\cdot u_{,\beta}) \end{aligned} \tag{1.112}$$

But

$$U_{,\alpha} = (\nabla_{\alpha}U^{\lambda} - B_{\alpha}^{\lambda}U^3)A_{\lambda} + (U^3_{,\alpha} + B_{\lambda\alpha}U^{\lambda})A_3 \tag{1.113}$$

$$U_{,\beta} = (\nabla_{\beta}U^{\rho} - B_{\beta}^{\rho}U^3)A_{\rho} + (U^3_{,\beta} + B_{\rho\beta}U^{\rho})A_3 \tag{1.114}$$

It follows that

$$\begin{aligned} E_{\alpha\beta} = & \tfrac{1}{2}((\nabla_{\alpha}U_{\beta} + \nabla_{\beta}U_{\alpha} - 2U_3B_{\alpha\beta}) \\ & + A^{\rho\gamma}(\nabla_{\beta}U_{\rho} - U_3B_{\beta\rho})(\nabla_{\alpha}U_{\gamma} - U_3B_{\alpha\gamma}) \\ & + (\nabla_{\beta}U_3 + B_{\beta\rho}U^{\rho})(\nabla_{\alpha}U_3 + B_{\alpha\gamma}U^{\gamma})) \end{aligned} \tag{1.115}$$

This is the Lagrangian expression of the strain tensor of the surface S. We can write $E_{\alpha\beta} = e_{\alpha\beta} + \ldots$ nonlinear terms where

$$e_{\alpha\beta} = \frac{1}{2}(\nabla_{\alpha}U_{\beta} + \nabla_{\beta}U_{\alpha} - 2U_3B_{\alpha\beta}) \tag{1.116}$$

The Eulerian expression of this linear term reads

$$e_{\alpha\beta} = \frac{1}{2}(\nabla_{\alpha}u_{\beta} + \nabla_{\beta}u_{\alpha} - 2u_3b_{\alpha\beta}) \tag{1.117}$$

We assume small strain or infinitesimal deformation, i.e. we shall consider in the linear analysis herein, only $(e_{\alpha\beta}) = e$.

The variation of the curvature tensor (or the change of the second fundamental form) is defined by

$$K_{\alpha\beta} = b_{\alpha\beta} - B_{\alpha\beta} \tag{1.118}$$

We have $b_{\alpha\beta} = a_{\alpha,\beta} \cdot a_3 = -a_{3,\beta} \cdot a_\alpha$, $B_{\alpha\beta} = A_{\alpha,\beta} \cdot A_3 = -A_{3,\beta} \cdot A_\alpha$. So

$$K_{\alpha\beta} = -a_{3,\beta} \cdot a_\alpha + A_{3,\beta} \cdot A_\alpha = -a_{3,\beta} \cdot (A_\alpha + U,_\alpha) + A_{3,\beta} \cdot A_\alpha \tag{1.119}$$

Let $A_3 - R = a_3$ where $R = R_\rho A^\rho + R_3 A^3 = R^\rho A_\rho + R^3 A_3$, then

$$K_{\alpha\beta} = -(A_3 - R),_\beta \cdot (A_\alpha + U,_\alpha) + A_{3,\beta} \cdot A_\alpha = -A_{3,\beta} \cdot U,_\alpha + R,_\beta \cdot A_\alpha + R,_\beta \cdot U,_\alpha \tag{1.120}$$

From $a_3 \cdot a_3 = 1 = (A_3 - R) \cdot (A_3 - R) = 1 - 2A_3 \cdot R + R \cdot R$, assuming $R \cdot R \simeq 0$ we deduce that $A_3 \cdot R = R^3 \approx 0 \cdots and \cdots R \simeq R_\gamma A^\gamma$. We have

$$a_3 \cdot a_\alpha = (A_3 - R) \cdot (A_\alpha + U,_\alpha) = 0 \tag{1.121}$$

$$A_3 \cdot U,_\alpha - R \cdot A_\alpha - R \cdot U,_\alpha = 0 \tag{1.122}$$

By neglecting $R \cdot U_{,\alpha}$, $R_{,\beta} \cdot U_{,\alpha}$ we deduce that $A_3 \cdot U,_\alpha \approx R \cdot A_\alpha$. Consider $U_{3,\alpha} + B_{\lambda\alpha} U^\lambda = R_\alpha$ and $K_{\alpha\beta} \simeq -A_{3,\beta} \cdot U,_\alpha + R,_\beta \cdot A_\alpha$. We denote $U_{3,\alpha} = \nabla_\alpha U_3$. We deduce that the rotation vector of the normal of the surface

$$R = R_\alpha A^\alpha = (\nabla_\alpha U_3 + B_{\lambda\alpha} U^\lambda) A^\alpha \tag{1.123}$$

is plane. We should express $K_{\alpha\beta}$ as a function of the displacement vector. We have

$$U,_\beta = (\nabla_\beta U_\rho - B_{\rho\beta} U^3) A^\rho + (U_3,_\beta + B^\lambda_\beta U_\lambda) A_3 \tag{1.124}$$

$$R,_\beta = (\nabla_\beta R_\alpha - B_{\alpha\beta} R^3) A^\alpha + (R_3,_\beta + B_{\beta\rho} R^\rho) A^3 \approx (\nabla_\beta R_\alpha) A^\alpha + (B_{\beta\rho} R^\rho) A^3 \tag{1.125}$$

$$K_{\alpha\beta} = \nabla_\alpha (\nabla_\beta U_3 + B^\rho_\beta U_\rho) + B^\rho_\alpha (\nabla_\beta U_\rho - U_3 B_{\beta\rho}) + ... \tag{1.126}$$

So, $K_{\alpha\beta} = K^l_{\alpha\beta} + ...$ nonlinear terms, with

$$K^l_{\alpha\beta} = \nabla_\alpha (\nabla_\beta U_3 + B^\rho_\beta U_\rho) + B^\rho_\alpha (\nabla_\beta U_\rho - U_3 B_{\beta\rho}) \tag{1.127}$$

or

$$
\begin{aligned}
K^{l}_{\alpha\beta} = & \quad \nabla_\alpha \nabla_\beta U_3 + \nabla_\alpha (B^{\rho}_{\beta} U_\rho) + B^{\rho}_{\alpha} \nabla_\beta U_\rho - U_3 B^{\rho}_{\beta} B_{\rho\alpha} \\
& = \nabla_\alpha \nabla_\beta U_3 + U_\rho \nabla_\alpha B^{\rho}_{\beta} + B^{\rho}_{\beta} \nabla_\alpha U_\rho + B^{\rho}_{\alpha} \nabla_\beta U_\rho - U_3 B^{\rho}_{\alpha} B_{\beta\rho}
\end{aligned}
\tag{1.128}
$$

We also have

$$
\begin{aligned}
K^{i}_{\beta\alpha} = & \quad \nabla_\beta (\nabla_\alpha U_3 + B^{\rho}_{\alpha} U_\rho) + B^{\rho}_{\beta} (\nabla_\alpha U_\rho - U_3 B_{\alpha\rho}) \\
& = \nabla_\beta \nabla_\alpha U_3 + U_\rho \nabla_\beta B^{\rho}_{\alpha} + B^{\rho}_{\alpha} \nabla_\beta U_\rho + B^{\rho}_{\beta} \nabla_\alpha U_\rho - U_3 B^{\rho}_{\beta} B_{\alpha\rho}
\end{aligned}
\tag{1.129}
$$

We deduce from the relations established by Mainardi-Codazzi, $\nabla_\beta B^{\rho}_{\alpha} = \nabla_\alpha B^{\rho}_{\beta}$, that

$$
K^{l}_{\alpha\beta} = K^{l}_{\beta\alpha} \tag{1.130}
$$

Only the linear part will be considered in the subsequent analysis

$$
\begin{aligned}
K_{\alpha\beta} &= \nabla_\alpha \nabla_\beta U_3 + U_\rho \nabla_\alpha B^{\rho}_{\beta} + B^{\rho}_{\beta} \nabla_\alpha U_\rho + B^{\rho}_{\alpha} \nabla_\beta U_\rho - U_3 B^{\rho}_{\alpha} B_{\beta\rho} \\
&= \quad \nabla_\alpha (\nabla_\beta U_3 + B^{\rho}_{\beta} U_\rho) + B^{\rho}_{\alpha} (\nabla_\beta U_\rho - U_3 B_{\beta\rho}) \ldots\ldots \\
&= \quad \nabla_\alpha R_\beta + B^{\rho}_{\alpha} s_{\rho\beta}
\end{aligned}
\tag{1.131}
$$

where $s_{\rho\beta}$ is strain. We remark that the change of curvature is made of two components, the variation of the rotation or angle vector and the ratio of the variation of a line element ($s_{\rho\beta}$) over the radius of curvature. The theory of strength of material only accounts for the first term, which is the second derivative of the transverse displacement.

Example: Let us consider a sphere with radius R subject to a uniform radial displacement w. Its curvature tensors before and after deformation are

$$
B^{\cdot}_{\cdot} = \begin{bmatrix} 1/R & 0 \\ 0 & 1/R \end{bmatrix}, \quad b^{\cdot}_{\cdot} = \begin{bmatrix} 1/(R+w) & 0 \\ 0 & 1/(R+w) \end{bmatrix} \tag{1.132}
$$

But $1/(R+w) = (R(1+w/R))^{-1} \cong (1-w/R)/R$ and $1/(R+w) \cong 1/R - w/R^2$. We deduce that

$$
K_{..} = b_{..} - B_{..} = \begin{bmatrix} -w/R^2 & 0 \\ 0 & -w/R^2 \end{bmatrix} \tag{1.133}
$$

The result is obtained by applying directly the formula. In the same way, we obtain

$$
(e_{\alpha\beta}) = \begin{bmatrix} -w/R & 0 \\ 0 & -w/R \end{bmatrix}
$$

Exercise: Show the Minardi-Codazzi relations.

Variation of the third fundamental form (or Gauss curvature tensor)
The third fundamental form is defined as follows:

$$C_{\alpha\beta} = B^{\rho}_{\alpha} B_{\rho\beta} = A^{\rho\gamma} B_{\alpha\gamma} B_{\rho\beta} = B_{\alpha\gamma} A^{\rho\gamma} B_{\rho\beta} = B_{\alpha\gamma} B^{\gamma}_{\beta} = B^{\gamma}_{\beta} B_{\alpha\gamma} = C_{\beta\alpha}$$

It is a symmetric tensor. The variation (or change) of the third fundamental form is defined by

$$Q_{\alpha\beta} = \frac{1}{2}\left(c_{\alpha\beta} - C_{\alpha\beta}\right) = \frac{1}{2}\left(b^{\rho}_{\alpha} b_{\rho\beta} - B^{\rho}_{\alpha} B_{\rho\beta}\right)$$

we have

$$b^{\rho}_{\alpha} = a^{\rho\gamma} b_{\alpha\gamma} = (A^{\rho\gamma} + E^{\rho\gamma})(B_{\alpha\gamma} + K_{\alpha\gamma}), \; b_{\rho\beta} = (B_{\rho\beta} + K_{\rho\beta}),$$

$$\begin{aligned} c_{\alpha\beta} &= b^{\rho}_{\alpha} b_{\rho\beta} = (A^{\rho\gamma} + E^{\rho\gamma})(B_{\alpha\gamma} + K_{\alpha\gamma})(B_{\rho\beta} + K_{\rho\beta}) \\ &= A^{\rho\gamma} B_{\alpha\gamma} B_{\rho\beta} + A^{\rho\gamma} K_{\alpha\gamma} B_{\rho\beta} + A^{\rho\gamma} B_{\alpha\gamma} K_{\rho\beta} + \ldots \\ &= C_{\alpha\beta} + B^{\gamma}_{\beta} K_{\alpha\gamma} + B^{\gamma}_{\alpha} K_{\beta\gamma} + .. \end{aligned} \tag{1.134}$$

and

$$Q_{\alpha\beta} = \frac{1}{2}\left(c_{\alpha\beta} - C_{\alpha\beta}\right) = \frac{1}{2}(B^{\gamma}_{\beta} \nabla_{\alpha} R_{\gamma} + B^{\gamma}_{\alpha} \nabla_{\beta} R_{\gamma}) + \ldots.. \tag{1.135}$$

Only the linear part will be used in the future analyses.

$$Q_{\alpha\beta} = \frac{1}{2}\left(B^{\delta}_{\alpha} \nabla_{\beta}(\nabla_{\delta} u_3 + B^{\rho}_{\delta} u_{\rho}) + B^{\delta}_{\beta} \nabla_{\alpha}(\nabla_{\delta} u_3 + B^{\rho}_{\delta} u_{\rho})\right) \tag{1.136}$$

We remark that this tensor, referred to hereafter as the Gauss deformation tensor, is the product of the curvature tensor and the variation of the rotational vector (Fig. 1.9).
Correction of the exercise on Mainardi-Codazzi relations
We have $A_{3,\alpha} = -B^{\rho}_{\alpha} A_{\rho}$ and

$$\begin{aligned} A_{3,\alpha\beta} &= -B^{\rho}_{\alpha,\beta} A_{\rho} - B^{\rho}_{\alpha} A_{\rho,\beta} \\ &= -B^{\rho}_{\alpha,\beta} A_{\rho} - B^{\rho}_{\alpha} \Gamma^{\gamma}_{\rho\beta} A_{\gamma} - B^{\rho}_{\alpha} B_{\rho\beta} A_3 \\ &= -(B^{\gamma}_{\alpha,\beta} + B^{\rho}_{\alpha} \Gamma^{\gamma}_{\rho\beta}) A_{\gamma} - B^{\rho}_{\alpha} B_{\rho\beta} A_3 \end{aligned} \tag{1.137}$$

Similarly

$$\begin{aligned} A_{3,\beta\alpha} &= -(B^{\rho}_{\beta} A_{\rho})_{,\alpha} = -B^{\rho}_{\beta,\alpha} A_{\rho} - B^{\rho}_{\beta} \Gamma^{\gamma}_{\rho\alpha} A_{\gamma} - B^{\rho}_{\beta} B_{\rho\alpha} A_3 \\ &= -(B^{\gamma}_{\beta,\alpha} + B^{\rho}_{\beta} \Gamma^{\gamma}_{\rho\alpha}) A_{\gamma} - B^{\rho}_{\beta} B_{\rho\alpha} A_3 \end{aligned} \tag{1.138}$$

From $A_{3,\alpha\beta} = A_{3,\beta\alpha}$, we deduce that

$$B^{\gamma}_{\alpha,\beta} + B^{\rho}_{\alpha}\Gamma^{\gamma}_{\rho\beta} = B^{\gamma}_{\beta,\alpha} + B^{\rho}_{\beta}\Gamma^{\gamma}_{\rho\alpha} \tag{1.139}$$

and adding on both sides $-B^{\gamma}_{\rho}\Gamma^{\rho}_{\alpha\beta}$ which is equal to $-B^{\gamma}_{\rho}\Gamma^{\rho}_{\beta\alpha}$, we obtain

$$B^{\gamma}_{\alpha,\beta} + B^{\rho}_{\alpha}\Gamma^{\gamma}_{\rho\beta} - B^{\gamma}_{\rho}\Gamma^{\rho}_{\alpha\beta} = B^{\gamma}_{\beta,\alpha} + B^{\rho}_{\beta}\Gamma^{\gamma}_{\rho\alpha} - B^{\gamma}_{\rho}\Gamma^{\rho}_{\beta\alpha} \tag{1.140}$$

so

$$\nabla_{\beta} B^{\rho}_{\alpha} = \nabla_{\alpha} B^{\rho}_{\beta} \tag{1.141}$$

We also deduce that $\nabla_{\beta} B_{\rho\alpha} = \nabla_{\alpha} B_{\rho\beta}$ because $B^{\rho}_{\beta} = A^{\rho\gamma} B_{\gamma\beta}$ and

$$\begin{aligned} \nabla_{\alpha} B^{\rho}_{\beta} &= \nabla_{\alpha}(A^{\rho\gamma} B_{\gamma\beta}) = \nabla_{\alpha}(A^{\rho\gamma}) B_{\gamma\beta} + (A^{\rho\gamma})\nabla_{\alpha} B_{\gamma\beta} \\ &= \quad (A^{\rho\gamma})\nabla_{\alpha} B_{\gamma\beta} = \nabla_{\beta} B^{\rho}_{\alpha} = (A^{\rho\gamma})\nabla_{\beta} B_{\gamma\alpha} \end{aligned} \tag{1.142}$$

It should be remarked that the same results are obtained by using $A_{\alpha,\beta\lambda} = A_{\alpha,\lambda\beta}$. Indeed we have

$$A_{\alpha,\beta\lambda} - A_{\alpha,\lambda\beta} = (R^{\rho}_{\alpha\lambda\beta} + B_{\alpha\lambda} B^{\rho}_{\beta} - B_{\alpha\beta} B^{\rho}_{\lambda}) A_{\rho} + (\nabla_{\lambda} B_{\alpha\beta} - \nabla_{\beta} B_{\alpha\lambda}) A_3, \tag{1.143}$$

$$R^{\rho}_{\alpha\lambda\beta} = \Gamma^{\rho}_{\alpha\beta,\lambda} - \Gamma^{\rho}_{\alpha\lambda,\beta} + \Gamma^{\rho}_{\lambda\delta}\Gamma^{\delta}_{\alpha\beta} - \Gamma^{\rho}_{\beta\delta}\Gamma^{\delta}_{\alpha\lambda} \tag{1.144}$$

From the above equality, we deduce that

$$\begin{aligned} &\nabla_{\lambda} B_{\alpha\beta} - \nabla_{\beta} B_{\alpha\lambda} = 0, \qquad R^{\rho}_{\alpha\lambda\beta} = B_{\alpha\beta} B^{\rho}_{\lambda} - B_{\alpha\lambda} B^{\rho}_{\beta} \\ &A_{\rho\gamma} R^{\rho}_{\alpha\lambda\beta} = R_{\gamma\alpha\lambda\beta} = B_{\alpha\beta} B_{\gamma\lambda} - B_{\alpha\lambda} B_{\gamma\beta}, \quad A_{\rho 1} R^{\rho}_{212} = R_{1212} = det B_{..} \end{aligned} \tag{1.145}$$

This relation shows that Gauss curvature

$$\bar{G} = det(B^{\alpha}_{\beta}) = \frac{det B_{..}}{A} = \frac{A_{\rho 1} R^{\rho}_{212}}{A} = \frac{A_{\rho 1}}{A}(\Gamma^{\rho}_{22,1} - \Gamma^{\rho}_{21,2} + \Gamma^{\rho}_{1\delta}\Gamma^{\delta}_{22} - \Gamma^{\rho}_{2\delta}\Gamma^{\delta}_{21}) \tag{1.146}$$

depends only on the metric tensor. In other words, it means that a surface can change shape with an invariant Gauss curvature provided its metric doesn't change. For example, a plane metal sheet can be transformed into a cylindrical surface without any strain (or change of metric). The Gauss curvature remains invariant before and after transformation (Fig. 1.10).

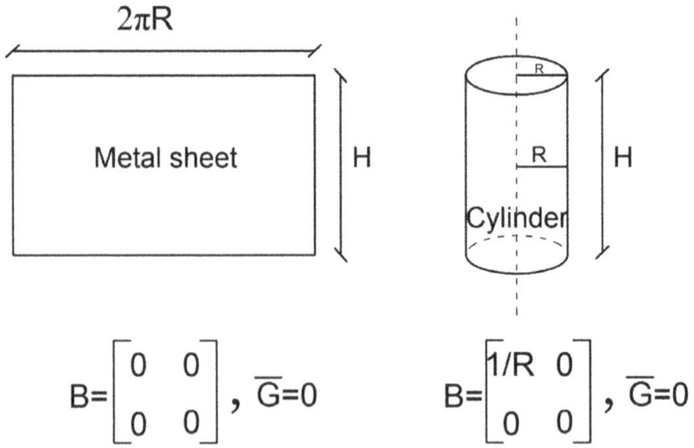

Fig. 1.9 Transformation of a plane surface into a cylindrical surface

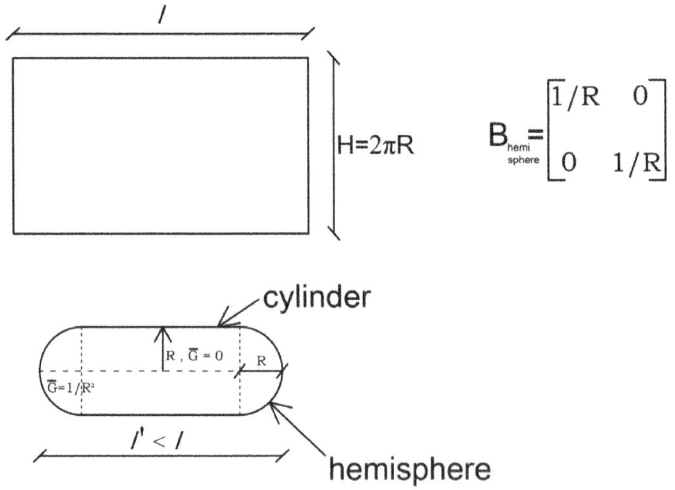

Fig. 1.10 Cylindrical tank with hemispherical tips

Examples

- *Cylindrical surface with radius R:* $\left\{\overrightarrow{om} = (Rcos\theta,\ Rsin\theta,\ x)\right\}$

The coordinates are $(x,\ R\theta) = (x_1,\ x_2)$

$$A_1 = \begin{pmatrix} 0 \\ 0 \\ 1 \end{pmatrix}\ A_2 = \begin{pmatrix} -sin\theta \\ cos\theta \\ 0 \end{pmatrix},\ A_3 = \begin{pmatrix} -cos\theta \\ -sin\theta \\ 0 \end{pmatrix}, \tag{1.147}$$

$$A_{1,1} = \begin{pmatrix} 0 \\ 0 \\ 0 \end{pmatrix}, \; A_{1,2} = \begin{pmatrix} 0 \\ 0 \\ 0 \end{pmatrix}, \; A_{2,1} = \begin{pmatrix} 0 \\ 0 \\ 0 \end{pmatrix}$$

$$A_{2,2} = \frac{1}{R} \begin{pmatrix} -cos\theta \\ -sin\theta \\ 0 \end{pmatrix}, \; \Gamma^1_{\alpha\beta} = \begin{bmatrix} 0 & 0 \\ 0 & 0 \end{bmatrix}, \; \Gamma^2_{\alpha\beta} = \begin{bmatrix} 0 & 0 \\ 0 & 0 \end{bmatrix}$$

$$B_{..cyl} = \begin{bmatrix} -\frac{1}{R} & 0 \\ 0 & \frac{1}{\infty} \end{bmatrix} = \begin{bmatrix} -\frac{1}{R} & 0 \\ 0 & 0 \end{bmatrix}, \; \nabla_\alpha u_\beta = u_{\beta,\alpha}, \tag{1.148}$$

$$(e_{\alpha\beta}) = \frac{1}{2} \begin{bmatrix} 2u_{1,x} & u_{2,1} + \frac{1}{R} u_{1,\theta} \\ & \\ * & \frac{2}{R}(u_{2,\theta} + u_3) \end{bmatrix},$$

$$(K_{\alpha\beta}) = \begin{bmatrix} u_{3,xx} & \frac{1}{R}(u_{3,x\theta} - u_{2,x}) \\ & \\ * & -\frac{1}{R^2}(2u_{2,\theta} + u_3 - u_{3,\theta\theta}) \end{bmatrix},$$

$$(Q_{\alpha\beta}) = \frac{1}{2} \begin{bmatrix} 0 & \frac{1}{R^2}(u_{2,x} - u_{3,x\theta}) \\ * & -\frac{2}{R^3}(u_{2,\theta} - u_{3,\theta\theta}) \end{bmatrix}$$

- *Spherical surface with radius R:*

$$\left\{\overrightarrow{om} = R\,(Sin\theta_1 Cos\theta_2,\; sin\theta_1 sin\theta_2,\; cos\theta_1)\right\}$$

The coordinates are $(\theta_1, \theta_2) = (x_1, x_2)$

$$A_1 = R \begin{pmatrix} cos\theta_1 \; cos\theta_2 \\ cos\theta_1 \; sin\theta_2 \\ - \; sin\theta_1 \end{pmatrix}, \; A_2 = R \begin{pmatrix} -sin\theta_1 sin\theta_2 \\ sin\theta_1 cos\theta_2 \\ 0 \end{pmatrix},$$

$$A_3 = \begin{pmatrix} sin\theta_1 \; cos\theta_2 \\ sin\theta_1 \; sin\theta_2 \\ cos\theta_1 \end{pmatrix};$$

$$(\Gamma^1_{\alpha\beta}) = \begin{bmatrix} 0 & 0 \\ 0 & -sin\theta_1 cos\theta_1 \end{bmatrix}, \; (\Gamma^2_{\alpha\beta}) = \begin{bmatrix} 0 & cotg\theta_1 \\ cotg\theta_1 & 0 \end{bmatrix}$$

$$B_{..sph} = \begin{bmatrix} -R & 0 \\ 0 & -Rsin^2\theta_1 \end{bmatrix}, \quad \bar{B}^{\cdot}_{\cdot sph} = \begin{bmatrix} -\dfrac{1}{R} & 0 \\ 0 & -\dfrac{1}{R} \end{bmatrix},$$

$$(\nabla_\alpha u_\beta) = \begin{bmatrix} u_{1,1} & u_{1,2} - u_2 cotg\theta_1 \\ u_{2,1} - u_2 cotg\theta_1 & u_{2,2} + u_1 sin\theta_1 cos\theta_1 \end{bmatrix},$$

$$(\nabla_{\alpha\beta} u_3) = \begin{bmatrix} u_{3,11} & u_{3,12} - u_{3,2} cotg\theta_1 \\ & u_{3,22} + u_{3,1} sin\theta_1 cos\theta_1 \end{bmatrix},$$

$$(e_{\alpha\beta}) = \frac{1}{2} \begin{bmatrix} 2u_{1,1} + 2Ru_3 & u_{2,1} + u_{1,2} - 2u_2 cotg\theta_1 \\ * & 2(u_{2,2} + u_3 Rsin^2\theta_1) \end{bmatrix},$$

$$(K_{\alpha\beta}) = \begin{bmatrix} u_{3,11} - u_3 - \dfrac{2}{R} u_{1,1} & \dfrac{2}{R} u_2 cotg\theta_1 - \dfrac{1}{R}(u_{2,1} + u_{1,2}) \\ & +u_{3,12} - u_{3,2} cotg\theta_1 \\ * & -\dfrac{2}{R}(u_{2,2} + u_1 sin\theta_1 cos\theta_1) + \\ & u_{3,22} + u_{3,1} sin\theta_1 cos\theta_1 - u_3 sin\theta_1^2 \end{bmatrix}$$

$$(Q_{\alpha\beta}) = \frac{1}{2} \begin{bmatrix} -\dfrac{u_{3,11}}{R} + \dfrac{u_{1,1}}{R^2} & \dfrac{1}{R^2}(u_{1,2} - 2u_2 cotg\theta_1 + u_{2,1}) \\ & -\dfrac{2}{R}(u_{3,12} - u_{3,2} cotg\theta_1) \\ * & \dfrac{1}{R^2}(u_{2,2} + u_1 sin\theta_1 cos\theta_1) \\ & -\dfrac{1}{R}(u_{3,22} + u_{3,1} sin\theta_1 cos\theta_1) \end{bmatrix}$$

By using the curvilinear abscissas $(R\theta_1,\ Rsin\theta_1) = (x_1,\ x_2)$ in its physical base (orthonormal base) $\{\bar{a}_\alpha,\ a_3\}$, we deduce as above the following relations:

$$\bar{e}_{\alpha\beta} = e_{\alpha\beta}/r_\alpha r_\beta,\ \bar{K}_{\alpha\beta} = K_{\alpha\beta}/r_\alpha r_\beta,\ \bar{Q}_{\alpha\beta} = Q_{\alpha\beta}/r_\alpha r_\beta,\ \bar{u}_\alpha = u_\alpha/r_\alpha,$$
$$\bar{u}_3 = u_3,\ d/d\theta_\alpha = r_\alpha d/dx_\alpha \qquad \bar{\Gamma}^\rho_{\alpha\beta} = \frac{r_\rho}{r_\alpha r_\beta}\Gamma^\rho_{\alpha\beta} \tag{1.149}$$

$$(\bar{e}_{\alpha\beta}) = \frac{1}{2}\begin{bmatrix} 2\bar{u}_{1,1} + \frac{2}{R}u_3 & \bar{u}_{2,1} + \bar{u}_{1,2} - \frac{2}{R}\bar{u}_2 cotg\theta_1 \\ * & 2(\bar{u}_{2,2} + \frac{u_3}{R}) \end{bmatrix},$$

$$(\bar{K}_{\alpha\beta}) = \begin{bmatrix} u_{3,11} - \frac{u_3}{R^2} - \frac{2}{R}\bar{u}_{1,1} & \begin{matrix} \frac{2}{R^2}\bar{u}_2 cotg\theta_1 - \frac{1}{R}(\bar{u}_{2,1} + \bar{u}_{1,2}) \\ +u_{3,12} - \frac{1}{R}u_{3,2} cotg\theta_1 \end{matrix} \\ * & \begin{matrix} -\frac{2}{R}(\bar{u}_{2,2} + \frac{1}{R}\bar{u}_1 cotg\theta_1) \\ +u_{3,22} + \frac{1}{R}u_{3,1} cotg\theta_1 - \frac{u_3}{R^2} \end{matrix} \end{bmatrix} \quad (1.150)$$

$$(\bar{Q}_{\alpha\beta}) = \frac{1}{2}\begin{bmatrix} -\frac{u_{3,11}}{R} + \frac{\bar{u}_{1,1}}{R^2} & \begin{matrix} \frac{1}{R^2}(\bar{u}_{1,2} - \frac{2}{R}\bar{u}_2 cotg\theta_1 + \bar{u}_{2,1}) \\ -\frac{2}{R}(u_{3,12} - \frac{1}{R}u_{3,2} cotg\theta_1) \end{matrix} \\ * & \begin{matrix} \frac{1}{R^2}(\bar{u}_{2,2} + \frac{1}{R}\bar{u}_1 cotg\theta_1) \\ -\frac{1}{R}(u_{3,22} + \frac{1}{R}u_{3,1} cotg\theta_1) \end{matrix} \end{bmatrix}$$

Chapter 2
Equilibrium Equations

2.1 Geometry of a Shell

2.1.1 Description and Covariant Base

Description of a shell

We define in an Euclidean space base of origin O, a shell with mid-surface S and thickness h, by $C = \left\{\overrightarrow{OM} = \overrightarrow{Om} + zA^3, \ldots m \in S, \ldots z \in \left]-\frac{h}{2}, \frac{h}{2}\right[\right\} = S \times \left]-\frac{h}{2}, \frac{h}{2}\right[$ The coordinates of a generic point m on the mid-surface S are $x = (x_1, x_2)$ and $z \in I = \left]-\frac{h}{2}, \frac{h}{2}\right[$. The system $\{A_\alpha, A_3\}$ constitutes the (Fig. 2.1) covariant base of S.

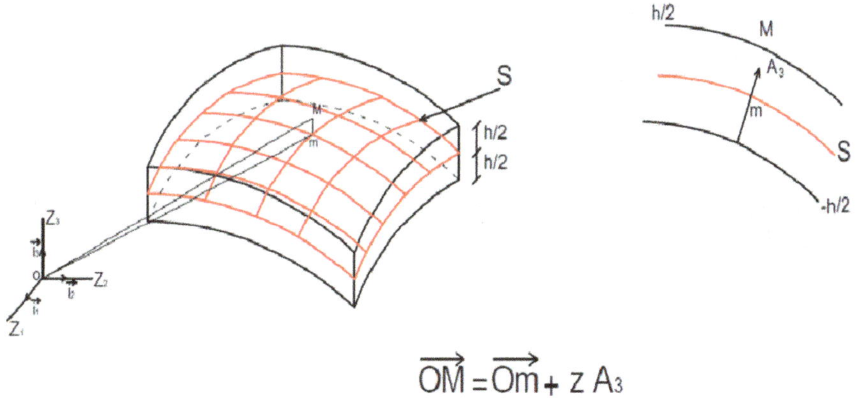

Fig. 2.1 Shell with mid-surface

R. Nzengwa, *Plate and Shell Models*,
https://doi.org/10.1007/978-981-97-2780-3_2

Covariant base of the shell:

We have

$$G_\alpha = \overrightarrow{OM},_\alpha = \overrightarrow{Om},_\alpha + zA_{3,\alpha} = A_\alpha - zB_\alpha^\lambda A_\lambda = \left(\delta_\alpha^\lambda - zB_\alpha^\lambda\right) A_\lambda = \mu_\alpha^\lambda A_\lambda,$$

$$G_3 = \overrightarrow{OM},_3 = \frac{\partial \overrightarrow{OM}}{\partial z} = A_3$$

The set $\{G_1, G_2, G_3\}$ constitutes a base if the vectors G_1, G_2 are independent, i.e. $G_1 \times G_2 \neq \overrightarrow{0}$ or $det\left(\mu_\alpha^\lambda\right) \neq 0$. By developing we have

$$\left(\mu_\alpha^\lambda\right) = \left(I - zB^\cdot_\cdot\right) = \begin{bmatrix} 1 - zB_1^1 & -zB_1^2 \\ -zB_2^1 & 1 - zB_2^2 \end{bmatrix} \tag{2.1}$$

and

$$\begin{aligned} det\left(\mu_\alpha^\lambda\right) = det\left(I - zB^\cdot_\cdot\right) &= \left(1 - zB_1^1\right)\left(1 - zB_2^2\right) - z^2 B_2^1 B_1^2 \\ &= 1 - z\left(B_1^1 + B_2^2\right) + z^2\left(B_1^1 B_2^2 - B_2^1 B_1^2\right) \\ &= 1 - ztr\,B + z^2 det\,B = 1 - 2z\bar{H} + z^2\bar{G} \end{aligned} \tag{2.2}$$

It is necessary that $det\left(\mu_\alpha^\lambda\right) > 0$. In the diagonal base of $B_{\cdot\cdot}$, we have

$$det\left(\mu_\alpha^\lambda\right) = det\begin{pmatrix} 1 - \frac{z}{R_1} & 0 \\ 0 & 1 - \frac{z}{R_2} \end{pmatrix} = \left(1 - \frac{z}{R_1}\right)\left(1 - \frac{z}{R_2}\right) \tag{2.3}$$

Let $\chi = \dfrac{max \mid z \mid}{min\{\mid R_1 \mid, \mid R_2 \mid\}} = h/2r$. If $\chi < 1$, then

$$det\left(\mu_\alpha^\lambda\right) = \left(1 - \frac{z}{R_1}\right)\left(1 - \frac{z}{R_2}\right) > (1 - \chi)^2 > 0 \tag{2.4}$$

and $\{G_1, G_2, A_3\}$ constitutes a covariant base of the shell. If $\chi < 0, 1$ (for example), the shell is said to be thin.

Metric tensor

We have

$$(G_{ij}) = \begin{bmatrix} G_{11} & G_{12} & 0 \\ G_{21} & G_{22} & 0 \\ 0 & 0 & 1 \end{bmatrix}, \quad G_{\alpha\beta} = \mu_\alpha^\nu A_\nu \cdot \mu_\beta^\rho A_\rho = A_{\alpha\beta} - 2zB_{\alpha\beta} + z^2 C_{\alpha\beta}, \tag{2.5}$$

$$det(G_{ij}) = G = det\left(\mu_\alpha^\nu\right)^2 A, \quad A = det(A_{\alpha\beta}) \tag{2.6}$$

The determinant being positive, we deduce that (G_{ij}) defines a metric. We define

$$G^\alpha = (\mu_\rho^\alpha)^{-1} A^\rho = (\mu^{-1})_\rho^\alpha A^\rho \tag{2.7}$$

The system $\{G^1, G^2, G^3 = A^3\}$ constitutes the contravariant base of the shell. From Cayley-Hamilton's theorem, $(\mu_\rho^\alpha)^{-1}$ is expanded in series as follows:

$$(\mu_\rho^\alpha)^{-1} = \sum_{n=0}^{\infty} z^n B^n = I + zB + z^2B^2 + ... + z^nB^n + ... \tag{2.8}$$

$B^0 = I, \ B^1 = B, \ B^n = BB^{n-1} = B^{n-1}B$, and deduce that the matrix commutes with B. We deduce that

$$G_{\alpha,\beta} = (\mu_\alpha^\nu A_\nu),_\beta = \mu_\alpha^\nu A_{\nu,\beta} + (\mu_\alpha^\nu),_\beta A_\nu = \mu_\alpha^\nu A_{\nu,\beta} - zB_{\alpha,\beta}^\nu A_\nu \tag{2.9}$$

$$G_{\alpha,3} = (\mu_\alpha^\nu A_\nu),_3 = (\mu_{\alpha,3}^\nu) A_\nu + \mu_\alpha^\nu A_{\nu,3} = -B_\alpha^\nu A_\nu \tag{2.10}$$

2.1.2 Relations Between 3D and 2D Christoffel Symbols

We have $G_{i,j} = \Gamma_{ij}^k G_k$, $\Gamma_{ij}^k = G_{i,j}.G^k$. We shall denote 2D Christoffel symbols (of the mid-surface) by $\bar{\Gamma}_{\alpha\beta}^\gamma$. Then

$$A_{\alpha,\beta} = \bar{\Gamma}_{\alpha\beta}^\gamma A_\gamma + B_{\alpha\beta} A_3$$

$$\Gamma_{\alpha\beta}^\gamma = G_{\alpha,\beta}.G^\gamma = \bar{\Gamma}_{\alpha\beta}^\gamma + \left(\mu^{-1}\right)_\nu^\gamma \nabla_\beta \mu_\alpha^\nu$$

$$\Gamma_{\beta 3}^\alpha = -\left(\mu^{-1}\right)_\lambda^\alpha B_\beta^\lambda, \ \Gamma_{\alpha\beta}^3 = \mu_\alpha^\nu B_{\nu\beta}, \ \Gamma_{3\beta}^3 = \Gamma_{33}^\alpha = \Gamma_{33}^3 = 0$$

Relations between vectors expressed in the bases $\{G_1, G_2, G_3\}$ and $\{A_1, A_2, A_3\}$

Let $T = T_i G^i = T_\alpha G^\alpha + T_3 A^3 = T^\alpha G_\alpha + T^3 A_3$ and $T = \bar{T}^\alpha A_\alpha + \bar{T}^3 A_3 = \bar{T}_\alpha A^\alpha + \bar{T}_3 A^3$. But $T_\alpha G^\alpha = T_\alpha \left(\mu^{-1}\right)^\alpha_\nu A^\nu = \overline{T}_\nu A^\nu$, $T^\alpha G_\alpha = T^\alpha \mu^\nu_\alpha A_\nu = \overline{T}^\nu A_\nu$ and $T^3 = T_3 = \overline{T}^3 = \overline{T}_3$. It follows that

$$T_\alpha = \mu^\nu_\alpha \overline{T}_\nu,\ T^\alpha = \left(\mu^{-1}\right)^\alpha_\nu \overline{T}^\nu,\ \overline{T}_\alpha = \left(\mu^{-1}\right)^\nu_\alpha T_\nu \text{ and } \overline{T}^\alpha = \mu^\alpha_\nu T^\nu$$

Relations between 3D and 2D covariant derivations

From $T_{i/j} = T_{i,j} - \Gamma^k_{ij} T_k$, we deduce the following relations:

$$T_{\alpha/\beta} = T_{\alpha,\beta} - \Gamma^\lambda_{\alpha\beta} T_\lambda - \Gamma^3_{\alpha\beta} T_3 = \mu^\nu_\alpha \left[\nabla_\beta \overline{T}_\nu - B_{\nu\beta} \overline{T}^3\right]$$

$$T^\alpha_{/\beta} = T^\alpha_{,\beta} + \Gamma^\alpha_{\beta\lambda} T^\lambda + \Gamma^\alpha_{\beta 3} T^3,\ T^\alpha_{/\beta} = \left(\mu^{-1}\right)^\alpha_\nu \left[\nabla_\beta \overline{T}^\nu - B^\nu_\beta \overline{T}^3\right]$$

$$T_{\alpha/3} = \mu^\nu_\alpha \overline{T}_{\nu,3};\ T_{3/\alpha} = \overline{T}_{3,\alpha} + B^\lambda_\alpha \overline{T}_\lambda$$

$$T^\alpha_{/3} = \left(\mu^{-1}\right)^\alpha_\nu \overline{T}^\nu_{,3};\ \ldots T^3_{/\alpha} = \overline{T}^3_{,\alpha} + B_{\alpha\lambda} \overline{T}^\lambda$$

$$T^3_{/3} = T_{3/3} = T_{3,3} = \overline{T}^3_{,3} = \overline{T}_{3,3}$$

Infinitesimal deformation tensor

We have $\left(\epsilon_{ij}\right) = \frac{1}{2}(g_{ij} - G_{ij})$. When developed, we obtain

$$\epsilon_{\alpha\beta} = \frac{1}{2}(g_{\alpha\beta} - G_{\alpha\beta}) = \frac{1}{2}(a_{\alpha\beta} - A_{\alpha\beta}) - z(b_{\alpha\beta} - B_{\alpha\beta}) + z^2 \frac{1}{2}(c_{\alpha\beta} - C_{\alpha\beta})$$

But $g_\alpha = G_\alpha + U_{,\alpha} = G_\alpha + U_{\rho/\alpha}G^\rho + U_{3/\alpha}A^3$ and $g_\beta = G_\beta + U_{,\beta} = G_\beta + U_{\rho/\beta}G^\rho + U_{3/\beta}A^3$ and when linearized we obtain $\epsilon = (\epsilon_{ij}) = \frac{1}{2}(U_{i/j} + U_{j/i})$, i.e.

$$\epsilon_{\alpha\beta} = \frac{1}{2}\left(U_{\alpha/\beta} + U_{\beta/\alpha}\right) = \frac{1}{2}\left[\mu_\alpha^\nu\left(\nabla_\beta \overline{U}_\nu - B_{\alpha\beta}\overline{U}_3\right) + \mu_\beta^\nu\left(\nabla_\alpha \overline{U}_\nu - B_{\nu\alpha}\overline{U}_3\right)\right]$$

$$\epsilon_{\alpha 3} = \frac{1}{2}\left(U_{\alpha/3} + U_{3/\alpha}\right) = \frac{1}{2}(\mu_\alpha^\nu \overline{U}_{\nu,3} + \overline{U}_{3,\alpha} + B_\alpha^\nu \overline{U}_\nu)$$

$$\epsilon_{33} = U_{3/3} = U_{3,3} = \overline{U}_{3,3}$$

Hypotheses: The constitutive material is homogenous, isotropic and the strain is plane

Anisotropic materials will be treated further. We recall that the kinematic (or displacement) that satisfies $(\epsilon_{i3} = 0)$ was rigorously deduced by Nzengwa and Tagne [87], from the 3D elastic shell problem, by applying the multi-scale dilatation and asymptotic analysis of mathematical theory, on the partial differential equations obtained, without any ad-hoc hypotheses. Then the stress-strain relation follows the classical constitutive law

$$\sigma^{ij} = \bar{\lambda}\epsilon_k^k G^{ij} + 2\bar{\mu}\epsilon^{ij} \text{ or } \epsilon^{ij} = \frac{1+\bar{\nu}}{E}\sigma^{ij} - \frac{\bar{\nu}}{E}\sigma_k^k G^{ij}$$

E is Young's modulus and $\bar{\nu}$ Poisson's ratio. We deduce from the above hypotheses that

$$\epsilon_{\alpha 3} = \frac{1}{2}\left[\mu_\alpha^\nu \overline{U}_{\nu,3} + \overline{U}_{3,\alpha} + B_\alpha^\nu \overline{U}_\nu\right] = 0;\ \epsilon_{33} = U_{3,3} = \overline{U}_{3,3} = 0$$

and

$$\begin{aligned}\sigma^{\alpha\beta} &= \tfrac{2\bar{\mu}\bar{\lambda}}{\bar{\lambda}+2\bar{\mu}}\epsilon_\rho^\rho G^{\alpha\beta} + 2\bar{\mu}\epsilon^{\alpha\beta} = \bar{\Lambda}\epsilon_\rho^\rho G^{\alpha\beta} + 2\bar{\mu}\epsilon^{\alpha\beta} \\ &= \frac{E}{(1-\bar{\nu}^2)}\epsilon_\rho^\rho G^{\alpha\beta} + \frac{E}{(1+\bar{\nu})}\epsilon^{\alpha\beta}\end{aligned} \tag{2.11}$$

$$or\ \ E = \bar{\mu}\frac{(3\bar{\lambda} + 2\bar{\mu})}{\bar{\lambda} + \bar{\mu}},\ \ \bar{\nu} = \frac{\bar{\lambda}}{2(\bar{\lambda} + \bar{\mu})},\ \ \bar{\mu} = \frac{E}{2(1+\bar{\nu})},\ \ \bar{\lambda} = \frac{E\bar{\nu}}{(1+\bar{\nu})(1-2\bar{\nu})}$$

From $U_3(x, z) = U_3(x)$, $\mu_\alpha^\nu \overline{U}_{\nu,3} + \overline{U}_{3,\alpha} + B_\alpha^\nu \overline{U}_\nu = 0$ it follows that $\bar{U}_3$ does not depend on z; therefore, $U_3(x) = \bar{U}_3(x) = u_3(x)$. We verify also that

$$\overline{U}_{\lambda,3} + \left(\mu^{-1}\right)_\lambda^\alpha B_\alpha^\gamma \overline{U}_\gamma = -\left(\mu^{-1}\right)_\lambda^\alpha \overline{U}_{3,\alpha}$$

$$\left(\left(\mu^{-1}\right)_\beta^\alpha \overline{U}_\alpha\right)_{,3} = -\left(\mu^{-1}\right)_\beta^\lambda \left(\mu^{-1}\right)_\lambda^\alpha \overline{U}_{3,\alpha}$$

Let $\left(\mu^{-1}\right)_\beta^\lambda = A_\beta^\lambda$; $w = \left(\overline{U}_1, \overline{U}_2\right) = \overline{U}_\alpha A^\alpha$.
Then,

$$\frac{d}{dz}(Aw) = -A^2 \nabla u_3 \tag{2.12}$$

because $AB = BA$ and $\frac{d}{dz}A = BA^2 = A^2 B$. The resolution of these equations leads to the following relations:

$$w_\gamma = \bar{U}_\gamma(x, z) = \mu_\gamma^\beta u_\beta(x) - z u_{3,\gamma}(x)$$

so

$$\begin{aligned} U_\alpha &= \mu_\alpha^\rho \bar{U}_\rho = u_\alpha - z\left(u_{3,\alpha} + 2B_\alpha^\tau u_\tau\right) + z^2\left(B_\alpha^\tau B_\tau^\nu u_\nu + B_\alpha^\tau u_{3,\tau}\right) \\ &= \quad u_\alpha - z\vartheta_\alpha + z^2 \varphi_\alpha \\ U_3 &= \quad u_3 \end{aligned} \tag{2.13}$$

With the displacement (U_α, U_3) so obtained , the strain tensor $(\epsilon_{ij}(U))$ reads

$$\epsilon_{\alpha\beta}(U) = e_{\alpha\beta}(u) - zK_{\alpha\beta}(u) + z^2 Q_{\alpha\beta}(u);$$

$$\epsilon_{\alpha 3} = 0; \quad \epsilon_{33} = 0; \quad u_i : x = (x^1, x^2) \to u_i(x) \in \mathbb{R}$$

where

$$e_{\alpha\beta}(u) = \frac{1}{2}(\nabla_\alpha u_\beta + \nabla_\beta u_\alpha - 2u_3 B_{\alpha\beta}) = e_{\beta\alpha}$$

$$K_{\alpha\beta}(u) = \nabla_\alpha(\nabla_\beta u_3 + B_\beta^\rho u_\rho) + B_\alpha^\rho(\nabla_\beta u_\rho - u_3 B_{\beta\rho}) = K_{\beta\alpha}$$

$$\begin{aligned} Q_{\alpha\beta}(u) &= \tfrac{1}{2}\left(B_\alpha^\nu \nabla_\beta(\nabla_\nu u_3 + B_\nu^\rho u_\rho) + B_\beta^\nu \nabla_\alpha(\nabla u_3 + B_\nu^\rho u_\rho)\right) \\ &= \quad \tfrac{1}{2}\left(B_\alpha^\nu \nabla_\beta \theta_\nu + B_\beta^\nu \nabla_\alpha \theta_\nu\right) = Q_{\beta\alpha} \end{aligned}$$

$$\theta_\nu = \nabla_\nu u_3 + B_\nu^\rho u_\rho$$

The kinematic of the N-T (Nzengwa-Tagne) model thus obtained shows the consistency between the change of the metric tensor $(g_{\alpha\beta} - G_{\alpha\beta})/2$ and the Gauss deformation tensor $(Q_{\alpha\beta})$. Contribution of the Gauss deformation in the strain energy is significant when 2χ is approximately beyond $\sqrt{1/10}$. In this model, the plane stress-strain constitutive relations still hold, i.e.

$$\sigma^{\alpha\beta} = \frac{2\bar{\mu}\bar{\lambda}}{\bar{\lambda} + 2\bar{\mu}} \epsilon^{\rho}_{\rho} G^{\alpha\beta} + 2\bar{\mu}\epsilon^{\alpha\beta} = \bar{\Lambda}\epsilon^{\rho}_{\rho} G^{\alpha\beta} + 2\bar{\mu}\epsilon^{\alpha\beta}$$

$$G^{\alpha\beta} = \left(\mu^{-1}\right)^{\alpha}_{\nu} \cdot \left(\mu^{-1}\right)^{\beta}_{\rho} A^{\nu\rho}$$

2.2 Euler's Equations and Variational Formulation

2.2.1 Variational Formulation of Equilibrium

Let $\mathbb{E}$ be the internal energy, C the kinetic energy and P_{ext} the external power. We have from the first principle of thermodynamics

$$\frac{d}{dt}(\mathbb{E} + \mathcal{C}) = P_{ext}$$

which is equivalent to

$$\frac{d}{dt}\mathcal{C} = -\frac{d\mathbb{E}}{dt} + P_{ext} \quad \text{or} \quad P_{inertia} = P_i + P_{ext}$$

$P_{inertia}$ is the inertia power to be considered in dynamic analysis and $-P_i$ the strain power. Dynamic analysis will be treated later. The equilibrium of a shell subjected to a body force f, a surface force on part of its border p and clamped on another part of its border (model problem) is characterized by the equations:

$$\begin{cases} \sigma^{ij}_{/j} + f^i = 0 \text{ in } \Omega \\ \sigma\vec{n} = p \;\; on \;\; \Gamma_- \cup \Gamma_+ \cup \bar{\gamma} \times [-\frac{h}{2}, \frac{h}{2}];\; \Gamma_- = S \times \left\{-\frac{h}{2}\right\} \\ \Gamma_+ = S \times \left\{\frac{h}{2}\right\};\; \gamma \text{ is a part of } \partial S \\ U = 0 \text{ on } \Gamma_0;\; \Gamma_0 = \bar{\gamma}_0 \times [-\frac{h}{2}, \frac{h}{2}], \\ \bar{\gamma}_0 \text{ is the complementary part of} \bar{\gamma} \text{ of } \partial S \end{cases} \tag{2.14}$$

It is the 3D elasticity of Euler's equation in curvilinear media. The variational equation reads

$$\int_{\Omega} \sigma^{ij}(U)\epsilon_{ij}(V)d\Omega = P_{ext}(V) \tag{2.15}$$

$$\begin{aligned} d\Omega &= (G_1,\ G_2,\ A_3)\, dx^1 dx^2 dz = \sqrt{G} dx dz \\ &= \left(det(\mu^{\alpha}_{\beta})\right) \sqrt{A} dx dz = \psi(x,z) dS dz; \end{aligned} \tag{2.16}$$

$$\epsilon_{\alpha\beta}(V) = e_{\alpha\beta}(v) - zK_{\alpha\beta}(v) + z^2 Q_{\alpha\beta}(v) \tag{2.17}$$

$$e_{\alpha\beta}(v) = \frac{1}{2}\left(\nabla_\alpha v_\beta + \nabla_\beta v_\alpha - 2v_3 B_{\alpha\beta}\right) = \bar{e}_{\alpha\beta}$$

$$K_{\alpha\beta}(v) = \nabla_\alpha B^{\rho}_{\beta} v_\rho + B^{\rho}_{\alpha}\nabla_\beta v_\rho + B^{\rho}_{\beta}\nabla_\alpha v_\rho + \nabla_\alpha\nabla_\beta v_3 - B^{\rho}_{\alpha} B_{\rho\beta} v_3 = \bar{K}_{\alpha\beta}$$

$$Q_{\alpha\beta}(v) = \frac{1}{2}\left(B^{\delta}_{\alpha}\nabla_\beta(\nabla_\delta v_3 + B^{\rho}_{\delta} v_\rho) + B^{\delta}_{\beta}\nabla_\alpha(\nabla_\delta v_3 + B^{\rho}_{\delta} v_\rho)\right) = \bar{Q}_{\alpha\beta}$$

$v_i: \quad x = (x_1,\ x_2) \rightarrow v_i(x) \in \mathbb{R}$. From

$$\sigma^{\alpha\beta} = \bar{\Lambda}\epsilon^{\rho}_{\rho} G^{\alpha\beta} + 2\bar{\mu}\epsilon^{\alpha\beta} = (\bar{\Lambda} G^{\alpha\beta} G^{\gamma\rho} + \bar{\mu}(G^{\alpha\rho}G^{\beta\gamma} + G^{\beta\rho}G^{\alpha\gamma}))\epsilon_{\rho\gamma}(U)$$

$$G^{\alpha\beta} = \left(\mu^{-1}\right)^{\alpha}_{\nu} \cdot \left(\mu^{-1}\right)^{\beta}_{\rho} A^{\nu\rho} \tag{2.18}$$

It follows that

$$\begin{aligned} \textstyle\int_{\Omega} \sigma^{ij}(U)\epsilon_{ij}(V)d\Omega = &\quad \textstyle\int_{\Omega} \sigma^{\alpha\beta}(U)\epsilon_{\alpha\beta}(V)d\Omega \\ = &\quad \textstyle\int_S dS \int_{\frac{-h}{2}}^{\frac{h}{2}} \psi\,(x,z)\,[\bar{\Lambda} G^{\alpha\beta} G^{\rho\gamma} \\ &+ \bar{\mu}\left(G^{\alpha\rho}G^{\beta\gamma} + G^{\beta\rho}G^{\alpha\gamma}\right)]\epsilon_{\rho\gamma}(U)\,\epsilon_{\alpha\beta}(V)\,dz \\ = &\quad A(u,v) \end{aligned} \tag{2.19}$$

Let $A_n^{\alpha\beta\rho\gamma} = \int_{\frac{-h}{2}}^{\frac{h}{2}} \psi z^n \left(\bar{\Lambda} G^{\alpha\beta} G^{\rho\gamma} + \bar{\mu}\left(G^{\alpha\rho}G^{\beta\gamma} + G^{\beta\rho}G^{\alpha\gamma}\right)\right) dz$; n=0,1,2,3,4.

Then

$$
\begin{aligned}
\int_\Omega \sigma^{\alpha\beta}(U)\epsilon_{\alpha\beta}(V)d\Omega = & \int_S \left[A_0^{\alpha\beta\rho\gamma} e_{\rho\gamma} - A_1^{\alpha\beta\rho\gamma} K_{\rho\gamma} + A_2^{\alpha\beta\rho\gamma} Q_{\rho\gamma}\right]\overline{e}_{\alpha\beta} \\
& + \left[-A_1^{\alpha\beta\rho\gamma} e_{\rho\gamma} + A_2^{\alpha\beta\rho\gamma} K_{\rho\gamma} - A_3^{\alpha\beta\rho\gamma} Q_{\rho\gamma}\right]\overline{K}_{\alpha\beta} \\
& + \left[A_2^{\alpha\beta\rho\gamma} e_{\rho\gamma} - A_3^{\alpha\beta\rho\gamma} K_{\rho\gamma} + A_4^{\alpha\beta\rho\gamma} Q_{\rho\gamma}\right]\overline{Q}_{\alpha\beta} dS \\
= & \int_S (N_0^{\alpha\beta} - N_1^{\alpha\beta} + N_2^{\alpha\beta})\overline{e}_{\alpha\beta} + (-N_1^{\alpha\beta} + N_2^{\alpha\beta} \\
- & \quad N_3^{\alpha\beta})\overline{K}_{\alpha\beta} + (N_2^{\alpha\beta} - N_3^{\alpha\beta} + N_4^{\alpha\beta})\overline{Q}_{\alpha\beta})dS \\
= & \int_\Omega N^{\alpha\beta}\overline{e}_{\alpha\beta} + M^{\alpha\beta}\overline{K}_{\alpha\beta} + M^{*\alpha\beta}\overline{Q}_{\alpha\beta})dS;
\end{aligned}
\tag{2.20}
$$

Therefore

$$
\int_\Omega \sigma : \epsilon d\Omega = A(u, v) = \int_S \left(N : \overline{e} + M : \overline{K} + M^* : \overline{Q}\right) dS; \tag{2.21}
$$

$N^{\alpha\beta} \equiv (N/m)$ is the membrane force; $M^{\alpha\beta} \equiv (mN/m)$ the bending moment; $M^{*\alpha\beta} \equiv \left(m^2 N/m\right)$ Gauss moment; $e_{\alpha\beta} \equiv (1/1)$, $K_{\alpha\beta} \equiv (1/m)$, $Q_{\alpha\beta} \equiv \left(1/m^2\right)$, respectively, the mid-surface strain, change in curvature and Gauss deformation tensors. It follows that the global internal virtual work is a sum of three terms due to membrane, flectional and Gaussian deformations. The contribution of the Gauss deformation is negligible in thin shells. However, it constitutes a significant energy reserve (or resistance) when the shell is sufficiently thick (see example later).
We can now evaluate the external power or virtual work in static state P_{ext}. We have

$$
\begin{aligned}
& \int_\Omega \left(f^\alpha V_\alpha + f^3 V_3\right) d\Omega + \int_{\Gamma_-} \sigma \vec{n}\, V dS + \int_{\Gamma_+} \sigma \vec{n}\, V dS + \int_{\Gamma_1} \sigma \vec{n}\, V dS \\
& = \int_\Omega \left(f^\alpha V_\alpha + f^3 V_3\right) d\Omega + \int_{\Gamma_-} p.V dS + \int_{\Gamma_+} p.V dS + \int_{\Gamma_1} p.V dS
\end{aligned}
\tag{2.22}
$$

with $V_\alpha = v_\alpha - z\left(v_{3,\alpha} + 2B_\alpha^\tau v_\tau\right) + z^2\left(B_\alpha^\tau B_\tau^\nu v_\nu + B_\alpha^\tau v_{3,\tau}\right)$; $V_3 = v_3$. We also have

$$
\begin{aligned}
\int_{\frac{-h}{2}}^{\frac{h}{2}} (f^\alpha V_\alpha + f^3 V_3)\psi dz = & \int_{\frac{-h}{2}}^{\frac{h}{2}} \left(f^\alpha v_\alpha - 2z f^\alpha B_\alpha^\tau v_\tau + z^2 f^\alpha B_\alpha^\nu B_\nu^\tau v_\tau\right)\psi dz \\
& + \int_{\frac{-h}{2}}^{\frac{h}{2}} \left(-z f^\alpha v_{3,\alpha} + z^2 f^\alpha B_\alpha^\tau v_3, \tau\right)\psi dz \\
& + \int_{\frac{-h}{2}}^{\frac{h}{2}} f^3 v_3 \psi dz
\end{aligned}
\tag{2.23}
$$

From these expressions, we obtain

$$
\begin{aligned}
P_{ext} = \quad & \int_{\Omega} f^{\alpha} V_{\alpha} d\Omega + \int_{\Omega} f^3 V_3 d\Omega + \int_{\Gamma_-} \left(p_-^{\alpha} V_{\alpha} + p_-^3 V_3\right) d\Gamma_- \\
& + \int_{\Gamma_+} \left(p_+^{\alpha} V_{\alpha} + p_+^3 V_3\right) d\Gamma_+ + \int_{\Gamma_1} \left(p^{\alpha} V_{\alpha} + p^3 V_3\right) d\Gamma_1 \\
= & \int_S \left(\bar{P}^{\alpha} v_{\alpha} + \bar{P}^3 v_3\right) dS + \int_{\bar{\gamma}_1} \left(q^{\alpha} v_{\alpha} + q^3 v_3\right) d\bar{\gamma} + \int_{\bar{\gamma}_1} m^{\alpha} \bar{\theta}_{\alpha} d\bar{\gamma};
\end{aligned}
\tag{2.24}
$$

with

$$
\bar{P}^{\alpha} = \int_{\frac{-h}{2}}^{\frac{h}{2}} f^{\tau} w_{\tau}^{\alpha}(z) \psi dz + p_+^{\tau} w_{\tau}^{\alpha}(h/2) + p_-^{\tau} w_{\tau}^{\alpha}(-h/2)
$$

$$
w_{\tau}^{\alpha}(z) = \delta_{\tau}^{\alpha} - 2z B_{\tau}^{\alpha} + z^2 B_{\upsilon}^{\alpha} B_{\tau}^{\upsilon}
$$

$$
\begin{aligned}
\bar{P}^3 = \quad & \int_{\frac{-h}{2}}^{\frac{h}{2}} f^3 \psi dz + p_+^3 (h/2) + p_-^3 (-h/2) \\
& -\partial_{\tau} \left[\int_{\frac{-h}{2}}^{\frac{h}{2}} f^{\alpha} \overline{w}_{\alpha}^{\tau}(z) \psi dz + p_+^{\alpha} \overline{w}_{\alpha}^{\tau}(h/2) + p_-^{\alpha} \overline{w}_{\alpha}^{\tau}(-h/2)\right];
\end{aligned}
$$

$$
\ldots \overline{w}_{\alpha}^{\tau}(z) = -z\delta_{\alpha}^{\tau} + z^2 B_{\alpha}^{\tau};
$$

$$
q^{\alpha} = \int_{\frac{-h}{2}}^{\frac{h}{2}} p^{\alpha} dz - \int_{\frac{-h}{2}}^{\frac{h}{2}} z B_{\tau}^{\alpha} p^{\tau} dz; \quad q^3 = \int_{\frac{-h}{2}}^{\frac{h}{2}} p^3 dz;
$$

$$
m^{\alpha} = \int_{\frac{-h}{2}}^{\frac{h}{2}} z p^{\alpha} dz - \int_{\frac{-h}{2}}^{\frac{h}{2}} z^2 B_{\tau}^{\alpha} p^{\tau} dz; \quad m = m^{\alpha} A_{\alpha} = m_{\alpha} A^{\alpha}
$$

$$
\bar{\theta}_{\alpha} = -\left(\nabla_{\alpha} v_3 + B_{\alpha}^{\gamma} v_{\gamma}\right)
$$

Let us recall that $\bar{\theta}_{\alpha}$ here defined on the border after deformation is opposite to the angle vector of rotation, of A_3 , the normal of the mid-surface. Indeed after deformation, the normal vector of the undeformed mid-surface A_3 and that of the deformed mid-surface $(A_3 + \delta A_3)$ define an angle $\bar{\theta}$,

$$
\bar{\theta} = A_3 \times (A_3 + \delta A_3) = A_3 \times (A_3 - R) = -A_3 \times R \tag{2.25}
$$

The variational equilibrium equation follows

$$\int_S [N(u) : e(v) + M(u) : K(v) + M^*(u) : Q(v)] dS$$
$$= \int_S \left(\bar{P}^\alpha v_\alpha + \bar{P}^3 v_3\right) dS + \int_{\bar{\gamma}_1} \left(q^\alpha v_\alpha + q^3 v_3\right) d\bar{\gamma} \tag{2.26}$$
$$+ \int_{\bar{\gamma}_1} m^\alpha \bar{\theta}_\alpha d\bar{\gamma}$$

The expressions of the force, bending moment and Gaussian moment tensors can be calculated exactly, if necessary (see approximations further), and lead to the variational equation,

$$\begin{cases} \text{Find } u \in U_{ad} = H^1_{\gamma_o}(S) \times H^1_{\gamma_o}(S) \times H^2_{\gamma_o}(S) \\ A(u, v) = L(v) \end{cases} \tag{2.27}$$

where $A(.,.)$ is a continuous , symmetric and strictly positive (coercive) bilinear form in the space of admissible functions $u,\ v$; $L(.)$, the virtual work of external forces is a continuous linear form, $H^1_{\gamma_o}(S) = \{w,\ w_{,\alpha} \in L^2(S),\ \ w = 0 \text{ on } \gamma_o\}$, $H^2(S) = \{w,\ w_{,\alpha},\ w_{,\alpha\beta} \in L^2(S)\}$, $H^2_{\gamma_o}(S) = \{w \in H^1_{\gamma_o}(S) \cap H^2(S),\ \partial_{\bar{\nu}} w_3 = 0 \text{ on } \gamma_o\}$.

Exercise: Proof that this variational equation has a unique solution in the space $U_{ad} = H^1_{\gamma_o}(S) \times H^1_{\gamma_o}(S) \times H^2_{\gamma_o}(S)$ if the forces $\bar{P}^i$, q^i, m^α are, respectively, in $L^2(S)$, $L^2(\gamma_1)$, $L^2(\gamma_1)$. The condition $\partial_{\bar{\nu}} u = 0 \ on \ \gamma_o$ means the shell is clamped at the border γ_o.

Analysis of the terms N, M, M and the kinematic*

Let us consider the kinematic obtained under the only assumption of plane strain

$$U = U_\alpha G^\alpha + U_3 A^3;$$
$$U_\alpha = u_\alpha - z\left(u_{3,\alpha} + 2B^\tau_\alpha u_\tau\right) + z^2\left(B^\tau_\alpha B^\nu_\tau u_\nu + B^\tau_\alpha u_{3,\tau}\right) \tag{2.28}$$
$$U_3 = u_3$$

Expressing the displacement U using the mid-surface base $\{A_\alpha,\ A_3\}$ we have

$$\begin{aligned}\bar{U}_\alpha(x,z) &= \mu^\rho_\alpha u_\rho(x) - z u_{3,\alpha}(x)\\ &= u_\alpha(x) - z(u_{3,\alpha}(x) + B^\rho_\alpha u_\rho(x))\\ &= u_\alpha(x) - z\theta_\alpha(x)\\ \bar{U}_3 &= u_3(x)\end{aligned} \tag{2.29}$$

where θ_α is the rotation of the normal of the mid-surface. This form is commonly known as the Reissner-Mindlin kinematic. It is obvious that if the rotation depends on u_i as calculated above then

$$\epsilon_{\alpha\beta}(U) = \epsilon_{\alpha\beta}(\bar{U}) = e_{\alpha\beta}(u) - zK_{\alpha\beta}(u) + z^2 Q_{\alpha\beta}(u);\ \epsilon_{\alpha 3} = 0;\ \epsilon_{33} = 0$$

Therefore, there is transverse strain only if the rotation does not depend on u_i. The correct expression of transverse strain deduced from the Reissner-Mindlin displacement reads

$$\begin{aligned}\epsilon^{RM}_{\alpha 3}(U) = \epsilon^{RM}_{\alpha 3}(\bar{U}) = \tfrac{1}{2}\left(-\theta_\alpha + u_{3,\alpha} + B^\rho_\alpha u_\rho\right),\\ \epsilon^{RM}_{33}(U) = \epsilon^{RM}_{33}(\bar{U}) = 0\end{aligned} \tag{2.30}$$

If the characteristic ratio $\chi = max(h/2r) \prec 0,1$ then we obtain the following approximations:

$$G^\alpha \approx A^\alpha,\ G_\alpha \approx A_\alpha,\ \mu^\beta_\alpha \approx \delta^\beta_\alpha \tag{2.31}$$

and

$$U_\alpha \simeq u_\alpha - z u_{3,\alpha} = \bar{U}_\alpha;\ U_3 = \bar{U}_3 = u_3$$

which is the Kirchhoff-Love displacement in the mid-surface base $\{A_\alpha,\ A_3\}$. In the real base of the shell $\{G^\alpha,\ A^3\}$, the K-L displacement reads

$$U_\alpha = u_\alpha(x) - z(u_{3,\alpha}(x) + B^\rho_\alpha u_\rho(x)) + z^2 B^\rho_\alpha u_{3,\rho}(x) \approx u_\alpha(x) - z\theta_\alpha(x);\ U_3 = u_3 \tag{2.32}$$

and

$$\epsilon^{KL}_{\alpha 3}(U) = \epsilon^{KL}_{\alpha 3}(\bar{U}) = B^\rho_\alpha u_\rho/2$$

$$\epsilon^{KL}_{33}(U) = \epsilon^{KL}_{33}(\bar{U}) = 0$$

By simplifying also $dét(\mu^\alpha_\beta)$ we obtain $\psi(x,z) = 1 - 2z\bar{H} + z^2\bar{G} \approx 1$, $\bar{H}$ and $\bar{G}$ being, respectively, the mean and Gauss curvatures;

$$\sigma^{\alpha\beta} = (\bar{\Lambda} A^{\alpha\beta} A^{\rho\gamma} + \bar{\mu}(A^{\alpha\rho} A^{\beta\gamma} + A^{\beta\rho} A^{\alpha\gamma}))\epsilon_{\rho\gamma}(u),\ \epsilon_{\rho\gamma} \approx e_{\rho\gamma}(u) - zK_{\rho\gamma}(u) \tag{2.33}$$

and

$$\int_{\frac{-h}{2}}^{\frac{h}{2}} \sigma^{\alpha\beta} \epsilon_{\alpha\beta} \psi dz = N : \overline{e} + M : \overline{K}; \quad N = \frac{Eh}{1-\bar{\nu}^2}\left[(1-\bar{\nu})\, e^{\alpha\beta} + \bar{\nu} e_{\lambda}^{\lambda} A^{\alpha\beta}\right]; \tag{2.34}$$

$$M = \frac{Eh^3}{12\left(1-\bar{\nu}^2\right)}\left[(1-\bar{\nu})\, K^{\alpha\beta} + \bar{\nu} K_{\lambda}^{\lambda} A^{\alpha\beta}\right] \tag{2.35}$$

By expanding the tensors $N^{\alpha\beta}$, $M^{\alpha\beta}$, $M^{*\alpha\beta}$ with respect to χ^n; For $(n = 0)$ we obtain

$$N^{\alpha\beta} = \frac{Eh}{1-\bar{\nu}^2}\left[(1-\bar{\nu})\, e^{\alpha\beta} + \bar{\nu} e_{\lambda}^{\lambda} A^{\alpha\beta}\right] + 0(\chi) = N_0^{\alpha\beta} + 0(\chi); \tag{2.36}$$

with $\frac{Eh}{1-\bar{\nu}^2}\left[(1-\bar{\nu})\, e^{\alpha\beta} + \bar{\nu} e_{\lambda}^{\lambda} A^{\alpha\beta}\right] = N_0^{\alpha\beta}$

$$M^{\alpha\beta} = \frac{Eh^3}{12\left(1-\bar{\nu}^2\right)}\left[(1-\bar{\nu})\, K^{\alpha\beta} + \bar{\nu} K_{\lambda}^{\lambda} A^{\alpha\beta}\right] + 0(\chi) = M_0^{\alpha\beta} + 0(\chi) \tag{2.37}$$

with $\frac{Eh^3}{12(1-\bar{\nu}^2)}\left[(1-\bar{\nu})\, K^{\alpha\beta} + \bar{\nu} K_{\lambda}^{\lambda} a^{\alpha\beta}\right] = M_0^{\alpha\beta}$. The expressions of N_0 and M_0 are the terms found in Kirchhoff-Love thin shell model discussed above. At the first order $(n = 1)$, we obtain

$$N^{\alpha\beta} = N_0^{\alpha\beta} + \frac{Eh^3}{12\left(1-\bar{\nu}^2\right)}\left[(1-\bar{\nu})\, Q^{\alpha\beta} + \bar{\nu} Q_{\lambda}^{\lambda} A^{\alpha\beta}\right] + 0(\chi^2) \tag{2.38}$$

$$M^{\alpha\beta} = M_0^{\alpha\beta} + ..0(\chi^2) \tag{2.39}$$

$$\begin{aligned} M^{*\alpha\beta} = \quad & \tfrac{Eh^3}{12(1-\bar{\nu}^2)}\left[(1-\bar{\nu})\, e^{\alpha\beta} + \bar{\nu} e_{\lambda}^{\lambda} A^{\alpha\beta}\right] \\ & + \tfrac{Eh^5}{80(1-\bar{\nu}^2)}\left[(1-\bar{\nu})\, Q^{\alpha\beta} + \bar{\nu} Q_{\lambda}^{\lambda} A^{\alpha\beta}\right] + 0(\chi^2) \end{aligned} \tag{2.40}$$

These tensors contain additional terms to those found in the classical Kirchhoff-Love model. Terms in h^{2n} disappear because the thickness is constant and integration with respect to z^{2n+1} is zero.

2.2.2 *Thick Shells Euler's Equations*

Recall that the variational equation is

$$\int\limits_S [N : e(v) + M : K(v) + M^* : Q(v)] dS$$

$$= \int\limits_S \left(\bar{P}^\alpha v_\alpha + \bar{P}^3 v_3\right) dS + \int\limits_{\bar{\gamma}_1} (q^\alpha v_\alpha + q^3 v_3) d\bar{\gamma} + \int\limits_{\bar{\gamma}_1} m^\alpha \bar{\theta}_\alpha (v)\, d\bar{\gamma} \tag{2.41}$$

$$\bar{\theta}_\alpha (v) = -\left(\nabla_\alpha v_3 + B_\alpha^\rho v_\rho\right) \tag{2.42}$$

The line densities of the forces q^α, q^3 and moment $m = m^\alpha A_\alpha$ on the free border $\bar{\gamma} \subset \partial S$ constitute the torsor of forces on the border Γ_1 relatively to the border of the mid-surface. Let us consider a direct orthonormal base on the border $\bar{\gamma}$ defined by $\left\{\vec{\nu}, \vec{t}, \vec{n} = A^3\right\}$, where $\vec{\nu}$, $\vec{t}$ are, respectively, the unit outer normal vector and the unit tangent vector on $\bar{\gamma}$ (Fig. 2.2).

We decompose the moment $m = m^\alpha A_\alpha$ in $m = m^t \vec{t} + m^\nu \vec{\nu}$ with $\vec{t} = t_\alpha A^\alpha$, $\vec{\nu} = \nu_\alpha A^\alpha$. Then

$$m^t = m^\alpha A_\alpha \cdot \vec{t} = m^\alpha t_\alpha, \quad m^\nu = m^\alpha \nu_\alpha$$

The components m^t and m^ν are, respectively, line densities of the bending moment of axis $\vec{t}$ and torque of axis $\vec{\nu}$ applied on $\bar{\gamma}_1$. Let

$$\vec{\nu} = \nu_\alpha A^\alpha + \nu_3 A^3, \quad \bar{\theta}_\nu = \bar{\theta}.\vec{\nu} = -\left(\nabla_\alpha v_3 + B_\alpha^\gamma v_\gamma\right) \nu^\alpha = -\vec{n}.\partial_{\vec{\nu}} v$$

$$\bar{\theta}_t = \bar{\theta}.\vec{t} = -\left(\nabla_\alpha v_3 + B_\alpha^\gamma v_\gamma\right) t^\alpha = -\vec{n}.\partial_{\vec{t}} v$$

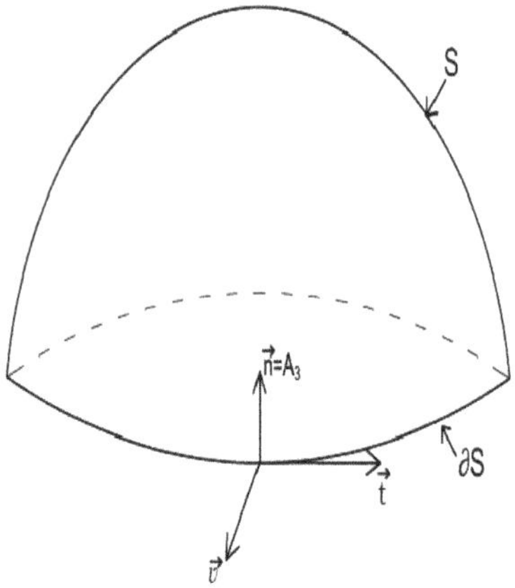

Fig. 2.2 Border local base

Then

$$m.\bar{\theta} = m^{\alpha}\bar{\theta}_{\alpha} = m^{t}\bar{\theta}_{\nu} - m^{\nu}\bar{\theta}_{t}$$

Indeed, on the free border

$$\bar{\theta} = \overrightarrow{n} \times \delta\overrightarrow{n};\ \delta\overrightarrow{n} = -\delta R = -\delta R_{\alpha}A^{\alpha} = -\left(\partial_{\alpha}v_{3} + B^{\rho}_{\alpha}v_{\rho}\right)A^{\alpha} \tag{2.43}$$

$$\begin{aligned} m^{\alpha}\bar{\theta}_{\alpha} &= m^{t}\overrightarrow{t}\cdot\left(\overrightarrow{n}\times\delta\overrightarrow{n}\right) + m^{\nu}\overrightarrow{\nu}\cdot\left(\overrightarrow{n}\times\delta\overrightarrow{n}\right) \\ &= m^{t}(\overrightarrow{t}\times\overrightarrow{n})\cdot\delta\overrightarrow{n} + m^{\nu}(\overrightarrow{\nu}\times\overrightarrow{n})\cdot\delta\overrightarrow{n} \\ &= m^{t}\overrightarrow{\nu}\cdot\delta\overrightarrow{n} - m^{\nu}\overrightarrow{t}\cdot\delta\overrightarrow{n} \\ &= m^{t}\bar{\theta}_{\nu} - m^{\nu}\bar{\theta}_{t} \end{aligned} \tag{2.44}$$

therefore

$$m^{\alpha}\bar{\theta}_{\alpha} = m^{t}\bar{\theta}_{\nu} + m^{\nu}\overrightarrow{n}\cdot\partial_{\overrightarrow{t}}v \tag{2.45}$$

and it can be remarked that the bending moment m^{t} (of axis $\overrightarrow{t}$) works on the angle due to variation in the normal direction $\overrightarrow{\nu}$, and the torque m^{ν} (of axis $\overrightarrow{\nu}$) works on the angle due to variation in the tangent direction $\overrightarrow{t}$. We then have

$$\begin{aligned} L(V) &= \int_{S}\left(\bar{P}^{\alpha}v_{\alpha} + \bar{P}^{3}v_{3}\right)dS + \int_{\bar{\gamma}_{1}}\left(q^{\alpha}v_{\alpha} + q^{3}v_{3}\right)d\bar{\gamma} \\ &+ \int_{\bar{\gamma}_{1}}(m^{t}\bar{\theta}_{\nu} + m^{\nu}\overrightarrow{n}\cdot\partial_{\overrightarrow{t}}v)d\bar{\gamma} \end{aligned} \tag{2.46}$$

But

$$\begin{aligned} \int_{\bar{\gamma}_{1}}(m^{t}\bar{\theta}_{\nu} + m^{\nu}\overrightarrow{n}\cdot\partial_{\overrightarrow{t}}v)d\bar{\gamma} &= \int_{\bar{\gamma}_{1}}(m^{t}\bar{\theta}_{\nu} - \partial_{\overrightarrow{t}}m^{\nu}\overrightarrow{n}\cdot v - m^{\nu}\partial_{\overrightarrow{t}}\overrightarrow{n}\cdot v)d\bar{\gamma} \\ &= \int_{\bar{\gamma}_{1}}(m^{t}\bar{\theta}_{\nu} - \partial_{\overrightarrow{t}}m^{\nu}v_{3} - m^{\nu}t^{\alpha}\partial_{\alpha}\overrightarrow{n}\cdot v)d\bar{\gamma} \\ &= \int_{\bar{\gamma}_{1}}(m^{t}\bar{\theta}_{\nu} - \partial_{\overrightarrow{t}}m^{\nu}v_{3} + m^{\nu}t^{\alpha}B^{\rho}_{\alpha}A_{\rho}\cdot v)d\bar{\gamma} \\ &= \int_{\bar{\gamma}_{1}}(m^{t}\bar{\theta}_{\nu} - \partial_{\overrightarrow{t}}m^{\nu}v_{3} + m^{\nu}t^{\alpha}B^{\rho}_{\alpha}v_{\rho})d\bar{\gamma} \end{aligned} \tag{2.47}$$

Therefore

$$L(V) = \int\limits_{S} \left(\bar{P}^{\alpha} v_{\alpha} + \bar{P}^{3} v_{3}\right) dS + \int\limits_{\bar{\gamma}_1} \left(q^{\alpha} v_{\alpha} + q^{3} v_{3}\right) d\bar{\gamma}$$
$$+ \int\limits_{\bar{\gamma}_1} (m^{t} \bar{\theta}_{\nu} - \partial_{\vec{t}} m^{\nu} v_{3} + m^{\nu} t^{\alpha} B_{\alpha}^{\rho} v_{\rho}) d\bar{\gamma} \tag{2.48}$$

The internal virtual work (or virtual power in dynamic analysis)

$$\int\limits_{S} (N^{\alpha\beta} \bar{e}_{\alpha\beta} + M^{\alpha\beta} \bar{K}_{\alpha\beta} + M^{*\alpha\beta} \bar{Q}_{\alpha\beta}) dS$$
$$= \int\limits_{S} (N^{\alpha\beta} \bar{e}_{\alpha\beta} + M^{\alpha\beta} \bar{K}_{\alpha\beta} + M^{*\alpha\beta} B_{\alpha}^{\lambda} \nabla_{\beta} (\nabla_{\lambda} v_{3} + B_{\lambda}^{\rho} v_{\rho}) dS \tag{2.49}$$

We have

$$N^{\alpha\beta} \bar{e}_{\alpha\beta} = N^{\alpha\beta} \left(\nabla_{\alpha} v_{\beta} - B_{\alpha\beta} v_{3}\right)$$

$$M^{\alpha\beta} \bar{K}_{\alpha\beta} = M^{\alpha\lambda} B_{\lambda}^{\beta} \left(\nabla_{\alpha} v_{\beta} - B_{\alpha\beta} v_{3}\right) + M^{\alpha\beta} \nabla_{\beta} \left(\nabla_{\alpha} v_{3} + B_{\alpha}^{\rho} v_{\rho}\right)$$

$$M^{*\alpha\beta} B_{\alpha}^{\lambda} \nabla_{\beta} (\nabla_{\lambda} v_{3} + B_{\lambda}^{\rho} v_{\rho}) = M^{*\lambda\beta} B_{\lambda}^{\alpha} \nabla_{\beta} \nabla_{\alpha} v_{3} + M^{*\alpha\beta} (B_{\alpha}^{\lambda} \nabla_{\beta} (B_{\lambda}^{\rho}) v_{\rho})$$
$$+ M^{*\alpha\beta} B_{\alpha}^{\lambda} B_{\lambda}^{\rho} \nabla_{\beta} v_{\rho} \tag{2.50}$$

$$N^{\alpha\beta} \bar{e}_{\alpha\beta} + M^{\alpha\beta} \bar{K}_{\alpha\beta} = \left(N^{\alpha\beta} + M^{\alpha\lambda} B_{\lambda}^{\beta}\right) \left(\nabla_{\alpha} v_{\beta} - B_{\alpha\beta} v_{3}\right)$$
$$+ M^{\alpha\beta} \nabla_{\beta} \left(\nabla_{\alpha} v_{3} + B_{\alpha}^{\rho} v_{\rho}\right) \tag{2.51}$$

$$N^{\alpha\beta} \bar{e}_{\alpha\beta} + M^{\alpha\beta} \bar{K}_{\alpha\beta} = \left(N^{\alpha\beta} + M^{\alpha\lambda} B_{\lambda}^{\beta}\right) \left(\nabla_{\alpha} v_{\beta} - B_{\alpha\beta} v_{3}\right)$$
$$+ M^{\alpha\beta} \nabla_{\beta} \nabla_{\alpha} v_{3} + M^{\alpha\beta} \nabla_{\beta} (B_{\alpha}^{\rho}) v_{\rho} \tag{2.52}$$
$$+ M^{\alpha\beta} B_{\alpha}^{\rho} \nabla_{\beta} v_{\rho}$$

Also

$$N^{\alpha\beta} \bar{e}_{\alpha\beta} + M^{\alpha\beta} \bar{K}_{\alpha\beta} + M^{*\alpha\beta} B_{\alpha}^{\lambda} \nabla_{\beta} (\nabla_{\lambda} v_{3} \tag{2.53}$$
$$+ B_{\lambda}^{\rho} v_{\rho}) = \left(N^{\alpha\beta} + M^{\alpha\lambda} B_{\lambda}^{\beta}\right) \left(\nabla_{\alpha} v_{\beta} - B_{\alpha\beta} v_{3}\right) + (M^{\alpha\beta} + M^{*\lambda\beta} B_{\lambda}^{\alpha}) \nabla_{\beta} \nabla_{\alpha} v_{3} \tag{2.54}$$
$$+ (M^{\alpha\beta} + M^{*\lambda\beta} B_{\lambda}^{\alpha}) \nabla_{\beta} (B_{\alpha}^{\rho}) v_{\rho} + (M^{\alpha\beta} B_{\alpha}^{\rho} + M^{*\rho\alpha} B_{\rho}^{\lambda} B_{\lambda}^{\beta}) \nabla_{\alpha} v_{\beta} \tag{2.55}$$
$$= (M^{\alpha\beta} + M^{*\lambda\beta} B_{\lambda}^{\alpha}) \nabla_{\beta} (B_{\alpha}^{\rho}) v_{\rho} - \left(N^{\alpha\beta} + M^{\alpha\lambda} B_{\lambda}^{\beta}\right) B_{\alpha\beta} v_{3} + \tag{2.56}$$
$$+ (N^{\alpha\beta} + 2M^{\alpha\lambda} B_{\lambda}^{\beta} + M^{*\rho\alpha} B_{\rho}^{\lambda} B_{\lambda}^{\beta}) \nabla_{\alpha} v_{\beta} + (M^{\alpha\beta} + M^{*\lambda\beta} B_{\lambda}^{\alpha}) \nabla_{\beta} \nabla_{\alpha} v_{3} \tag{2.57}$$

Let $L^{\alpha\beta} = \left(N^{\alpha\beta} + 2M^{\alpha\lambda}B^{\beta}_{\lambda} + M^{*\rho\alpha}B^{\lambda}_{\rho}B^{\beta}_{\lambda}\right)$. Then

$$L^{\alpha\beta}\nabla_{\alpha}v_{\beta} = \nabla_{\alpha}(L^{\alpha\beta}v_{\beta}) - \nabla_{\alpha}(L^{\alpha\beta})v_{\beta} \tag{2.58}$$

and

$$\begin{aligned}\int_S L^{\alpha\beta}\nabla_{\alpha}v_{\beta}dS &= \int_S \nabla_{\alpha}(L^{\alpha\beta}v_{\beta})dS - \int_S \nabla_{\alpha}(L^{\alpha\beta})v_{\beta}dS \\ &= -\int_S \nabla_{\alpha}(L^{\alpha\beta})v_{\beta}dS + \int_{\bar{\gamma}} (L^{\alpha\beta})\nu_{\alpha}v_{\beta}d\bar{\gamma}\end{aligned}$$

The outer unit vector on the border $\vec{\nu}$ has ν_{α} as components. Therefore

$$\begin{aligned}\int_S L^{\alpha\beta}\nabla_{\alpha}v_{\beta}dS &= -\int_S \nabla_{\alpha}\left(N^{\alpha\beta} + 2M^{\alpha\lambda}B^{\beta}_{\lambda} + M^{*\rho\alpha}B^{\lambda}_{\rho}B^{\beta}_{\lambda}\right)v_{\beta}dS \\ &+\int_{\bar{\gamma}}\left(N^{\alpha\beta} + 2M^{\alpha\lambda}B^{\beta}_{\lambda} + M^{*\rho\alpha}B^{\lambda}_{\rho}B^{\beta}_{\lambda}\right)\nu_{\alpha}v_{\beta}d\bar{\gamma}\end{aligned} \tag{2.59}$$

Let

$$\mathcal{L}^{\alpha\beta} = (M^{\alpha\beta} + M^{*\lambda\beta}B^{\alpha}_{\lambda}) \tag{2.60}$$

Then

$$\mathcal{L}^{\alpha\beta}\nabla_{\beta}\nabla_{\alpha}v_3 = +\nabla_{\alpha}\nabla_{\beta}\mathcal{L}^{\alpha\beta}v_3 + \nabla_{\beta}\left(\mathcal{L}^{\alpha\beta}\nabla_{\alpha}v_3\right) - \nabla_{\alpha}(\nabla_{\beta}\mathcal{L}^{\alpha\beta}v_3) \tag{2.61}$$

and

$$\begin{aligned}\int_S \mathcal{L}^{\alpha\beta}\nabla_{\beta}\nabla_{\alpha}v_3dS &= \int_S \nabla_{\alpha}\nabla_{\beta}\mathcal{L}^{\alpha\beta}v_3dS + \int_S \nabla_{\beta}\left(\mathcal{L}^{\alpha\beta}\nabla_{\alpha}v_3\right)dS \\ &\quad -\int_S \nabla_{\alpha}(\nabla_{\beta}\mathcal{L}^{\alpha\beta}v_3)dS \\ &= \int_S \nabla_{\alpha}\nabla_{\beta}\mathcal{L}^{\alpha\beta}v_3dS + \int_{\bar{\gamma}}\left(\mathcal{L}^{\alpha\beta}\nu_{\beta}\nabla_{\alpha}v_3\right)d\bar{\gamma} \\ &\quad -\int_{\bar{\gamma}}(\nabla_{\beta}\mathcal{L}^{\alpha\beta}v_3)\nu_{\alpha}d\bar{\gamma}\end{aligned} \tag{2.62}$$

Therefore

$$\int_S (M^{\alpha\beta} + M^{*\lambda\beta} B_\lambda^\alpha) \nabla_\beta \nabla_\alpha v_3 dS = \int_S \nabla_\alpha \nabla_\beta (M^{\alpha\beta} + M^{*\lambda\beta} B_\lambda^\alpha) v_3 dS$$

$$+ \int_{\bar{\gamma}} \left((M^{\alpha\beta} + M^{*\lambda\beta} B_\lambda^\alpha) \nu_\beta \nabla_\alpha v_3 \right) d\bar{\gamma} \tag{2.63}$$

$$- \int_{\bar{\gamma}} (\nabla_\beta (M^{\alpha\beta} + M^{*\lambda\beta} B_\lambda^\alpha) v_3) \nu_\alpha v_\beta d\bar{\gamma}$$

We regroup the internal virtual work as follows:

$$\int_S [N : e(v) + M : k(v) + M^* : Q(v)] dS = \int_S [(M^{\alpha\rho} + M^{*\lambda\rho} B_\lambda^\alpha) \nabla_\rho (B_\alpha^\beta) \tag{2.64}$$

$$- \nabla_\alpha (N^{\alpha\beta} + 2M^{\alpha\lambda} B_\lambda^\beta + M^{*\rho\alpha} B_\rho^\lambda B_\lambda^\beta)] v_\beta dS - \int_S [(N^{\alpha\beta} + M^{\alpha\lambda} B_\lambda^\beta) B_{\alpha\beta} - \nabla_\alpha \nabla_\beta (M^{\alpha\beta} \tag{2.65}$$

$$+ M^{*\lambda\beta} B_\lambda^\alpha)] v_3 dS + \int_{\bar{\gamma}} (N^{\alpha\beta} + 2M^{\alpha\lambda} B_\lambda^\beta + M^{*\rho\alpha} B_\rho^\lambda B_\lambda^\beta) \nu_\alpha v_\beta d\bar{\gamma} + \int_{\bar{\gamma}} ((M^{\alpha\beta} \tag{2.66}$$

$$+ M^{*\lambda\beta} B_\lambda^\alpha) \nu_\beta \nabla_\alpha v_3) d\bar{\gamma} - \int_{\bar{\gamma}} (\nabla_\beta (M^{\alpha\beta} + M^{*\lambda\beta} B_\lambda^\alpha) \nu_\alpha v_3 d\bar{\gamma} \tag{2.67}$$

We rearrange terms on the border below:
Let $M^{\alpha\beta}\nu_\alpha = M\overrightarrow{\nu} = M^{\nu\beta}$, $M^{\alpha\beta}\nu_\alpha\nu_\beta = M\overrightarrow{\nu}\cdot\overrightarrow{\nu} = M^{\nu\nu}$, $M^{\alpha\beta}t_\alpha = M^{t\beta}$, $M\overrightarrow{\nu}\cdot\overrightarrow{t} = M^{\alpha\beta}\nu_\alpha t_\beta = M^{\nu\tau}$ and $T^\alpha\nu_\alpha = T^\nu$. The vector of resistant moment on the border $M\overrightarrow{\nu}$ shall be decomposed in $M^{\nu\nu}\overrightarrow{\nu} + M^{\nu\tau}\overrightarrow{t}$. The components $M^{\nu\nu}$ and $M^{\nu\tau}$ are, respectively, line densities of resistant bending moment of axis $\overrightarrow{t}$ and torque of axis $\overrightarrow{\nu}$. So, terms on the border are rewritten:

$$\int_{\bar{\gamma}} \left(N^{\alpha\beta} + M^{\alpha\lambda} B_\lambda^\beta \right) \nu_\alpha v_\beta d\bar{\gamma}$$

$$- \int_{\bar{\gamma}} (\nabla_\beta (M^{\alpha\beta} + M^{*\lambda\beta} B_\lambda^\alpha) \nu_\alpha v_3 d\bar{\gamma} + \int_{\bar{\gamma}} \left((M^{\alpha\beta} + M^{*\lambda\beta} B_\lambda^\alpha) \nu_\beta ((\nabla_\alpha v_3 + B_\alpha^\rho v_\rho) \right) d\bar{\gamma}$$

Let $\left(N^{\alpha\beta} + M^{\alpha\lambda} B_\lambda^\beta \right) = \mathcal{N}^{\alpha\beta}$; $\nabla_\beta (M^{\alpha\beta} + M^{*\lambda\beta} B_\lambda^\alpha) = T^\alpha$; $(M^{\alpha\beta} + M^{*\lambda\beta} B_\lambda^\alpha) = \mathcal{M}^{\alpha\beta}$ then border terms now read

$$\begin{aligned} & \int_{\bar{\gamma}} \mathcal{N}^{\alpha\beta} \nu_\alpha v_\beta d\bar{\gamma} - \int_{\bar{\gamma}} T^\alpha \nu_\alpha v_3 d\bar{\gamma} \\ + \quad & \int_{\bar{\gamma}} \left(\mathcal{M}^{\alpha\beta} \nu_\beta (\nabla_\alpha v_3 + B_\alpha^\rho v_\rho) \right) d\bar{\gamma} \\ = & \int_{\bar{\gamma}} \mathcal{N}^{\nu\beta} v_\beta d\bar{\gamma} - \int_{\bar{\gamma}} T^\nu v_3 d\bar{\gamma} - \int_{\bar{\gamma}} \mathcal{M}^{\alpha\nu} \bar{\theta}_\alpha d\bar{\gamma} \end{aligned}$$

Or

$$\mathcal{M}^{\alpha\nu}\bar{\theta}_\alpha = \mathcal{M}\overrightarrow{v}\cdot\bar{\theta} = \left(\mathcal{M}^{\nu\nu}\overrightarrow{v} + \mathcal{M}^{\nu\tau}\overrightarrow{t}\right)\cdot\bar{\theta}$$
$$= \mathcal{M}^{\nu\nu}\bar{\theta}_\nu + \mathcal{M}^{\nu\tau}\bar{\theta}_t$$
$$= \mathcal{M}^{\nu\nu}\bar{\theta}_\nu - \mathcal{M}^{\nu\tau}\overrightarrow{n}.\partial_{\overrightarrow{t}}v$$

And become

$$\int_{\bar{\gamma}} \mathcal{N}^{\nu\beta}v_\beta d\bar{\gamma} - \int_{\bar{\gamma}} T^\nu v_3 d\bar{\gamma} - \int_{\bar{\gamma}} (\mathcal{M}^{\nu\nu}\bar{\theta}_\nu - \mathcal{M}^{\nu\tau}\overrightarrow{n}.\partial_{\overrightarrow{t}}v)d\bar{\gamma}$$

The term

$$\int_{\bar{\gamma}}(\mathcal{M}^{\nu\nu}\bar{\theta}_\nu - \mathcal{M}^{\nu\tau}\overrightarrow{n}.\partial_{\overrightarrow{t}}v)d\bar{\gamma} = \int_{\bar{\gamma}}(\mathcal{M}^{\nu\nu}\bar{\theta}_\nu + \partial_{\overrightarrow{t}}\mathcal{M}^{\nu\tau}\overrightarrow{n}.v + \mathcal{M}^{\nu\tau}\partial_{\overrightarrow{t}}\overrightarrow{n}.v)d\bar{\gamma}$$
$$= \int_{\bar{\gamma}}(\mathcal{M}^{\nu\nu}\bar{\theta}_\nu + \partial_{\overrightarrow{t}}\mathcal{M}^{\nu\tau}v_3 - \mathcal{M}^{\nu\tau}t^\alpha B_\alpha^\rho v_\rho)d\bar{\gamma}$$

Therefore, border terms are rewritten

$$\int_{\bar{\gamma}} \mathcal{N}^{\nu\beta}v_\beta d\bar{\gamma} - \int_{\bar{\gamma}} T^\nu v_3 d\bar{\gamma} - \int_{\bar{\gamma}} (\mathcal{M}^{\nu\nu}\bar{\theta}_\nu + \partial_{\overrightarrow{t}}\mathcal{M}^{\nu\tau}v_3 - \mathcal{M}^{\nu\tau}t^\alpha B_\alpha^\rho v_\rho)d\bar{\gamma}$$

Finally, the variational formulation is equivalent to

$$\begin{aligned}
&\int_S[N : e(v) + M : k(v) + M^* : Q(v)]dS = \int_S((M^{\alpha\rho} + M^{*\lambda\rho}B_\lambda^\alpha)\nabla_\rho(B_\alpha^\beta)\\
-\ &\nabla_\alpha(N^{\alpha\beta} + 2M^{\alpha\lambda}B_\lambda^\beta + M^{*\rho\alpha}B_\rho^\lambda B_\lambda^\beta))v_\beta dS - \int_S[(N^{\alpha\beta} + M^{\alpha\lambda}B_\lambda^\beta)B_{\alpha\beta} - \nabla_\alpha\nabla_\beta(M^{\alpha\beta}\\
+\ &M^{*\lambda\beta}B_\lambda^\alpha)]v_3 dS + \int_{\bar{\gamma}}\mathcal{N}^{\nu\beta}v_\beta d\bar{\gamma} - \int_{\bar{\gamma}}T^\nu v_3 d\bar{\gamma} - \int_{\bar{\gamma}}(\mathcal{M}^{\nu\nu}\bar{\theta}_\nu + \partial_{\overrightarrow{t}}\mathcal{M}^{\nu\tau}v_3 - \mathcal{M}^{\nu\tau}t^\alpha B_\alpha^\rho v_\rho)d\bar{\gamma}\\
=\ &L(v)\\
=\ &\int_S\left(\bar{P}^\alpha v_\alpha + \bar{P}^3 v_3\right)dS + \int_{\bar{\gamma}_1}(q^\alpha v_\alpha + q^3 v_3)d\bar{\gamma}\\
+\ &\int_{\bar{\gamma}_1}(m^t\bar{\theta}_\nu - \partial_{\overrightarrow{t}}m^\nu v_3 + m^\nu t^\alpha B_\alpha^\rho v_\rho)d\bar{\gamma}
\end{aligned}$$

We thus deduce the indefinite Euler equations in S:

$$\begin{cases}(M^{\alpha\rho} + M^{*\lambda\rho}B_\lambda^\alpha)\nabla_\rho(B_\alpha^\beta) - \nabla_\alpha\left(N^{\alpha\beta} + 2M^{\alpha\lambda}B_\lambda^\beta + M^{*\rho\alpha}B_\rho^\lambda B_\lambda^\beta\right) = \bar{P}^\beta\\[2ex] \left(N^{\alpha\beta} + M^{\alpha\lambda}B_\lambda^\beta\right)B_{\alpha\beta} - \nabla_\alpha\nabla_\beta(M^{\alpha\beta} + M^{*\lambda\beta}B_\lambda^\alpha) = -\bar{P}^3\end{cases} \tag{2.68}$$

and the boundary conditions on the free border $\bar{\gamma}$:

$$\begin{cases} \left(N^{\alpha\beta} + M^{\alpha\lambda} B_\lambda^\beta\right) \nu_\beta - (M^{\gamma\beta} + M^{*\lambda\beta} B_\lambda^\gamma) \nu_\beta t_\gamma t^\rho B_\rho^\alpha = q^\alpha; \\ -\nabla_\beta (M^{\alpha\beta} + M^{*\lambda\beta} B_\lambda^\alpha) \nu_\alpha - \partial_{\vec{t}}((M^{\gamma\beta} + M^{*\lambda\beta} B_\lambda^\gamma) \nu_\beta t_\gamma) + \partial_{\vec{t}} m^\nu = q^3 \\ (M^{\alpha\beta} + M^{*\lambda\beta} B_\lambda^\alpha) \nu_\alpha \nu_\beta + m^t = 0 \end{cases} \tag{2.69}$$

on the clamped border $\bar{\gamma}_0$

$$U_i = 0 \text{ and } \bar{\theta}_\nu = 0$$

also equivalent to

$$u_\alpha = 0, u_3 = 0 \text{ and } \partial_{\vec{\nu}} u_3 = 0$$

These equations show how the moments $M^{*\lambda\beta} B_\lambda^\gamma$ contribute to the resistance of the shell by their gradient in the mid-surface and their curvilinear derivative on the free border, like the classical moments $M^{\lambda\beta}$. The equations also contain all the terms found in Kirchhoff-Love-Euler's equations if the moments $M^{*\lambda\beta} B_\lambda^\gamma$ are neglected and the force and bending moment N and M tensors are expanded only at the zero order.

Let us recall that

$$N = (N^{\alpha\beta}) = A_0 e(u) - A_1 K(u) + A_2 Q(u) \tag{2.70}$$

$$M = (M^{\alpha\beta}) = -A_1 e(u) + A_2 K(u) - A_3 Q(u) \tag{2.71}$$

$$M^* = (M^{*\alpha\beta}) = A_2 e(u) - A_3 K(u) + A_4 Q(u) \tag{2.72}$$

$$e_{\alpha\beta}(u) = \frac{1}{2}\left(\nabla_\alpha u_\beta + \nabla_\beta u_\alpha - 2B_{\alpha\beta} u_3\right) \tag{2.73}$$

$$K_{\alpha\beta}(u) = \nabla_\alpha\left(B_\beta^\nu u_\nu\right) + B_\alpha^\nu \nabla_\beta u_\nu + \nabla_\alpha \nabla_\beta u_3 - B_\alpha^\rho B_{\rho\beta} u_3 \tag{2.74}$$

$$\begin{aligned} Q_{\alpha\beta}(u) &= \tfrac{1}{2}\left(B_\alpha^\nu \nabla_\beta (\nabla_\nu u_3 + B_\nu^\rho u_\rho) + B_\beta^\nu \nabla_\alpha (\nabla_\nu u_3 + B_\nu^\rho u_\rho)\right) \\ &= \tfrac{1}{2}\left(B_\alpha^\nu \nabla_\beta \theta_\nu + B_\beta^\nu \nabla_\alpha \theta_\nu\right) \end{aligned} \tag{2.75}$$

$$\theta_\nu = \nabla_\nu u_3 + B_\nu^\rho u_\rho \tag{2.76}$$

We have

$$\begin{aligned} \sigma^{\alpha\beta} &= \left(\bar{\Lambda} G^{\alpha\beta} G^{\rho\delta} + \bar{\mu}\left(G^{\alpha\gamma} G^{\beta\delta} + G^{\alpha\delta} G^{\beta\gamma}\right)\right)\left(e_{\gamma\delta} - zK_{\gamma\delta} + z^2 Q_{\gamma\delta}\right) \\ &= \bar{A}e - z\bar{A}K + z^2\bar{A}Q \end{aligned} \tag{2.77}$$

Therefore

$$\sigma^{\alpha\beta} = \begin{bmatrix} \bar{A} & -z\bar{A} & z^2\bar{A} \end{bmatrix} \begin{bmatrix} e \\ K \\ Q \end{bmatrix}$$

But

$$\begin{pmatrix} N \\ M \\ M^* \end{pmatrix} = \begin{bmatrix} N_0 & -N_1 & N_2 \\ -N_1 & N_2 & -N_3 \\ N_2 & -N_3 & N_4 \end{bmatrix} \begin{pmatrix} e \\ K \\ Q \end{pmatrix} = \mathbb{A} \begin{pmatrix} e \\ K \\ Q \end{pmatrix} \tag{2.78}$$

where $\mathbb{A}$ is a symmetric positive defined matrix. We deduce that

$$\mathbb{A}^{-1} \begin{pmatrix} N \\ M \\ M^* \end{pmatrix} = \begin{pmatrix} e \\ K \\ Q \end{pmatrix}$$

and

$$\left[\sigma^{\alpha\beta}\right] = \begin{bmatrix} \bar{A} & -z\bar{A} & z^2\bar{A} \end{bmatrix} \mathbb{A}^{-1} \begin{pmatrix} N \\ M \\ M^* \end{pmatrix} \tag{2.79}$$

2.2.3 Calculations of Transverse Stresses

We deduce from the equilibrium equations ($div\sigma + f = 0$) and the boundary conditions that the transverse stresses satisfy the equation

$$\begin{cases} \frac{d}{dz}\sigma^{\alpha 3} + 2\Gamma^{\alpha}_{\beta 3}\sigma^{\beta 3} + \Gamma^{\beta}_{\beta 3}\sigma^{\alpha 3} & = -\left(\sigma,^{\alpha\beta}_{\beta} + \Gamma^{\alpha}_{\beta\tau}\sigma^{\tau\beta} + \Gamma^{\beta}_{\beta\tau}\sigma^{\alpha\tau}\right) - f^{\alpha} \\ \sigma^{\alpha 3}(-h/2) = -p^{\alpha}_{-},\ \sigma^{\alpha 3}(h/2) & = p^{\alpha}_{+} \end{cases} \tag{2.80}$$

$$\begin{cases} \frac{d}{dz}\sigma^{33} + \Gamma^{\alpha}_{\alpha 3}\sigma^{33} & = -\left(\sigma,^{3\alpha}_{\alpha} + \Gamma^{3}_{\alpha\tau}\sigma^{\tau\alpha} + \Gamma^{\beta}_{\beta\tau}\sigma^{3\tau}\right) - f^{3} \\ \sigma^{33}(-h/2) = -p^{3}_{-},\ \sigma^{33}(h/2) & = p^{3}_{+} \end{cases} \tag{2.81}$$

$\Gamma^{\alpha}_{\beta 3} = -\left(\mu^{-1}\right)^{\alpha}_{\lambda} B^{\lambda}_{\beta}$; $\Gamma^{\beta}_{\beta 3} = -\left(\mu^{-1}\right)^{\beta}_{\lambda} B^{\lambda}_{\beta}$; $\Gamma^{\alpha}_{\beta\tau} = \bar{\Gamma}^{\alpha}_{\beta\tau} + \left(\mu^{-1}\right)^{\alpha}_{\nu} \nabla_{\beta}\mu^{\nu}_{\tau}$; $\Gamma^{3}_{\alpha\tau} = \mu^{\nu}_{\alpha} B_{\nu\tau}$. It is proven in Nzengwa and Tagne [87] that these special differential equations have unique solutions. Their solutions with the initial conditions satisfy the final conditions.

2.2.4 Best First-Order Model for Thick Shells

The variational formulation

$$\int_S (N(u) : e(v) + M(u) : K(v) + M^*(u) : Q(v)) = L(v) \tag{2.82}$$

can be rewritten in the form

$$\mathrm{A}(u, v) = \mathrm{L}(v) \text{ with} \mathrm{A}(u, v) = A_1(u, v) + \sum_{n \geq 2} B^n A^n(u, v) = \mathrm{L}(v) \tag{2.83}$$

By neglecting terms greater than one as calculated above, we obtain

$$N^{\alpha\beta} = \frac{Eh}{1 - \bar{\nu}^2}\left[(1 - \bar{\nu}) e^{\alpha\beta}(u) + \bar{\nu} e^{\rho}_{\rho}(u) A^{\alpha\beta}\right]$$

$$M^{\alpha\beta} = \frac{Eh^3}{12\left(1 - \bar{\nu}^2\right)}\left[(1 - \bar{\nu}) K^{\alpha\beta}(u) + \bar{\nu} K^{\rho}_{\rho}(u) A^{\alpha\beta}\right]$$

$$\bar{N}^{\alpha\beta} = \frac{Eh^3}{12\left(1 - \bar{\nu}^2\right)}\left[(1 - \bar{\nu}) Q^{\alpha\beta}(u) + \bar{\nu} Q^{\rho}_{\rho}(u) A^{\alpha\beta}\right]$$

$$\bar{M}^{\alpha\beta} = \frac{Eh^3}{12\left(1 - \bar{\nu}^2\right)}\left[(1 - \bar{\nu}) e^{\alpha\beta}(u) + \bar{\nu} e^{\rho}_{\rho}(u) A^{\alpha\beta}\right]$$

$$\overline{\overline{M}}^{\alpha\beta} = \frac{Eh^5}{80\left(1 - \bar{\nu}^2\right)}\left[(1 - \bar{\nu}) Q^{\alpha\beta}(u) + \bar{\nu} Q^{\rho}_{\rho}(u) A^{\alpha\beta}\right]$$

Terms with even power on h disappear because the integration of z to the odd power in the interval $-h/2$ to $h/2$ is zero. The corresponding variational equation which also has a unique solution Nzengwa and Tagne [87] reads

$$\begin{cases} Find\ u \in U_{ad} = H^1_{\gamma_o}(S) \times H^1_{\gamma_o}(S) \times H^2_{\gamma_o}(S) \\ A_1(u, v) = \int\limits_S \left[\left(N + \overline{N}\right) : e(v) + M : K(v) + \left(\overline{M} + \overline{\overline{M}}\right) : Q(v)\right] dS \\ \quad = \int\limits_S [\mathcal{N} : e(v) + M : K(v) + \mathcal{M} : Q(v)]\, dS = L(v) \end{cases} \tag{2.84}$$

with $\mathcal{N} = N + \overline{N},\ \mathcal{M} = \overline{M} + \overline{\overline{M}}$

$$\sigma^{\alpha\beta} = \frac{1}{h} N^{\alpha\beta} - \frac{12z}{h^3} M^{\alpha\beta} + \frac{80}{h^5} z^2 \overline{\overline{M}}^{\alpha\beta} \tag{2.85}$$

σ^{i3} are calculated by integrating the differential equations established above for transverse stresses.

Chapter 3
Dynamic Evolution of Shells

3.1 Dynamic Equilibrium Equation of the N-T Model

3.1.1 Variational Equation

The kinematics derived from the static analysis of this model is defined by

$$
\begin{aligned}
U_\alpha(x,z,t) \quad &= \quad u_\alpha(x,t) - z(u_{3,\alpha}(x,t) + 2B_\alpha^\rho u_\rho(x,t)) \\
&\quad + z^2\left(B_\alpha^\tau B_\tau^\nu u_\nu(x,t) + B_\alpha^\tau u_{3,\tau}(x,t)\right) \qquad (3.1)
\end{aligned}
$$

$$U_3(x,z,t) = u_3(x,t)$$

and the 3D variational equation reads

$$\int\limits_\Omega \rho \frac{\partial^2 U}{\partial t^2} V d\Omega + \int\limits_\Omega \sigma\left(U\right) : \epsilon\left(V\right) d\Omega = L\left(V\right) \qquad (3.2)$$

By using the expressions of U, V in u and v, it becomes

$$B\left(\frac{d^2}{dt^2}u, v\right) + A\left(u\left(x,t\right), v\left(x\right)\right) = L\left(v\right) \qquad (3.3)$$

with initial conditions $u(x,o)$; $\dot{u}(x,o)$ and

R. Nzengwa, *Plate and Shell Models*,
https://doi.org/10.1007/978-981-97-2780-3_3

$$\begin{cases} B\left(\frac{d^2}{dt^2}u, v\right) = \int\limits_{\Omega} \rho \frac{\partial^2 U}{\partial t^2} V d\Omega = \frac{d^2}{dt^2} B(u, v); \\ A(u(x,t), v(x)) = \int\limits_{S} [N(u(x,t)) : e(v) + M(u(x,t)) : K(v) \\ \qquad +M^*(u(x,t)) : Q(v)] dS \\ \frac{d^2}{dt^2} B(u, v) = \int\limits_{s} \left(J^{\beta}(\ddot{u}) v_{\beta} + J^3(\ddot{u}) v_3\right) dS \end{cases} \tag{3.4}$$

$$\begin{aligned} L(v) = \int\limits_{S} \left(\bar{P}^{\alpha}(x,t) v_{\alpha}(x) + \bar{P}^3(x,t) v_3(x)\right) dS + \int\limits_{\gamma_1} (q^{\alpha}(x,t) v_{\alpha}(x) \\ +q^3(x,t) v_3(x)) d\bar{\gamma} + \int\limits_{\gamma_1} m^{\alpha}(x,t) \bar{\theta}_{\alpha}(x) d\bar{\gamma} \end{aligned} \tag{3.5}$$

3.1.2 First-Order N-T Model Dynamic Equation

As in the static analysis, we consider only first-order terms in the strain energy. We thus obtain

$$\frac{d^2}{dt^2} B_1(u(x,t), v) + A_1(u(x,t), v) = L(v) \tag{3.6}$$

with

$$A_1(u(x,t), v) = \int\limits_{S} [\mathcal{N} : e(v) + M : K(v) + \mathcal{M} : Q(v)] dS$$

$$\mathcal{N} = N + \overline{N}, \ \mathcal{M} = \overline{M} + \overline{\overline{M}}$$

$$\bar{M}^{\alpha\beta} = \frac{Eh^3}{12(1-\bar{\nu}^2)} \left[(1-\bar{\nu}) e^{\alpha\beta}(u(x,t)) + \bar{\nu} e^{\rho}_{\rho}(u(x,t)) A^{\alpha\beta}\right]$$

$$\overline{\overline{M}}^{\alpha\beta} = \frac{Eh^5}{80(1-\bar{\nu}^2)} \left[(1-\bar{\nu}) Q^{\alpha\beta}(u(x,t)) + \bar{\nu} Q^{\rho}_{\rho}(u(x,t)) A^{\alpha\beta}\right]$$

$$\bar{N}^{\alpha\beta} = \frac{Eh^3}{12(1-\bar{\nu}^2)} \left[(1-\bar{\nu}) Q^{\alpha\beta}(u(x,t)) + \bar{\nu} Q^{\rho}_{\rho}(u(x,t)) A^{\alpha\beta}\right]$$

$$M^{\alpha\beta} = \frac{Eh^3}{12\left(1-\bar{\nu}^2\right)}\left[(1-\bar{\nu})\,K^{\alpha\beta}\,(u(x,t)) + \bar{\nu}K^{\rho}_{\rho}\,(u(x,t))\,A^{\alpha\beta}\right]$$

$$N^{\alpha\beta} = \frac{Eh}{1-\bar{\nu}^2}\left[(1-\bar{\nu})\,e^{\alpha\beta}\,(u(x,t)) + \bar{\nu}e^{\rho}_{\rho}\,(u(x,t))\,A^{\alpha\beta}\right]$$

$$\frac{d^2}{dt^2}B_1\,(u(x,t),v) = \int_s \left(I^{\beta}\,(\ddot{u})\,v_\beta + I^3\,(\ddot{u})\,v_3\right)dS; \tag{3.7}$$

$$\begin{aligned} I^{\beta}\,(\ddot{u}) = \quad & \rho h A^{\alpha\beta}\ddot{u}_\alpha + \rho\{\tfrac{h^3}{12}\{A^{\alpha\tau}B^{\beta}_{\nu}B^{\nu}_{\tau}\ddot{u}_\alpha \\ & +2A^{\alpha\tau}B^{\beta}_{\tau}\left(\partial_\alpha\ddot{u}_3 + 2B^{\nu}_{\alpha}\ddot{u}_\nu\right) + A^{\alpha\beta}(B^{\nu}_{\tau}B^{\tau}_{\alpha}\ddot{u}_\nu \\ & +B^{\tau}_{\alpha}\partial_\tau\ddot{u}_3)\} + \tfrac{h^5}{80}A^{\delta\alpha}B^{\beta}_{\gamma}B^{\gamma}_{\delta}(B^{\nu}_{\tau}B^{\tau}_{\alpha}\ddot{u}_\nu + B^{\tau}_{\alpha}\partial_\tau\ddot{u}_3)\} \end{aligned}$$

$$\begin{aligned} I^3\,(\ddot{u}) = \rho h\ddot{u}_3 - \rho\tfrac{h^3}{12}\partial_\beta\left(A^{\alpha\tau}B^{\beta}_{\tau}\ddot{u}_\alpha + A^{\alpha\beta}\left(\partial_\alpha\ddot{u}_3 + 2B^{\tau}_{\alpha}\ddot{u}_\tau\right)\right) \\ -\rho\tfrac{h^5}{80}\partial_\beta\left(A^{\alpha\tau}B^{\beta}_{\tau}\left(B^{\nu}_{\delta}B^{\delta}_{\alpha}\ddot{u}_\nu + B^{\nu}_{\alpha}\partial_\nu\ddot{u}_3\right)\right) \end{aligned}$$

The inertia I^3 contains the classical term $\rho h\ddot{u}_3$ and the term proposed by Morozov (1967)

$$-\rho\frac{h^3}{12}\partial_\beta\left(A^{\alpha\beta}\partial_\alpha\ddot{u}_3\right) = \rho\frac{h^3}{12}\,\Delta\,\ddot{u}_3 \tag{3.8}$$

in a physical base ($A^{\alpha\beta} = \delta^{\alpha\beta}$). Euler's equations are obtained by adding to the static equations inertial terms and initial conditions. The equations now read

$$I^{\alpha}\,(\ddot{u}) + (M^{\alpha\rho} + M^{*\lambda\rho}B^{\alpha}_{\lambda})\nabla_\rho(B^{\beta}_{\alpha}) - \nabla_\alpha\left(N^{\alpha\beta} + 2M^{\alpha\lambda}B^{\beta}_{\lambda} + M^{*\rho\alpha}B^{\lambda}_{\rho}B^{\beta}_{\lambda}\right) = \bar{P}^{\beta}; \tag{3.9}$$

$$I^3\,(\ddot{u}) + (N^{\alpha\beta} + M^{\alpha\lambda}B^{\beta}_{\lambda})B_{\alpha\beta} - \nabla_\alpha\nabla_\beta(M^{\alpha\beta} + M^{*\lambda\beta}B^{\alpha}_{\lambda}) = -\bar{P}^3 \tag{3.10}$$

to be completed with boundary and initial conditions.

3.1.3 Transverse Stress Equations

Transverse stresses are obtained by solving the equations

$$\begin{cases} \frac{d}{dz}\sigma^{\alpha 3} + 2\Gamma^{\alpha}_{\beta 3}\sigma^{\beta 3} + \Gamma^{\beta}_{\beta 3}\sigma^{\alpha 3} & = -\left(\sigma,^{\alpha\beta}_{\beta} + \Gamma^{\alpha}_{\beta\tau}\sigma^{\tau\beta} + \Gamma^{\beta}_{\beta\tau}\sigma^{\alpha\tau}\right) - f^{\alpha} + \rho\frac{\partial^2}{\partial t^2}U_{\alpha} \\ \sigma^{\alpha 3}(-h/2) = -p^{\alpha}_{-}, \\ \sigma^{\alpha 3}(h/2) = p^{\alpha}_{+} \end{cases} \tag{3.11}$$

$$\begin{cases} \frac{d}{dz}\sigma^{33} + \Gamma^{\alpha}_{\alpha 3}\sigma^{33} & = -\left(\sigma,^{3\alpha}_{\alpha} + \Gamma^{3}_{\alpha\tau}\sigma^{\tau\alpha} + \Gamma^{\beta}_{\beta\tau}\sigma^{3\tau}\right) - f^{3} + \rho\frac{\partial^2}{\partial t^2}U_{3} \\ \sigma^{33}(-h/2) = -p^{3}_{-}, \\ \sigma^{33}(h/2) = p^{3}_{+} \end{cases} \tag{3.12}$$

Exercise: Proof that the dynamic variational equations and its first-order version have unique solutions in $H^2(0, T; \; H^1_{\gamma_o}(S) \times H^1_{\gamma_o}(S) \times H^2_{\gamma_o}(S))$. Use properties of maximal monotone operators (see H. Brezis 1983).

3.2 Free Vibrations

3.2.1 Free Vibrations with Total Inertia

The free vibration equations under plane strain are

$$\begin{cases} \rho\frac{\partial^2}{\partial t^2}U(x, z, t) - div\sigma = 0 \text{ in } \Omega \\ \sigma\vec{n} = 0 \text{ on } \Gamma = \partial\Omega \setminus \Gamma_0 \\ U(x, z, t) = 0 \text{ on } \Gamma_0 \end{cases} \tag{3.13}$$

$$\begin{aligned} U_{\alpha}(x, z, t) \quad &= \quad u_{\alpha}(x, t) - z(u_{3,\alpha}(x, t) + 2B^{\rho}_{\alpha}u_{\rho}(x, t)) \\ &+z^2\left(B^{\tau}_{\alpha}B^{\nu}_{\tau}u_{\nu}(x, t) + B^{\tau}_{\alpha}u_{3,\tau}(x, t)\right) \end{aligned} \tag{3.14}$$

$$U_3(x, z, t) = u_3(x, t)$$

$$\sigma^{\alpha\beta}(U(x,z,t)) = \left(\bar{\Lambda}G^{\alpha\beta}G^{\gamma\delta} + \bar{\mu}\left(G^{\alpha\gamma}G^{\beta\delta} + G^{\alpha\delta}G^{\beta\gamma}\right)\right)\epsilon_{\gamma\delta}(U(x,z,t))$$

$$= \left(\bar{\Lambda}G^{\alpha\beta}G^{\gamma\delta} + \bar{\mu}\left(G^{\alpha\gamma}G^{\beta\delta} + G^{\alpha\delta}G^{\beta\gamma}\right)\right)(e_{\gamma\delta}(x,t) - zK_{\gamma\delta}(x,t) + z^2 Q_{\gamma\delta}(x,t)) \quad (3.15)$$

$$e_{\alpha\beta}(x,t) = \frac{1}{2}\left(\nabla_\alpha u_\beta(x,t) + \nabla_\beta u_\alpha(x,t) - 2B_{\alpha\beta}u_3(x,t)\right) \quad (3.16)$$

$$K_{\alpha\beta}(x,t) = \nabla_\alpha\left(B^\nu_\beta u_\nu(x,t)\right) + B^\nu_\alpha\nabla_\beta u_\nu(x,t) + \nabla_\alpha\nabla_\beta u_3(x,t) - B^\rho_\alpha B_{\rho\beta}u_3(x,t) \quad (3.17)$$

$$Q_{\alpha\beta}(x,t) = \tfrac{1}{2}(B^\nu_\alpha\nabla_\beta(\nabla_\nu u_3(x,t) + B^\rho_\nu u_\rho(x,t)) + B^\nu_\beta\nabla_\alpha(\nabla u_3(x,t) + B^\rho_\nu u_\rho(x,t))) \quad (3.18)$$

$$= \tfrac{1}{2}\left(B^\nu_\alpha\nabla_\beta\theta_\nu + B^\nu_\beta\nabla_\alpha\theta_\nu\right)$$

$$\theta_\nu = \nabla_\nu u_3(x,t) + B^\rho_\nu u_\rho(x,t)$$

Let $U(x,z,t) = cos(\sqrt{\Lambda}\ t)U(x,z)$ or $U(x,z,t) = sin(\sqrt{\Lambda}\ t)U(x,z)$. After simplification the equation reads

$$-\,div\sigma(U(x,z)) = \Lambda\rho U(x,z) \text{ in } \Omega \quad (3.19)$$

or in its variational form

$$\int_\Omega \sigma\left(U(x,z)\right) : \epsilon\left(V(x,z)\right) d\Omega = \Lambda\int_\Omega \rho U(x,z).V(x,z)d\Omega \quad (3.20)$$

equivalent to

$$A\left(u\left(x\right), v\left(x\right)\right) = \Lambda\int_S \left(J^\beta\left(u\right)v_\beta + J^3\left(u\right)v_3\right)dS = \Lambda B\left(u,v\right) \quad (3.21)$$

Exercise: Calculate the terms $J^\beta\left(u\right),\ J^3\left(u\right)$

3.2.2 First-Order Free Vibrations

At the first order, we have

$$A_1(u(x), v(x)) = \Lambda \int_S \left(I^\beta(u) v_\beta + I^3(u) v_3\right) dS = \Lambda B_1(u, v)$$

The inertia $(J^\beta(u), J^3(u))$ or $(I^\beta(u), I^3(u))$ contain the terms $pha^{\alpha\beta}u_\alpha$ and phu_3 which are very significant and predominant in certain structures during vibration. We then define

$$\begin{aligned} B_2(u, v) &= \int_S \left(\rho h A^{\alpha\beta} u_\alpha v_\beta + \rho h u_3 v^3\right) dS \\ &= \int_S \left(\rho h u_\alpha v^\alpha + \rho h u_3 v^3\right) dS \\ &= (u, v) \end{aligned}$$

By using the approximate inertia, the 2D variational full and first-order equations are, respectively,

$$A(u, v) = \Lambda B_2(u, v) = \Lambda(u, v)$$
$$A_1(u, v) = \Lambda B_2(u, v) = \Lambda(u, v)$$

Let us consider the operator T defined in $U_{ad} = H^1_{\gamma_o}(S) \times H^1_{\gamma_o}(S) \times H^2_{\gamma_o}(S))$ as follows: $u \longmapsto Tu$; such that $A(Tu, v) = B_2(u, v) = (u, v)$. We have $(Tu, v) = (u, Tv)$ and there exists a constant $c > 0$ such that $(Tu, u) = A(Tu, Tu) \geqslant c \parallel Tu \parallel^2$. It follows that the operator T is positive defined and satisfies all the properties of symmetric compact operators. Consequently T has a sequence of eigenvalues and orthonormal eigenvectors (ϖ^n, u^n), $n \mapsto \infty$, $\varpi^n > 0$ and $(u^n, u^m) = \delta^{nm}$.

Exercise: Show that the operators T and T_1 corresponding to the variational problems defined by A and A_1 respectively, are compact. By replacing the inertia operator by B and B_1 for A and A_1, respectively, are the operators T and T_1 still compact?

We deduce from the properties of the operator T that the eigenvectors (or free modes) satisfy the equations

$$A\left(u^n, v\right) = \frac{1}{\varpi^n} B_2\left(u^n, v\right) = \frac{1}{\varpi^n}\left(u^n, v\right) = \Lambda^n\left(u^n, v\right), \quad \Lambda^n = \frac{1}{\varpi^n}$$

$$0 < \Lambda^1 \leq \Lambda^2 \leq \Lambda^3 ...... \leq \Lambda^n, \cdots \Lambda^n \rightarrow \infty \text{ when n} \mapsto \infty$$

$$\begin{aligned} U^n_\alpha(x,z) &= u^n_\alpha(x) - z(u^n_{3,\alpha}(x) + 2B^\rho_\alpha u^n_\rho(x)) \\ &\quad + z^2\left(B^\nu_\tau B^\tau_\alpha u^n_\nu(x) + B^\tau_\alpha u^n_{3,\tau}(x)\right) \end{aligned} \tag{3.22}$$

$$\begin{aligned} U^n_3(x,z) &= u^n_3(x) \\ \sigma^{\alpha\beta}_n &= \left(\bar{\Lambda} G^{\alpha\beta} G^{\rho\delta} + \bar{\mu}\left(G^{\alpha\gamma} G^{\beta\delta} + G^{\alpha\delta} G^{\beta\gamma}\right)\right)(e_{\gamma\delta}(u^n) \\ &\quad - zK_{\gamma\delta}(u^n) + z^2 Q_{\gamma\delta}(u^n)) \end{aligned} \tag{3.23}$$

$$\begin{cases} \frac{d}{dz}\sigma^{\alpha3}_n + 2\Gamma^\alpha_{\beta3}\sigma^{\beta3}_n + \Gamma^\beta_{\beta3}\sigma^{\alpha3}_n = -\left(\sigma^{\alpha\beta}_{n,\beta} + \Gamma^\alpha_{\beta\tau}\sigma^{\tau\beta}_n\right) - \rho\Lambda^n\left(U^n\right)^\alpha \\ \sigma^{\alpha3}_n(-h/2) = 0, \\ \sigma^{\alpha3}_n(h/2) = 0 \end{cases} \tag{3.24}$$

$$\begin{cases} \frac{d}{dz}\sigma^{33}_n + \Gamma^\alpha_{\alpha3}\sigma^{33}_n = -\left(\sigma^{3\alpha}_{n,\alpha} + \Gamma^3_{\alpha\tau}\sigma^{\tau\alpha}_n + \Gamma^\beta_{\beta\tau}\sigma^{3\tau}_n\right) - \rho\Lambda^n\left(U^n\right)^3 \\ \sigma^{33}_n(-h/2) = 0, \\ \sigma^{33}_n(h/2) = 0 \end{cases}$$

Eigenvalues are calculated by using Rayleigh coefficients

$$R(v) = \frac{A_1(v,v)}{(v,v)}, \quad v \neq 0, \quad \Lambda^1 = min\{R(v),\ v \in U_{ad},\ v \neq 0\},$$

$$\begin{aligned} \Lambda^n &= min\{R(v),\ v \in U_{ad},\ v \neq 0, \left(v, u^k\right) = 0, \\ &\qquad 0 \le k \le n-1,\ n \geqslant 2\} \\ &= min\{max R(v),\ v \in W,\ v \neq 0, W \in F^n\} \end{aligned} \tag{3.25}$$

F^n denotes the family of all B_1-dimension sub-spaces of U_{ad}. During the vibration of certain structures, the transverse inertia is predominant, i.e.

$$\int_S \left(\rho h \ddot{u}_\alpha v^\alpha + \rho h \ddot{u}_3 v^3\right) dS \approx \int_S \rho h \ddot{u}_3 v^3 dS \tag{3.26}$$

3.2.3 Free Vibrations with Simplified Inertia

We define

$$B_3\,(u,v) = (u,v)_3 = \int\limits_S \rho h u_3 v^3 dS$$

and the operator

$$u \longmapsto Tu \ \ , A\,(Tu,v) = B_3\,(u,v) = (u,v)_3$$

or the equivalent first-order operator $u \longmapsto Tu;\ A_1\,(Tu,v) = B_3\,(u,v) = (u,v)_3$. The operators T thus defined with A or A_1 are symmetric, positive defined and compact. Let us calculate

$$min\left\{(Tu,u)_3\,,u \in U_{ad},\ u \neq 0,\,(u,u)=1,\left(u,u^k\right)=0,\,1 \le k \le n-1,\ n \geqslant 2\right\}$$

If we consider the extremum of the function f with ϖ the Lagrange multiplier

$$f\,(u,\varpi) = (Tu,u)_3 - \varpi\,((u,u)-1) \tag{3.27}$$

Then

$$(Tu,\delta u)_3 - \varpi\,(u,\delta u) = 0 \text{ and } \delta\varpi\,((u,u)-1) = 0 \tag{3.28}$$

By considering $\delta u = (u_1, u_2, 0)$ we deduce from

$$\varpi\,(u,\delta u) = \varpi \int\limits_S \left(\rho h u_\alpha \delta u^\alpha + \rho h u_3 \delta u^3\right) dS = 0$$

that $u = (0,0,u_3)$ and $(Tu - \varpi u,\ \delta u)_3 = 0$.
It follows that ϖ^n is the n-th eigenvalue of the operator T and the eigenvector $u^n = (0,0,u_3^n),\ u_3^n \in H^2_{\gamma_o}(S)$. Let us denote $u_3^n = \eta$. The free vibration problem with the simplified inertia operator B_3 now reads

$$A\,(u^n,v) = \int\limits_S \left(N(\eta) : e(v^3) + M(\eta) : K(v^3) + M^*(\eta) : Q(v^3)\right) dS$$

$$= \Lambda^n\,(u^n,v)_3 = \Lambda^n \int\limits_S \rho h \eta v^3 dS, \tag{3.29}$$

$$u^n = \eta, \quad v^3 \in H^2_{\gamma_o}(S)$$

The corresponding first-order variational formulation also follows:

$$A_1(u^n, v) = \int_S \left[\mathcal{N}(\eta) : e\left(v^3\right) + M(\eta) : K\left(v^3\right) + \mathcal{M}(\eta) : Q\left(v^3\right)\right] dS$$
$$= \Lambda^n \int_S \rho h \eta v^3 dS \tag{3.30}$$

$$\mathcal{N} = N + \overline{N}, \cdots \mathcal{M} = \overline{M} + \overline{\overline{M}}$$

$$N_n^{\alpha\beta} = \frac{Eh}{1-\bar{\nu}^2}\left[(1-\bar{\nu})\, e^{\alpha\beta}(\eta) + \bar{\nu} e_\rho^\rho(\eta)\, A^{\alpha\beta}\right],$$

$$\bar{N}_n^{\alpha\beta} = \frac{Eh^3}{12\left(1-\bar{\nu}^2\right)}\left[(1-\bar{\nu})\, Q^{\alpha\beta}(\eta) + \bar{\nu} Q_\rho^\rho(\eta)\, A^{\alpha\beta}\right]$$

$$M_n^{\alpha\beta} = \frac{Eh^3}{12\left(1-\bar{\nu}^2\right)}\left[(1-\bar{\nu})\, K^{\alpha\beta}(\eta) + \bar{\nu} K_\rho^\rho(\eta)\, A^{\alpha\beta}\right]$$

$$\overline{\overline{M}}_n^{\alpha\beta} = \frac{Eh^5}{80\left(1-\bar{\nu}^2\right)}\left[(1-\bar{\nu})\, Q^{\alpha\beta}(\eta) + \bar{\nu} Q_\rho^\rho(\eta)\, A^{\alpha\beta}\right]$$

$$\bar{M}_n^{\alpha\beta} = \frac{Eh^3}{12\left(1-\bar{\nu}^2\right)}\left[(1-\nu)\, e^{\alpha\beta}(\eta) + \bar{\nu} e_\rho^\rho(\eta)\, A^{\alpha\beta}\right]$$

$$e^n_{\alpha\beta}(\eta) = -\eta B_{\alpha\beta} \quad K^n_{\alpha\beta}(\eta) = \nabla_\alpha \nabla_\beta \eta - \eta B_\alpha^\rho B_{\beta\rho};$$

$$Q^n_{\alpha\beta}(\eta) = \frac{1}{2}\left(B_\alpha^\nu \nabla_\beta \nabla_\nu \eta + B_\beta^\nu \nabla_\alpha \nabla_\nu \eta\right)$$

$$e_{\alpha\beta}(v^3) = -v^3 B_{\alpha\beta};$$

$$K_{\alpha\beta}(v^3) = \nabla_\alpha \nabla_\beta v^3 - v^3 B_\alpha^\rho B_{\beta\rho};$$

$$Q_{\alpha\beta}(v^3) = \tfrac{1}{2}\left(B_\alpha^\nu \nabla_\beta \nabla_\nu v^3 + B_\beta^\nu \nabla_\alpha \nabla_\nu v^3\right)$$

$$U^n_\alpha(x, z) = -z\eta_{,\alpha}(x) + z^2 B_\alpha^\tau \eta_{,\tau}(x); \quad U^n_3(x, z) = \eta(x) \tag{3.31}$$

Note that the different variational problems addressed above with different inertia operators have numerical solutions.

Exercise: Write down Euler's equations (or strong formulations) of the above different eigenvalue problems.

3.3 The Model "N" of Thick Shells

3.3.1 Existence of a Transverse Strain Potential

Under the assumption that there exists a transverse strain potential, we define here a kinematic which accounts for through the thickness deformation. Thickness variation and section warping under torsion are thus addressed. Indeed, during the deformation of a thick shell, layers tend to slide relative to each other non-uniformly and to pull apart from each other. To maintain plane deformation, transverse pressure, σ^{i3}, should be applied to secure equilibrium. These transverse stresses were calculated above by solving appropriate differential equations. Transverse stresses thus calculated appear to be reactions because they are not deduced from a constitutive law. We remark that a simple non-uniform variation of thickness during deformation creates transverse strain which produces transverse stress. A torque around the normal axis of the mid-surface S also creates through the thickness transverse strain and significant warping of the extreme surfaces ($z = -h/2$ and $z = h/2$) which modifies the thickness. These observations suggest to find a kinematic which can reproduce the behaviour of a solid subject to torsion. Let

$$\begin{gathered}\phi_\alpha = 2\epsilon_{\alpha 3} = \left(U_{\alpha/3} + U_{3/\alpha}\right) = \left[\mu_\alpha^\nu \overline{U}_{\nu,3} + \overline{U}_{3,\alpha} + B_\alpha^\nu \overline{U}_\nu\right] \\ \phi_3 = \epsilon_{33} = U_{3/3} = U_{3,3} = \overline{U}_{3,3}\end{gathered} \tag{3.32}$$

We assume $rot\,(\phi) = 0$; $rot\,(\phi)_i = \phi_{i+1,i+2} - \phi_{i+2,i+1}$ mod [3]. Then there exists a potential $q(x, z)$ such that

$$\begin{gathered}\epsilon_{\alpha 3} = \tfrac{1}{2}\left(U_{\alpha/3} + U_{3/\alpha}\right) = \tfrac{1}{2}\left[\mu_\alpha^\nu \overline{U}_{\nu,3} + \overline{U}_{3,\alpha} + B_\alpha^\nu \overline{U}_\nu\right] = \tfrac{1}{2}\partial_\alpha q \\ \epsilon_{33} = \overline{U}_{3,3} = \partial_3 q = \partial_z q\end{gathered} \tag{3.33}$$

Let us denote

$$\left(U_{3,1},\, U_{3,2}\right) = \nabla U_3,\;\; A_\beta^\alpha = \left(\mu^{-1}\right)_\beta^\alpha \text{ and } \left(A^{-1}\right)_\beta^\alpha = \mu_\beta^\alpha \tag{3.34}$$

Then we have

$$\frac{d}{dz}\left(A\overline{U}\right) + A^2\nabla\left(U_3 - q\right) = 0,\;\; \frac{d}{dz}U_3 = \partial_z q,\, \bar{U} = \left(\bar{U}_\alpha\right) \tag{3.35}$$

Let $w = (U_3 - q)$. We obtain

$$\frac{d}{dz}\left(A\overline{U}\right) + A^2\nabla w = 0,\;\; \frac{d}{dz}w = 0 \tag{3.36}$$

We deduce from these equations that $U_3(x,z) = q(x,z) + u_3(x)$ and $\overline{U} = A^{-1}\overline{u} - z\nabla u_3$, $\bar{u} = (\bar{u}_\alpha(x))$. Indeed, because

$$\frac{d}{dz}w = 0 \ , u_3 = w$$

and

$$\begin{aligned}\frac{d}{dz}\left(A\left(A^{-1}\overline{u} - z\nabla w\right) + A^2\nabla w\right) = & \frac{d}{dz}(\overline{u} - zA\nabla w) + A^2\nabla w \\ = & -zA^2B\nabla w - A\nabla w + A^2\nabla w \\ = & -A^2\left(zB + A^{-1}\right)\nabla w + A^2\nabla w \\ = & -A^2\nabla w + A^2\nabla w = 0\end{aligned} \tag{3.37}$$

Therefore, the kinematic is

$$U_\alpha = u_\alpha - z\left(\partial_\alpha u_3 + 2B_\alpha^\rho u_\rho\right) + z^2\left(B_\alpha^\rho B_\rho^\tau u_\tau + B_\alpha^\rho \partial_\rho u_3\right), \ U_3 = u_3 + q \tag{3.38}$$

We remark that q is the *stretching* function and

$$\begin{aligned}\epsilon_{\alpha\beta} &= e_{\alpha\beta} - zK_{\alpha\beta} + z^2 Q_{\alpha\beta} + q\Upsilon_{\alpha\beta} \\ \epsilon_{\alpha 3} &= \frac{1}{2}\partial_\alpha q \\ \epsilon_{33} &= \partial_z q\,(x,z) \\ \Upsilon_{\alpha\beta} &= -\frac{1}{2}\left(\mu_\alpha^\rho B_{\rho\beta} + \mu_\beta^\rho B_{\rho\alpha}\right)\end{aligned} \tag{3.39}$$

The tensor $(q\Upsilon_{\alpha\beta})$ describes section warping. The kinematic thus obtained (Nzengwa 2005) and referred herein as the model "N", established after solving an equation of mechanical origin, is not heuristic. It contains, in addition to the N-T kinematics, a term which accounts for *stretching through the thickness*. Certain authors have guessed some kinematics without a mechanical explanation or foundation. Some of them reproduce $\epsilon_{\alpha 3} \neq 0$ and maintain $\epsilon_{33} = 0$. The strain tensor is therefore made of two components

$$
\begin{aligned}
\epsilon &= \epsilon(u,q) = \epsilon(u) + \epsilon(q) \\
\epsilon_{\alpha\beta}(u) &= e_{\alpha\beta}(u) - zK_{\alpha\beta}(u) + z^2 Q_{\alpha\beta}(u) \\
\epsilon_{i3}(u) &= 0, \qquad \epsilon_{\alpha\beta}(q) = q(x,z)\Upsilon_{\alpha\beta} \\
\epsilon_{\alpha 3}(q) &= \tfrac{1}{2}\partial_\alpha q, \qquad \epsilon_{33}(q) = \partial_z q
\end{aligned}
\tag{3.40}
$$

The constitutive law now reads

$$
\sigma = \bar{\lambda}\epsilon_l^l G^{\cdot\cdot} + 2\bar{\mu}\epsilon = \sigma(\epsilon(u)) + \sigma(\epsilon(q)) \tag{3.41}
$$

$$
\sigma^{ij}(u,q) = \bar{\lambda}\left(\epsilon_\alpha^\alpha(u) + \epsilon_l^l(q)\right)G^{ij} + 2\bar{\mu}\left(\epsilon^{ij}(u) + \epsilon^{ij}(q)\right) = C^{ijkl}\epsilon_{kl}(u,q) \tag{3.42}
$$

and the moduli tensor $C = \left(C^{ijkl}\right)$ satisfies all material symmetries and the ellipticity condition. We have

$$
\sigma^{33} = \bar{\lambda}\left(\epsilon_\alpha^\alpha(u) + \epsilon_\rho^\rho(q)\right) + \left(\bar{\lambda} + 2\bar{\mu}\right)\epsilon_3^3(q) \tag{3.43}
$$

Assume that $\sigma^{33} = 0$, then

$$
\epsilon_3^3 = -\frac{\bar{\lambda}}{\left(\bar{\lambda} + 2\bar{\mu}\right)}\left(\epsilon_\alpha^\alpha(u) + \epsilon_\alpha^\alpha(q)\right); \quad \sigma^{\alpha 3} = 2\bar{\mu}\epsilon^{\alpha 3}(q); \tag{3.44}
$$

$$
\sigma^{\alpha\beta} = \frac{2\bar{\mu}\bar{\lambda}}{\left(\bar{\lambda} + 2\bar{\mu}\right)}\left(\epsilon_\alpha^\alpha(u) + \epsilon_\rho^\rho(q)\right)G^{\alpha\beta} + 2\bar{\mu}\left(\epsilon^{\alpha\beta}(u) + \epsilon^{\alpha\beta}(q)\right) \tag{3.45}
$$

It shows that we can obtain plane stress only if $\partial_\alpha q = 0$. The equilibrium equation of a thick shell, in our model problem, under this kinematic reads

$$
\rho\ddot{U} - div\sigma = f \quad \text{in } \Omega \times [0,T], \ddot{U} = \frac{\partial^2 U}{\partial t^2} \tag{3.46}
$$

$$
q(.,t) = 0, \;\; U(.,t) = 0 \text{ on } \Gamma_0, \sigma\vec{n} = \text{p on } \Gamma = \Gamma_1 \cup \Gamma_- \cup \Gamma_+ \tag{3.47}
$$

$$
U_\alpha(.,0) = U_{\alpha 0}, \;\; u_3(.,0) = u_{30}, \;\; q(.,0) = q_0, \;\; \dot{U}_\alpha = \partial U_\alpha/\partial t = U_{\alpha 1} \tag{3.48}
$$

$$
\dot{u}_3 = \frac{\partial u_3}{\partial t} = u_{31}, \;\; \dot{q} = \frac{\partial q}{\partial t} = q_1 \; in \; \Omega, \tag{3.49}
$$

$$\sigma^{\cdot\cdot} = \bar{\lambda}\epsilon_l^l G^{\cdot\cdot} + 2\bar{\mu}\epsilon^{\cdot\cdot}$$

with

$$\begin{aligned} U_\alpha(x,z,t) &= u_\alpha(x,t) - z\left(\partial_\alpha u_3(x,t) + 2B_\alpha^\delta u_\delta(x,t)\right) \\ &\quad + z^2\left(B_\alpha^\delta B_\delta^\tau u_\tau(x,t) + B_\alpha^\delta \partial_\delta u_3(x,t)\right) \\ U_3(x,z,t) &= \quad u_3(x,t) + q(x,z,t) \end{aligned} \tag{3.50}$$

3.3.2 Choice of a Transverse Distribution Function

We consider a through the thickness Taylor expansion of the stretching function q across the mid-surface S. We have

$$\begin{aligned} q(x,z,t) = &\quad q(x,0,t) + z\partial_z q(x,0,t) + \frac{z^2}{2}\partial_{z^2}^2 q(x,0,t) \\ &\quad ... + z^{n+1}0(x,z^{n+1},t) \\ &= q_0(x,t) + z\bar{q}_1(x,t) + z^2\bar{q}_2(x,t) + ... + z^n\bar{q}_n(x,t) \\ &\quad + z^{n+1}0(x,z^{n+1},t) \end{aligned} \tag{3.51}$$

The term $q(x,0,t)$ is similar to $u_3(x,t)$. In order to define a 2D model we can consider a truncated expansion

$$q(x,z,t) = z\partial_z q(x,0,t) = z\bar{q}(x,t) = w(z)\,\overline{q}(x,t)$$

Many authors have proposed heuristically different distribution functions $w(z)$, in the form of polynomials or none polynomials, which all satisfy the conditions $w(0) = 0$. Recall that this condition is necessary to secure the transformation of the reference mid-surface into the mid-surface of the deformed shell. If a polynomial of degree n is chosen, then consistently, one should have to deal with unknown functions $\bar{q}_i, i = 1, \cdots n$. We assume $w(z)$ satisfies $w(0) = 0$. Then the variational formulation reads

$$B^w(\cdot\cdot U, V) + A^w(U, V) = L^w(V) \tag{3.52}$$

$U = U(u, \bar{q}), \quad V = V(v, \bar{y})\ ; U = (U_\alpha, U_3)\ ;$

$$U_\alpha(x,z,t) = u_\alpha(x,t) - z\left(\partial_\alpha u_3(x,t) + 2B_\alpha^\delta u_\delta(x,t)\right)$$
$$+z^2\left(B_\alpha^\delta B_\delta^\tau u_\tau(x,t) + B_\alpha^\delta \partial_\delta u_3(x,t)\right) \quad (3.53)$$
$$U_3(x,z,t) = u_3(x,t) + w(z)\bar{q}(x,t)$$

$V = (V_\alpha,\ V_3)$;

$$V_\alpha(x,z) = v_\alpha(x) - z\left(\partial_\alpha v_3(x) + 2B_\alpha^\delta v_\delta(x)\right)$$
$$+z^2\left(B_\alpha^\delta B_\delta^\tau v_\tau(x) + B_\alpha^\delta \partial_\delta v_3(x)\right) \quad (3.54)$$
$$V_3(x,z) = v_3(x) + w(z)\bar{y}(x)$$

which can also be written, for $U_{ad} = H^2(0,T;\ H^1_{\gamma_o}(S) \times H^1_{\gamma_o}(S) \times H^2_{\gamma_o}(S) \times H^1_{\gamma_o}(S))$

$$\begin{cases} \text{find } (u(x,t), \bar{q}(x,t)) \in U_{ad} \\ B^w\left((\ddot{u},\ddot{\bar{q}});(v,\bar{y})\right) + A^w\left((u,\bar{q});(v,\bar{y})\right) = L^w(v,\bar{y}); \end{cases} \quad (3.55)$$

$$B^w\left((\ddot{u},\ddot{\bar{q}});(v,\bar{y})\right) = \int_S \left(J^\alpha(\ddot{u})v_\alpha + J^3(\ddot{u},\ddot{\bar{q}})v_3 + J^4(\ddot{u},\ddot{\bar{q}})\bar{y}\right)dS; \quad (3.56)$$

$$J^3(\ddot{u},\ddot{\bar{q}}) = J^3(\ddot{u}) + \ddot{\bar{q}}\int_{-\frac{h}{2}}^{\frac{h}{2}} w\psi dz$$
$$J^4(\ddot{u},\ddot{\bar{q}}) = \ddot{\bar{q}}\int_{\frac{-}{h}2}^{\frac{h}{2}} w^2\psi dz + \ddot{u}_3\int_{\frac{-}{h}2}^{\frac{h}{2}} w\psi dz$$

$$A^w\left((u,\bar{q});(v,\bar{y})\right) = \int_S\int_{-\frac{h}{2}}^{\frac{h}{2}} \sigma^{ij}(u,q)\epsilon_{ij}(v,w\bar{y})\,\psi dz dS$$
$$= \int_S\int_{-\frac{h}{2}}^{\frac{h}{2}} [\bar{\lambda}\left(G^{\delta\tau}\epsilon_{\delta\tau}(u) + w\bar{q}G^{\delta\tau}\Upsilon_{\delta\tau} + w'\bar{q}\right)G^{\alpha\beta}$$
$$+2\bar{\mu}\left(G^{\alpha\delta}G^{\tau\beta}\epsilon_{\delta\tau}(u) + w\bar{q}G^{\alpha\delta}G^{\tau\beta}\Upsilon_{\delta\tau}\right)] \quad (3.57)$$
$$(e_{\alpha\beta}(v) - zK_{\alpha\beta}(v) + z^2 Q_{\alpha\beta}(v) + w\bar{y}\Upsilon_{\alpha\beta})\psi dz dS$$
$$+\int_S\int_{-\frac{h}{2}}^{\frac{h}{2}} [\bar{\mu}G^{\alpha\beta}w^2\partial_\beta\bar{q}\partial_\alpha\bar{y} + \bar{\lambda}(G^{\alpha\beta}\epsilon_{\alpha\beta}(u)$$
$$+w\bar{q}G^{\alpha\beta}\Upsilon_{\alpha\beta})w'\bar{y} + (\bar{\lambda}+2\bar{\mu})\,\bar{q}\,(w')^2\,\bar{y}]\psi dz dS;$$

$$q(x, z, t) = w(z)\bar{q}(x, t) \tag{3.58}$$

Let us denote $A^{\alpha\beta\delta\tau} = \bar{\lambda}G^{\alpha\beta}G^{\delta\tau} + 2\bar{\mu}G^{\alpha\delta}G^{\tau\beta}$, then

$$\begin{aligned}
A^{w}((u, \bar{q}); (v, \bar{y})) = & \int\limits_S \int_{-\frac{h}{2}}^{\frac{h}{2}} \sigma^{ij}(u, q)\epsilon_{ij}(v, w\bar{y})\,\psi dz dS \\
= & \int\limits_S \int_{-\frac{h}{2}}^{\frac{h}{2}} [A^{\alpha\beta\delta\tau}\epsilon_{\delta\tau}(u)\,\epsilon_{\alpha\beta}(v) + wA^{\alpha\beta\delta\tau}\Upsilon_{\delta\tau}\epsilon_{\alpha\beta}(u)\,\bar{y} \\
& +\bar{\lambda}w'G^{\alpha\beta}\epsilon_{\alpha\beta}(u)\,\bar{y} + (\bar{\lambda}G^{\alpha\beta}w' + wA^{\alpha\beta\delta\tau}\Upsilon_{\delta\tau}) \\
& \bar{q}\epsilon_{\alpha\beta}(v)]\psi dz dS + \int\limits_S \int_{-\frac{h}{2}}^{\frac{h}{2}} [(w^2 A^{\alpha\beta\delta\tau}\Upsilon_{\delta\tau}\Upsilon_{\alpha\beta} \\
& +\bar{\lambda}ww'G^{\alpha\beta}\Upsilon_{\alpha\beta})\bar{q}\bar{y} + \left(\bar{\lambda} + 2\bar{\mu}\right)\left(w'\right)^2\bar{q}\bar{y} \\
& +\bar{\mu}G^{\alpha\beta}w^2\partial_\beta\bar{q}\partial_\alpha\bar{y}]\psi dz dS \\
= & \int\limits_S (N^{\alpha\beta}(u)e_{\alpha\beta}(v) + M^{\alpha\beta}(u)K_{\alpha\beta}(v) \\
& +M^{*\alpha\beta}(u)Q_{\alpha\beta}(v))dS + \int\limits_S (D^{\alpha\beta}_{0w}e_{\alpha\beta}(u) - D^{\alpha\beta}_{1w}K_{\alpha\beta}(u) \\
& +D^{\alpha\beta}_{2w}Q_{\alpha\beta}(u))\bar{y}dS + \int\limits_S (E^{\alpha\beta}_{0w}e_{\alpha\beta}(u) - E^{\alpha\beta}_{1w}K_{\alpha\beta}(u) \\
& +E^{\alpha\beta}_{2w}Q_{\alpha\beta}(u))\bar{y}dS + \int\limits_S \bar{q}(F^{\alpha\beta}_{0w}(\Upsilon)e_{\alpha\beta}(v) - F^{\alpha\beta}_{1w}(\Upsilon)K_{\alpha\beta}(v) \\
& +F^{\alpha\beta}_{2w}(\Upsilon)Q_{\alpha\beta}(v))dS + \int\limits_S (\bar{q}\bar{y}(F^{33}_{w0}(\Upsilon) + F^{33}_{w1}(\Upsilon) + F^{33}_{w2})dS \\
& +\int\limits_S (I^{\alpha\beta}_{ww}\partial_\beta\bar{q}\partial_\alpha\bar{y}dS
\end{aligned} \tag{3.59}$$

The terms $N^{\alpha\beta}(u)$, $M^{\alpha\beta}(u)$, $M^{*\alpha\beta}(u)$ are defined as earlier with $\bar{\lambda}$ in place of $\bar{\Lambda} = 2\bar{\mu}\bar{\lambda}/(\bar{\lambda} + 2\bar{\mu})$. The additional coefficients are defined as follows:

$$D^{\alpha\beta}_{0w} = \int_{-\frac{h}{2}}^{\frac{h}{2}} \left[wA^{\alpha\beta\delta\tau}\Upsilon_{\delta\tau}\right]\psi dz \quad D^{\alpha\beta}_{1w} = \int_{-\frac{h}{2}}^{\frac{h}{2}} z\left[wA^{\alpha\beta\delta\tau}\Upsilon_{\delta\tau}\right]\psi dz;$$

$$D^{\alpha\beta}_{2w} = \int_{-\frac{h}{2}}^{\frac{h}{2}} z^2\left[wA^{\alpha\beta\delta\tau}\Upsilon_{\delta\tau}\right]\psi dz;$$

$$E^{\alpha\beta}_{0w} = \int_{-\frac{h}{2}}^{\frac{h}{2}} \left[\bar{\lambda} G^{\alpha\beta} w'\right] \psi dz, \;\; E^{\alpha\beta}_{1w} = \int_{-\frac{h}{2}}^{\frac{h}{2}} z \left[\bar{\lambda} G^{\alpha\beta} w'\right] \psi dz$$

,

$$E^{\alpha\beta}_{2w} = \int_{-\frac{h}{2}}^{\frac{h}{2}} z^2 \left[\bar{\lambda} G^{\alpha\beta} w'\right] \psi dz$$

$$F^{\alpha\beta}_{0w}(\Upsilon) = \int_{-\frac{h}{2}}^{\frac{h}{2}} \left[w A^{\alpha\beta\delta\tau} \Upsilon_{\delta\tau} + \bar{\lambda} w' G^{\alpha\beta}\right] \psi dz$$

$$F^{\alpha\beta}_{1w}(\Upsilon) = \int_{-\frac{h}{2}}^{\frac{h}{2}} z \left[w A^{\alpha\beta\delta\tau} \Upsilon_{\delta\tau} + \bar{\lambda} w' G^{\alpha\beta}\right] \psi dz$$

$$F^{\alpha\beta}_{2w}(\Upsilon) = \int_{-\frac{h}{2}}^{\frac{h}{2}} z^2 \left[w A^{\alpha\beta\delta\tau} \Upsilon_{\delta\tau} + \bar{\lambda} w' G^{\alpha\beta}\right] \psi dz$$

$$F^{33}_{w0}(\Upsilon) = \int =_{-\frac{h}{2}}^{\frac{h}{2}} \left[(w^2 A^{\alpha\beta\delta\tau} \Upsilon_{\delta\tau} \Upsilon_{\alpha\beta} + \bar{\lambda} w w' G^{\alpha\beta} \Upsilon_{\alpha\beta})\right] \psi dz,$$

$$F^{33}_{w1}(\Upsilon) = \int_{-\frac{h}{2}}^{\frac{h}{2}} \left[\bar{\lambda} w G^{\alpha\beta} \Upsilon_{\alpha\beta} w'\right] \psi dz,$$

$$F^{33}_{w2}(\Upsilon) = \int_{-\frac{h}{2}}^{\frac{h}{2}} \left[(\bar{\lambda} + 2\bar{\mu}) \left(w'\right)^2\right] \psi dz$$

$$I^{\alpha\beta}_{ww} = \int_{-\frac{h}{2}}^{\frac{h}{2}} \left[\bar{\mu} G^{\alpha\beta} w^2\right] \psi dz$$

$$L^w(v, \bar{y}) = \int_S \left(\bar{p}^i v_i + \bar{p}^4 \bar{y}\right) dS + \int_{\gamma_1} \left(q^i v_i + q^4 \bar{y}\right) d\gamma + \int_{\gamma_1} m^\alpha \bar{\theta}_\alpha(v) d\gamma$$

$$\bar{p}^4 = \int_{-\frac{h}{2}}^{\frac{h}{2}} f^3 w \psi dz + w(h/2) p^3_+ - w(-h/2) p^3_-, \quad q^4 = \int_{-\frac{h}{2}}^{\frac{h}{2}} p^3 w dz$$

It should be remarked that for a fixed distribution function w the virtual strain power A^w is a symmetric positive defined bilinear form, the virtual external power L^w and the virtual kinetic power B^w are linear forms. We deduce from the boundary conditions that the space of the four unknown admissible functions $(u(x,t), \bar{q}(x,t))$ denoted by $U_{ad} = H^2(0, T; \quad H^1_{\gamma_o}(S) \times H^1_{\gamma_o}(S) \times H^2_{\gamma_o}(S) \times H^1_{\gamma_o}(S))$ is a Banach space. The "N" model for 3D thick shell thus presented is a four-parameter model. The different forms are continuous in this space and the operator A^w is coercive. We deduce from a classical result on maximal monotone operators that the variational problem has a unique solution in U_{ad}. Inertia terms are ignored in static analysis.

This variational equation can be rearranged in order to highlight the impact of the transverse deformation on the resultant forces and moments. We rewrite the equation as follows:

$$\int_S ((N^{\alpha\beta}(u) + N_w^{\alpha\beta}(u,\bar{q}))e_{\alpha\beta}(v) + (M^{\alpha\beta}(u) + M_w^{\alpha\beta}(u,\bar{q}))K_{\alpha\beta}(v)$$

$$+(M^{*\alpha\beta}(u) + M_w^{*\alpha\beta}(u,\bar{q}))Q_{\alpha\beta}(v))dS + \int_S (T_w^{\alpha}(u,\bar{q})\partial_\alpha \bar{y}$$

$$+T_w^3(u,\bar{q})\bar{y})dS = L^w(v,\bar{y})$$

$$N_w^{\alpha\beta}(u,\bar{q}) = \bar{q}(F_{0w}^{\alpha\beta}(\Upsilon), \;\; M_w^{\alpha\beta}(u,\bar{q}) = -\bar{q}F_{1w}^{\alpha\beta}(\Upsilon),$$

$$M_w^{*\alpha\beta}(u,\bar{q}) = \bar{q}F_{2w}^{\alpha\beta}(\Upsilon)$$

Exercise:
Show that the different forms are continuous in their respective domain of definition and that A^w is coercive.
Show that the first-order operator of A^w and the approximate inertia operator of B^w have the same properties.
Write down Euler's equations of the different variational problems.

Chapter 4
Thin Shells

4.1 Theory of Thin Shells

The theory of thin shells is the most common in the engineering literature. Based on the works of Kirchhoff and Love, the model obtained under their hypotheses, referred to herein as the *Kirchhoff-Love, or K-L* model, is realistic if the characteristic ratio of the shell, $\chi = max(h/2r) \ll 1$. The notion of thin shell is intimately related to this ratio which ranges in an interval whose upper bound χ_0 diverges according to different authors. Certain authors recommend the model for $2\chi \leq 0, 1$ and others, around 0, 5. It appears that strain energies of this model and that of the N-T thick shell model diverge from $2\chi = \sqrt{1/10}$.

So, by assuming the shell $\Omega = S \times]-\frac{h}{2}; \frac{h}{2}[$ is thin, we have

$$\begin{cases} U_\alpha(x, z) = u_\alpha(x) - z\partial_\alpha u_3(x); \\ \\ U_3 = u_3(x); \end{cases} \tag{4.1}$$

$$\left(\mu^{-1}\right)^\alpha_\rho A^\rho = G^\alpha \approx A^\alpha \ and \ det\left(\mu^{-1}\right) \approx 1; \quad d\Omega = dSdz;$$

$$\epsilon_{\alpha\beta}(u) = e_{\alpha\beta}(u) - zK_{\alpha\beta}(u), \ \epsilon_{i3} = 0; \tag{4.2}$$

$$e_{\alpha\beta}(u) = \frac{1}{2}\left(\nabla_\alpha u_\beta + \nabla_\beta u_\alpha - 2B_{\alpha\beta}u_3\right); \tag{4.3}$$

$$K_{\alpha\beta}(u) = \nabla_\alpha\left(\nabla_\beta u_3 + B^\rho_\beta u_\rho\right) + B^\rho_\alpha\left(\nabla_\beta u_\rho - B_{\rho\beta}u_3\right). \tag{4.4}$$

R. Nzengwa, *Plate and Shell Models*,
https://doi.org/10.1007/978-981-97-2780-3_4

$$N^{\alpha\beta} = \frac{Eh}{1-\bar{\nu}^2}\left(\bar{\nu} e^{\rho}_{\rho} A^{\alpha\beta} + (1-\bar{\nu})\, e^{\alpha\beta}\right); \tag{4.5}$$

$$M^{\alpha\beta} = \frac{Eh^3}{12\left(1-\bar{\nu}^2\right)}\left(\bar{\nu} K^{\rho}_{\rho} A^{\alpha\beta} + (1-\bar{\nu})\, K^{\alpha\beta}\right) \tag{4.6}$$

The variational formulation becomes

$$\begin{aligned} &\frac{Eh}{1-\bar{\nu}^2}\int_S \left(\bar{\nu} e^{\rho}_{\rho}(u)\, A^{\alpha\beta} + (1-\bar{\nu})\, e^{\alpha\beta}(u)\right) e_{\alpha\beta}(v)\, dS \\ +\ &\frac{Eh^3}{12\left(1-\bar{\nu}^2\right)}\int_S (\bar{\nu} K^{\rho}_{\rho}(u)\, A^{\alpha\beta} + (1-\bar{\nu})\, K^{\alpha\beta}(u)) K_{\alpha\beta}(v)\, dS \\ =\ &\int_S \left(P^{\alpha} v_{\alpha} + P^3 v_3\right) dS + \int_{\gamma_1} (q^{\alpha} v_{\alpha} + q^3 v_3) d\gamma \\ +\ &\int_{\gamma_1} m^{\alpha} \bar{\theta}_{\alpha}(v)\, d\gamma = L(v) \end{aligned} \tag{4.7}$$

This variational equation characterizes the stationary point, in the space of admissible displacements, of the total potential energy

$$I(u) = J(u) - L(u);\ J(u) = \frac{1}{2}\int_S N : e(u)\, dS + \frac{1}{2}\int_S M : K(u)\, dS \tag{4.8}$$

Stress components are calculated by $\sigma^{\alpha\beta} = \frac{1}{h} N^{\alpha\beta} - \frac{12z}{h^3} M^{\alpha\beta}$. The strain energy $J(u)$ has two parts: the membrane energy

$$\frac{1}{2}\int_S N : e(u)\, dS = \frac{Eh}{2(1-\bar{\nu}^2)}\int_S ((1-\bar{\nu}) e^{\alpha\beta}(u)\, e_{\alpha\beta}(u) + \bar{\nu} e^{\rho}_{\rho}(u)\, e^{\rho}_{\rho}(u)) dS \tag{4.9}$$

which characterizes expansion (or change of metric) of the mid-surface and bending or flectional energy

$$\begin{aligned} \tfrac{1}{2}\int_S M : K(u)\, dS = \tfrac{Eh^3}{24(1-\bar{\nu}^2)}\int_S &((1-\bar{\nu}) K^{\alpha\beta}(u)\, K_{\alpha\beta}(u) \\ &+\bar{\nu} K^{\rho}_{\rho}(u)\, K^{\rho}_{\rho}(u)) dS \end{aligned} \tag{4.10}$$

which characterizes change of curvature. The model problem is still the genuine mixed boundary condition problem. The space of admissible displacements is still the same as earlier. Particular boundary conditions can be addressed by considering additional energy related to the appropriate Lagrangian multiplier. In certain cases the membrane energy can be relatively more significant than the flectional energy or vice versa. It is advisable to consider only the most significant energy contribution as the total strain energy. We then have to deal with simplified theories: the *membrane theory* or the *flectional theory*.

4.2 The Membrane Theory of Thin Shells

We assume constant stress through the thickness, i.e. $\sigma^{\alpha\beta}(x, z) = \sigma^{\alpha\beta}(x, 0)$ and no or negligible bending moment $M^{\alpha\beta}$ almost everywhere. Some localized bending moments (on supports for example) may exist but will be treated separately. We assume the membrane energy is very significant, i.e. flectional energy is negligible, then the equilibrium equations read

$$\nabla_\alpha N^{\alpha\beta} + P^\beta = 0, \;\; N^{\alpha\beta} B_{\alpha\beta} + P^3 = 0, \;\; N^{\alpha\beta}\nu_\beta = q^\alpha \; on \; \gamma_1; \tag{4.11}$$

$$N^{\alpha\beta} = \frac{Eh}{1-\bar{\nu}^2}\left(\bar{\nu} e^\rho_\rho A^{\alpha\beta} + (1-\bar{\nu})\, e^{\alpha\beta}\right),$$

$$N^{\alpha\beta} = C^{\alpha\beta\gamma\delta} e_{\gamma\delta}, \;\; or \;\; e_{\gamma\delta} = S_{\gamma\delta\alpha\beta} N^{\alpha\beta}$$

4.2.1 Axisymmetric Structures

Shell with mid-surface of revolution (Fig. 4.1).

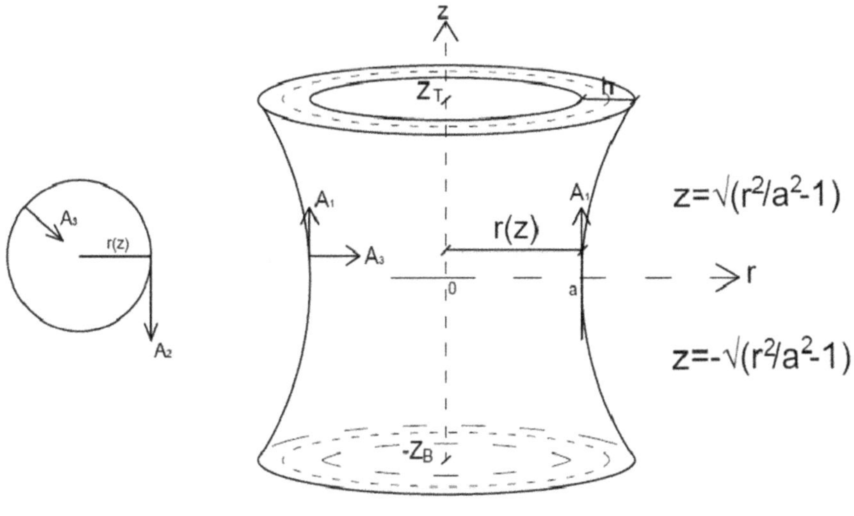

Fig. 4.1 Hyperbolic tower

We consider the function $z = z(r) = \pm\sqrt{(r^2/a^2) - 1}$ or $r = \pm a\sqrt{z^2 + 1}$ defined in the space of $(r,\ z)$, $r \geq a$, $a \succ 0$. Let C_{BT} denote the arc delimited by $z = -Z_B$ and $z = +Z_T$. The surface S_R, obtained by rotating the arc C_{BT} around the axis Oz, is the mid-surface of the shell of thickness h. We have

$$M = \begin{pmatrix} r cos\theta \\ r sin\theta \\ z\,(r) \end{pmatrix} = M\,(r, \theta) = M\,(z, \theta) \tag{4.12}$$

Let S be the curvilinear coordinate of a point on the surface. We shall also refer to the point as $M\,(S, \theta)$. Let ϕ denote the angle between the horizontal line and the tangent vector of the curve C_{BT}. We have

$$dS^2 = dz^2 + dr^2 = \left(z'\,(r)\,dr\right)^2 + dr^2 = \left(z'^2 + 1\right) dr^2 \tag{4.13}$$

$$\frac{dS}{dr} = \sqrt{1 + z'^2} = \frac{1}{cos\phi}, \quad \frac{dS}{dz} = \frac{1}{sin\phi}, \quad \frac{dz}{dr} = \frac{sin\phi}{cos\phi} \tag{4.14}$$

$$A_1 = \frac{dM}{dS} = \frac{dM}{dr}\frac{dr}{dS} = \frac{dM}{dz}\frac{dz}{dS} = A_S = cos\phi \begin{pmatrix} cos\theta \\ sin\theta \\ z' \end{pmatrix} \tag{4.15}$$

$$A_2 = \frac{dM}{d\theta} = A_\theta = \begin{pmatrix} -r sin\theta \\ r cos\theta \\ 0 \end{pmatrix} \tag{4.16}$$

$$A_3 = \frac{A_1 \times A_2}{\mid A_1 \times A_2 \mid} = \frac{1}{\sqrt{1 + (z')^2}} \begin{pmatrix} -z' cos\theta \\ -z' sin\theta \\ 1 \end{pmatrix} = cos\phi \begin{pmatrix} -z' cos\theta \\ -z' sin\theta \\ 1 \end{pmatrix} \tag{4.17}$$

$$(A_{\alpha\beta}) = \begin{bmatrix} 1 & 0 \\ 0 & r^2 \end{bmatrix}, \quad \left(A^{\alpha\beta}\right) = \begin{bmatrix} 1 & 0 \\ 0 & \frac{1}{r^2} \end{bmatrix} \tag{4.18}$$

From $A^\alpha = A^{\alpha\beta} A_\beta$, we deduce that $A^1 = A_1$: $A^2 = \frac{1}{r^2} A_2$

$$B_{11} = A_{1,1} \cdot A_3 = \tfrac{d^2M}{dS^2} \cdot A_3 = \tfrac{d}{dS}\left(cos\phi\begin{pmatrix} cos\theta \\ sin\theta \\ z'(r) \end{pmatrix}\right) \cdot A_3$$

$$\begin{aligned} &= cos^2\phi\tfrac{d}{dr}\begin{pmatrix} cos\theta \\ sin\theta \\ z'(r) \end{pmatrix} \cdot A_3 \\ &= cos^2\phi\begin{pmatrix} 0 \\ 0 \\ z''(r) \end{pmatrix} \cdot A_3 = cos^2\phi\frac{z''(r)}{\sqrt{1+(z')^2}} \\ &= \frac{-\, cos^2\phi}{a^2z^3\sqrt{1+(z')^2}} = \frac{-\, cos^3\phi}{a^2z^3} \end{aligned} \tag{4.19}$$

The coordinate $\phi \to \pi/2$ as $z \to 0$. By letting $\phi = \pi/2 - z$ we deduce that $\dfrac{-\, cos^3\phi}{a^2z^3} \to \dfrac{-\,1}{a^2}$. Also

$$B_{22} = A_{2,2} \cdot A_3 = \begin{pmatrix} -rcos\theta \\ -rsin\theta \\ 0 \end{pmatrix} \cdot cos\phi\begin{pmatrix} -z'cos\theta \\ -z'sin\theta \\ 1 \end{pmatrix} = rcos\phi\frac{dz}{dr} = rsin\phi,$$

$$B_{12} = B_{21} = 0$$

$$\left(B_{\alpha\beta}\right) = \begin{bmatrix} \dfrac{-\, cos^3\phi}{a^2z^3} & 0 \\ 0 & rsin\phi \end{bmatrix} = \begin{bmatrix} \dfrac{-\,1}{R_1} & 0 \\ 0 & rsin\phi \end{bmatrix},\ R_1 > 0 \tag{4.20}$$

It can be remarked that $rsin\phi$ is not a curvature because the base vector A^2 is not a unit vector unlike A^1 and the normal unit vector A^3 is directed inside the surface, i.e. outside the osculating circle tangent to the generic curve C_{BT} of the surface. This is why the curvature is negative. Because the normal vector is directed inside the circle which is also the osculating circle tangent to A^2, the second radius $rsin\phi$ is positive. A base is said to be physical if it is orthonormal. In such a base the tensor components $B_{\alpha\beta}$ and B^{α}_{β} are equal. Let $\{\bar{A}^1,\ \bar{A}^2,\ A^3\}$ and $\{\bar{A}_1,\ \bar{A}_2,\ A_3\}$ be two dual physical bases. The tensor $(B_{\alpha\beta})$, in the two bases (one physical and the other non-physical), reads

$$(B_{\alpha\beta}) = \begin{bmatrix} -\frac{1}{R_1} & 0 \\ & \\ 0 & rsin\phi \end{bmatrix} = B_{\alpha\beta}A^\alpha \otimes A^\beta = \bar{B}_{\alpha\beta}\bar{A}^\alpha \otimes \bar{A}^\beta \tag{4.21}$$

Let $\| A_\alpha \| = r_\alpha$, then (without the repeated index convention)

$$A^\alpha = \frac{A_\alpha}{r_\alpha^2}, \quad \bar{A}_\alpha = A_\alpha / r_\alpha, \quad \bar{A}^\alpha = A_\alpha / r_\alpha, \quad A^\alpha = \bar{A}^\alpha / r_\alpha \tag{4.22}$$

$$B_{\alpha\beta}A^\alpha \otimes A^\beta = B_{\alpha\beta}\frac{\bar{A}^\alpha}{r_\alpha} \otimes \frac{\bar{A}^\beta}{r_\beta} \quad \text{and} \quad \bar{B}_{\alpha\beta} = B_{\alpha\beta}\frac{1}{r_\alpha r_\beta} \tag{4.23}$$

It follows that $\bar{B}_{11} = -1/R_1, \bar{B}_{22} = \dfrac{rsin\phi}{r^2} = \dfrac{sin\phi}{r} = \frac{1}{R_2}$. This result coincides with $B_2^2 = A^{22}B_{22} = \dfrac{rsin\phi}{r^2}$ which is the curvature in the A_2 direction.

In the physical base the curvature tensor is now

$$(\bar{B}_{\alpha\beta}) = \begin{bmatrix} -\frac{1}{R_1} & 0 \\ & \\ 0 & -\frac{1}{R_2} \end{bmatrix} = \begin{bmatrix} \dfrac{-cos^3\phi}{a^2z^3} & 0 \\ 0 & \dfrac{sin\phi}{r} \end{bmatrix} \tag{4.24}$$

Let us calculate the symbols $\Gamma^\rho_{\alpha\beta}$. Because A_1 is a unit vector $\frac{d}{dS}A_1 \cdot A_1 = 0$, i.e. $\frac{d}{dS}A_1$ is orthogonal to the surface. It follows that

$$\Gamma^1_{11} = \Gamma^2_{11} = 0$$

$$\begin{aligned} \Gamma^2_{12} = \Gamma^2_{21} = cos\phi \begin{pmatrix} -sin\theta \\ cos\theta \\ 0 \end{pmatrix} \cdot A^2 = cos\phi \begin{pmatrix} -sin\theta \\ cos\theta \\ 0 \end{pmatrix} \cdot \tfrac{1}{r^2}A_2 \\ = \quad cos\phi \begin{pmatrix} -sin\theta \\ cos\theta \\ 0 \end{pmatrix} \cdot \tfrac{1}{r^2} \begin{pmatrix} -rsin\theta \\ rcos\theta \\ 0 \end{pmatrix} = \tfrac{cos\phi}{r} \end{aligned} \tag{4.25}$$

$$\Gamma^1_{22} = A_{2,2} \cdot A^1 = -rcos\phi$$

$$\Gamma^1_{12} = \Gamma^1_{21} = A_{1,2} \cdot A^1 = cos\phi \begin{pmatrix} -sin\theta \\ cos\theta \\ 0 \end{pmatrix} \cdot cos\phi \begin{pmatrix} cos\theta \\ sin\theta \\ z'(r) \end{pmatrix} = 0 \tag{4.26}$$

$$\Gamma^2_{22} = A_{2,2} \cdot A^2 = \begin{pmatrix} -rcos\theta \\ -rsin\theta \\ 0 \end{pmatrix} \cdot \frac{1}{r^2} \begin{pmatrix} -rsin\theta \\ rcos\theta \\ 0 \end{pmatrix} = 0 \tag{4.27}$$

Christoffel symbols are written in a matrix format as follows:

$$\left(\Gamma^1_{\alpha\beta}\right) = \begin{bmatrix} 0 & 0 \\ 0 & -rcos\phi \end{bmatrix}, \ \left(\Gamma^2_{\alpha\beta}\right) = \begin{bmatrix} 0 & \frac{cos\phi}{r} \\ \frac{cos\phi}{r} & 0 \end{bmatrix} \tag{4.28}$$

Let us consider the base $\{A^1,\ A^2,\ A^3\}$, the physical base $\left\{\bar{A}^1,\ \bar{A}^2,\ A^3\right\}$ and the tensor of membrane forces expressed in the two bases

$$N^{\alpha\beta} A_\alpha \otimes A_\beta = N^{\alpha\beta} r_\alpha r_\beta \bar{A}_\alpha \otimes \bar{A}_\beta = \bar{N}^{\alpha\beta} \bar{A}_\alpha \otimes \bar{A}_\beta \tag{4.29}$$

We deduce the relations $\bar{N}^{\alpha\beta} = N^{\alpha\beta} r_\alpha r_\beta$. Therefore, $\bar{N}^{11} = N^{11}$, $\bar{N}^{12} = rN^{12}$ and $\bar{N}^{22} = r^2 N^{22}$. We assume that the tower of thickness h undergoes membrane strain under its own weight $-\rho g \overrightarrow{k}$, which can also be decomposed as $-\rho gh \overrightarrow{k} = P^1 \bar{A}_1 + P^2 \bar{A}_2 + P^3 A_3$. Then $P^i = -\rho gh \overrightarrow{k} \cdot \bar{A}^i$

$$P^1 = -\rho gh \begin{pmatrix} 0 \\ 0 \\ 1 \end{pmatrix} \cdot cos\phi \begin{pmatrix} cos\theta \\ sin\theta \\ z' \end{pmatrix} = -\rho ghsin\phi, \tag{4.30}$$

$$P^2 = -\rho gh \begin{pmatrix} 0 \\ 0 \\ 1 \end{pmatrix} \cdot \begin{pmatrix} -sin\theta \\ cos\theta \\ 0' \end{pmatrix} = 0 \tag{4.31}$$

$$P^3 = -\rho gh \begin{pmatrix} 0 \\ 0 \\ 1 \end{pmatrix} \cdot cos\phi \begin{pmatrix} -z'cos\theta \\ -z'sin\theta \\ 1 \end{pmatrix} = -\rho ghcos\phi \tag{4.32}$$

The equilibrium equations are

$$\begin{gathered} \nabla_\alpha N^{\alpha\beta} + P^\beta = 0, \quad N^{\alpha\beta} B_{\alpha\beta} + P^3 = 0 \text{ on } S \\ N\overrightarrow{\nu} = 0 \text{ at } z = +Z_T;\ u = v = w = w' = 0 \text{ at } z = -Z_B \end{gathered} \tag{4.33}$$

From the vertical equilibrium equation of a cap of the tower at the height z and assuming the membrane force $\bar{N}^{11}$ is in the outer normal direction, we obtain the equation (Fig. 4.2)

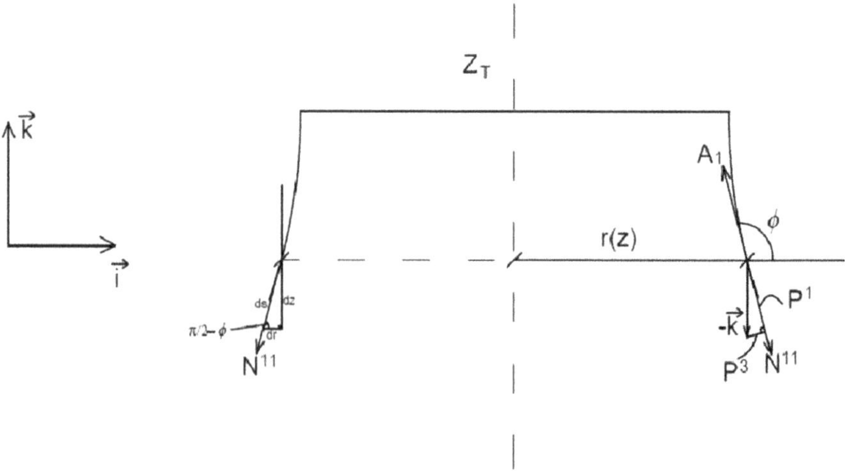

Fig. 4.2 Vertical equilibrium of a cap

$$\begin{cases} +2\pi r(z)\bar{N}^{11}sin\phi + 2\pi\rho gh\int_z^{Z_T} r(z)dS = 0\,and \\ \bar{N}^{11} = -\frac{\rho gh}{r(z)sin\phi}\int_z^{Z_T} r(z)dS \\ \quad = -\frac{\rho gha}{r(z)sin\phi}\int_z^{Z_T}\sqrt{(1+z^2+a^2z^2)}dz \end{cases} \tag{4.34}$$

$$\begin{cases} \bar{N}^{\alpha\beta}\bar{B}_{\alpha\beta} + P^3 = 0 = \bar{N}^{11}\bar{B}_{11} + \bar{N}^{22}\bar{B}_{22} + P^3 \\ \qquad = -\frac{1}{R_1}\bar{N}^{11} + \frac{1}{R_2}\bar{N}^{22} + P^3 = 0 \text{ and} \\ \bar{N}^{22} = R_2(\frac{1}{R_1}\bar{N}^{11} - P^3) = \frac{r}{sin\phi}(-\frac{cos^3\phi}{a^2z^3}\frac{\rho gh}{r(z)sin\phi}\int_z^{Z_T} r(z)dS + \rho ghcos\phi) \end{cases} \tag{4.35}$$

The equation

$$\nabla_1 N^{11} + \nabla_2 N^{21} + P^1 = 0 \tag{4.36}$$

$$\nabla_1 N^{12} + \nabla_2 N^{22} + P^2 = 0 \tag{4.37}$$

is equivalent to

$$\begin{aligned} N^{11}_{,1} + \Gamma^1_{11}N^{11} + \Gamma^1_{12}N^{21} + \Gamma^1_{11}N^{11} + \Gamma^1_{12}N^{12} + N^{21}_{,2} \\ + \Gamma^2_{21}N^{11} + \Gamma^2_{22}N^{21} + \Gamma^1_{21}N^{21} + \Gamma^1_{22}N^{22} + P^1 \quad &= 0 \end{aligned} \tag{4.38}$$

$$\begin{aligned} N^{12}_{,1} + \Gamma^1_{11} N^{12} + \Gamma^1_{12} N^{22} + \Gamma^2_{11} N^{11} + \Gamma^2_{12} N^{12} + N^{22}_{,2} \\ + \Gamma^2_{21} N^{12} + \Gamma^2_{22} N^{22} + \Gamma^2_{21} N^{21} + \Gamma^2_{22} N^{22} + P^2 &= 0 \end{aligned} \tag{4.39}$$

$$N^{11}_{,1} + N^{21}_{,2} + \Gamma^2_{21} N^{11} + \Gamma^1_{22} N^{22} + P^1 = 0 \tag{4.40}$$

$$N^{12}_{,1} + 3\Gamma^2_{12} N^{12} + N^{21}_{,2} + P^2 = 0 \tag{4.41}$$

We also have

$$N^{11}_{,1} + N^{21}_{,2} + \frac{cos\phi}{r} N^{11} - r cos\phi N^{22} + P^1 = 0 \tag{4.42}$$

$$N^{12}_{,1} + 3\frac{cos\phi}{r} N^{12} + N^{21}_{,2} + P^2 = 0 \tag{4.43}$$

$$r N^{11}_{,1} + r N^{21}_{,2} + cos\phi N^{11} - r^2 cos\phi N^{22} + r P^1 = 0 \tag{4.44}$$

$$r N^{12}_{,1} + 3cos\phi N^{12} + r N^{21}_{,2} + r P^2 = 0 \tag{4.45}$$

i.e.

$$(r\bar{N}^{11})_{,1} + \bar{N}^{21}_{,2} - cos\phi \bar{N}^{22} + r\bar{P}^1 = 0 \tag{4.46}$$

$$r^2(\bar{N}^{12})_{,1} + 2r cos\phi \bar{N}^{12} + r^2 \bar{N}^{21}_{,2} + r^2 \bar{P}^2 = 0 \tag{4.47}$$

The membrane force satisfies the equation $\bar{N}^{21}_{,2} = 0$. Therefore, the first equation is unnecessary, and it follows that

$$(r^2 \bar{N}^{12})_{,1} + r^2 \bar{P}^2 = 0 \tag{4.48}$$

In this example $r^2 \bar{P}^2 = 0$. Therefore $r^2 \bar{N}^{12} = Cste$. It can be remarked that the membrane force component $\bar{N}^{11}$ is a compression, while $\bar{N}^{22}$ can be a compression or a traction depending on the altitude. From the constitutive law

$$\bar{N}^{\alpha\beta} = \bar{N}_{\alpha\beta} = \frac{Eh}{1-\nu^2}\left((1-\bar{\nu})\, e_{\alpha\beta} + \bar{\nu} e^\rho_\rho \delta_{\alpha\beta}\right) \tag{4.49}$$

We calculate the strain components as follows:

$$e_{\alpha\beta} = \frac{(1+\bar{\nu})}{hE}\bar{N}_{\alpha\beta} - \frac{\bar{\nu}}{hE}\bar{N}^{\rho}_{\rho}\delta_{\alpha\beta} \tag{4.50}$$

4.2.2 Spherical Dome

We consider a spherical dome (Fig. 4.3), of radius R, thickness h and open angle 2α, which is subject to its weight. The normal vector is directed outside the centre of the sphere, and the coordinate system is $(dX^1, dX^2) = (d\psi, d\varphi)$. A position vector is

$$\overrightarrow{OM} = (Rsin\psi cos\varphi, Rsin\psi sin\varphi, Rcos\psi)$$

We have

$$A_1 = \begin{pmatrix} Rcos\psi cos\varphi \\ Rcos\psi sin\varphi \\ -Rsin\psi \end{pmatrix}, \ A_2 = \begin{pmatrix} -Rsin\psi sin\varphi \\ Rsin\psi cos\varphi \\ 0 \end{pmatrix}, \ A_3 = \begin{pmatrix} sin\psi cos\varphi \\ sin\psi sin\varphi \\ cos\psi \end{pmatrix}, \tag{4.51}$$

Fig. 4.3 A self-weighted spherical tank

$$A_{\alpha\beta} = \begin{bmatrix} R^2 & 0 \\ 0 & R^2 sin^2\psi \end{bmatrix}, \; A^{\alpha\beta} = \begin{bmatrix} 1/R^2 & 0 \\ 0 & 1/R^2 sin^2\psi \end{bmatrix}, \tag{4.52}$$

$$B_{..} = \begin{bmatrix} -R & 0 \\ 0 & -R sin^2\psi \end{bmatrix}$$

$$A^1 = \frac{1}{R}\begin{pmatrix} cos\psi cos\varphi \\ cos\psi sin\varphi \\ -sin\psi \end{pmatrix}, \; A^2 = \frac{1}{R sin\psi}\begin{pmatrix} -sin\varphi \\ cos\varphi \\ 0 \end{pmatrix}, \tag{4.53}$$

$$\Gamma^1_{\alpha\beta} = \begin{bmatrix} 0 & 0 \\ 0 & -cos\psi sin\psi \end{bmatrix}, \; \Gamma^2_{\alpha\beta} = \begin{bmatrix} 0 & 1/tg\psi \\ 1/tg\psi & 0 \end{bmatrix} \tag{4.54}$$

From the vertical equilibrium equation of a cap of angle 2ψ (see Fig. 4.4) and assuming N^{11} is directed outward, we obtain the equation

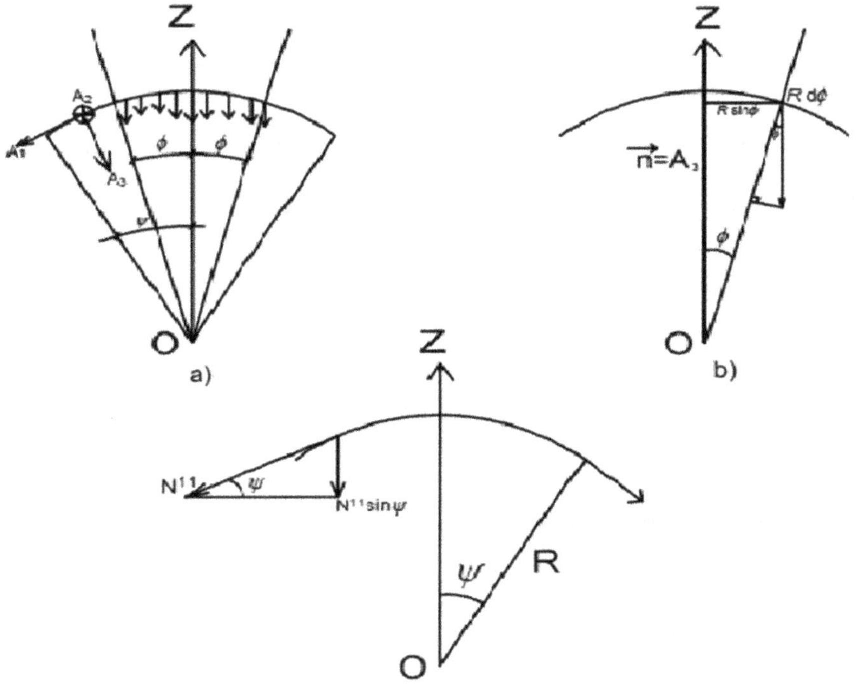

Fig. 4.4 Vertical equilibrium of a spherical cap

$$0 = 2\pi N^{11} R sin^2\psi + 2\pi\rho g h \int_0^{\psi} R^2 sin\phi d\phi \tag{4.55}$$

which yields

$$N^{11} = -\rho g h R \frac{1 - cos\psi}{sin^2\psi} = -\rho g h R \frac{1}{1 + cos\psi} \tag{4.56}$$

From

$$\begin{aligned} P^1 &= -\rho g h \vec{k} \cdot A^1 = \rho g h sin\psi \quad P^2 = -\rho g h \vec{k} \cdot A^2 = 0 \\ P^3 &= -\rho g h \vec{k} \cdot A^3 = -\rho g h \vec{k} \cdot \vec{e}_r = -\rho g h cos\psi \end{aligned} \tag{4.57}$$

and the third equation of (4.33), it follows that

$$N^{22} = \left(-P^3 - N^{11}\right) R = -\rho g h R \left(cos\psi - \frac{1}{1 + cos\psi}\right) \tag{4.58}$$

while the stress component $\sigma^{11} = \frac{1}{h} N^{11}$ is always a compression independent of half the open angle $\alpha < \frac{\pi}{2}$, $\sigma^{22}(\psi) = \frac{1}{h} N^{22}$ is a compression only if $cos\psi - \frac{1}{1+cos\psi}$ is positive, i.e. $\alpha \leqq \theta_l < \frac{\pi}{2}$, solution of $cos\theta_l - \frac{1}{1+cos\theta_l} = 0$. The solution is $\theta_l = 51°68'$. Traction appears in the structure if $\alpha > \theta_l$. In practice for $\alpha < 51°$ there is compression in both directions, which means a roof can be constructed only with masonry, without any reinforcement, provided that half the open angle is inferior to θ_l. The stress N^{12} is calculated by solving the equations

$$N^{11}_{,1} + N^{12}_{,2} + \frac{1}{tg(\psi)} N^{11} - cos(\psi) sin(\psi) N^{22} + P^1 = 0 \tag{4.59}$$

$$N^{12}_{,1} + N^{22}_{,2} + \frac{3}{tg(\psi)} N^{12} + P^2 = 0 \tag{4.60}$$

4.2.3 Equilibrium of a Sphere

We consider a shell of radius R and thickness h, subject to a uniform pressure (see Fig. 4.5). The third equation reads

$$N^{\alpha\beta} B_{\alpha\beta} + p = 0, \quad B_{..} = \begin{bmatrix} \frac{1}{R} & 0 \\ 0 & \frac{1}{R} \end{bmatrix} \tag{4.61}$$

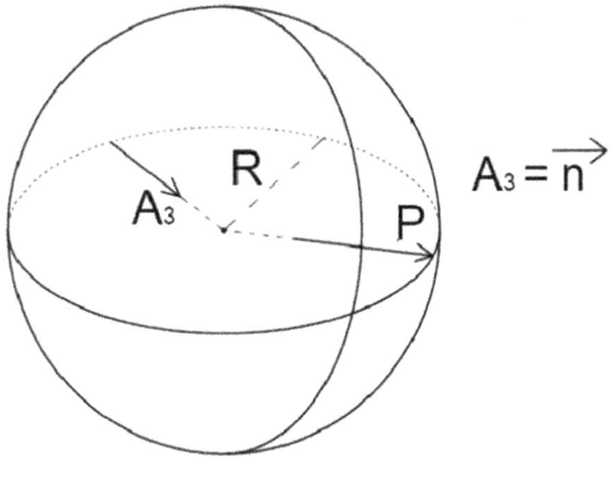

Fig. 4.5 A pressurized spherical tank

The pressure p has an algebraic value depending on the orientation of the normal vector (Fig. 4.5). We assume the normal vector is directed toward the centre. In this case $R > 0$ and the pressure is $\overrightarrow{P} = -p\overrightarrow{n}$, $p > 0$. If the normal vector is directed outside the centre, $R < 0$ and $\overrightarrow{P} = p\overrightarrow{n}$, $p > 0$. It follows from the symmetry of the problem that $N^{11} = N^{22}$ and the third equation reads

$$\frac{N^{11}}{R} + \frac{N^{22}}{R} - p = 0, \quad 2\frac{N^{11}}{R} - p = 0, \quad N^{11} = N^{22} = \frac{pR}{2} \tag{4.62}$$

$$\sigma^{11} = \sigma^{22} = \frac{N^{11}}{h} = \frac{pR}{2h}, \quad N^{12} = 0 \tag{4.63}$$

We find here the famous welding and metal fabrication formula used to determine the thickness h of the elastic metal of a pressure tank, whose elastic limit is f_e. The condition to fulfil is

$$max\left(|\,\sigma^{11}\,|, |\,\sigma^{22}\,|, |\,\sigma^{12}\,|\right) = \frac{pR}{2h} < \frac{f_e}{\gamma_s}, \text{ i.e. h} > \frac{\gamma_s pR}{2f_e} \tag{4.64}$$

For $p = 0,5\,\text{MPa}$, $R = 1,5\,\text{m}$, $f_e = 235\,\text{MPa}$, $\gamma_s = 1,15$, $E = 210\,000\,\text{MPa}$, $\bar{\nu} = 0,38$, the thickness $h > \frac{\gamma_s pR}{2f_e}$ must be greater than 2 mm. The strain tensor is

$$e_{..} = \frac{1+\bar{\nu}}{Eh}N_{..} - \frac{\bar{\nu}}{Eh}N^{\alpha}_{\alpha}A. \tag{4.65}$$

$$
\begin{aligned}
e_{11} = e_{22} &= \tfrac{1+\bar{\nu}}{Eh}\left(\tfrac{pR}{2}\right) - \tfrac{\bar{\nu}}{Eh}pR = \left(\tfrac{pR}{Eh}\right)\left(\tfrac{1+\bar{\nu}}{2} - \bar{\nu}\right) \\
&= \left(\tfrac{pR}{Eh}\right)\left(\tfrac{1-\bar{\nu}}{2}\right) \\
&= \tfrac{1}{2}\left(\nabla_1 u_1 + \nabla_1 u_1 - 2B_{11}u_3\right) \qquad (4.66)\\
&= -u_3/R \\
e_{12} &= 0
\end{aligned}
$$

It follows that the radial displacement (the normal vector is directed toward the centre) $u_3 = -\left(\frac{1-\bar{\nu}}{2}\right)\frac{pR^2}{Eh} = -0.8\,\text{mm}$ is an expansion.

4.2.4 Equilibrium of a Cylindrical Tank

We consider a cylindrical tank of radius R and thickness h, under uniform pressure (Fig. 4.6). We have

$$
B_{..} = \begin{bmatrix} \frac{1}{R} & 0 \\ 0 & 0 \end{bmatrix};\; S = R\theta = x^1,\; z = x^2,\; R > 0,\; p > 0,\; \frac{N^{11}}{R} - p = 0 \qquad (4.67)
$$

We deduce that $N^{11} = pR$. If the tank has spherical tips of the same radius on both edges, then, from the above results, at the junction, we have

$$
N^{11}_{sphere} = \frac{pR}{2},\; N^{11}_{cylinder} = pR,\; N^{22}_{sphere} = \frac{pR}{2} = N^{22}_{cylinder} \qquad (4.68)
$$

Therefore, there is a jump at the junction, between $N^{11}_{cylinder}$ and N^{11}_{sphere}, while there is continuity on N^{22} even if the spherical tip is replaced by a circular disc of the same

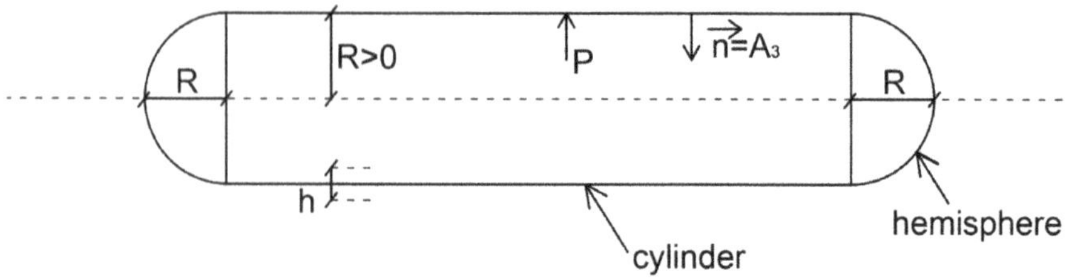

Fig. 4.6 A pressurized cylindrical tank with hemispherical tips

radius. We calculate $N^{22}_{cylinder}$ in a section whose coordinate $x^2 = z_0$ by considering the equilibrium on the z-axis of the portion of the tank $z \geq z_0$. This example shows the limit of application of the membrane theory. Such a discrepancy should be addressed technically or by refining the theory. The choice of the membrane theory or bending theory depends on the ratio of the two strain energies. In order to compare the two energies in this case, we assume membrane displacements are negligible, i.e. $u_\alpha = 0$. Let us denote $u_3 = w$. Then

$$\begin{aligned} K_{\alpha\beta} = & \quad \nabla_\alpha \left(\partial_\beta w + B^\rho_\beta u_\rho \right) + B^\rho_\alpha \left(\nabla_\beta u_\rho - B_{\beta\rho} w \right) \\ &= \nabla_\alpha \partial_\beta w - B^\rho_\alpha B_{\beta\rho} w = w,_{\alpha\beta} - w,_\rho \Gamma^\rho_{\alpha\beta} - B^\rho_\alpha B_{\beta\rho} w \\ &= \quad -B^\rho_\alpha B_{\beta\rho} w \end{aligned} \tag{4.69}$$

since w is constant. We deduce that $K_{11} = K_{22} = -\frac{w}{R^2}$, $K_{12} = 0$. It should be remarked that the same result can be obtained from the expansion of $K_{11} = K_{22} = b_{11} - B_{11} = \frac{1}{R+w} - \frac{1}{R} = \frac{1}{R}(\frac{1}{1+w/R} - 1) \approx \frac{1}{R}(1 - \frac{w}{R} - 1) \approx -\frac{w}{R^2}$. Also, $e_{11} = e_{22} = -\frac{w}{R}$, $e_{12} = 0$; and the energy density ratio

$$\begin{aligned} \chi = \frac{\text{Bending energiy}}{\text{Membrane energy}} = & \frac{\dfrac{Eh^3}{12\left(1-\bar{\nu}^2\right)}\left[\bar{\nu}\,(tr K)^2 + (1-\bar{\nu})\,K^2\right]}{\dfrac{Eh}{\left(1-\bar{\nu}^2\right)}\left[\bar{\nu}\,(tre)^2 + (1-\bar{\nu})\,e^2\right]} \\ = & \quad \frac{Eh^3}{12\left(1-\bar{\nu}^2\right)} \cdot \frac{w^2}{R^4} \times \frac{(1-\bar{\nu}^2)R^2}{Ehw^2} = \frac{1}{12}\left(\frac{h}{R}\right)^2 \end{aligned} \tag{4.70}$$

where $K^2 = K : K$; $e^2 = e : e$ justifies the application of the membrane theory because if $\frac{h}{R} = 0.1$, χ is of the order of a thousandth. Under some simplified hypotheses as stated earlier, it may appear that the ratio of the two energies does not show a clear preponderance. Both energies are considered and this leads to the simplified mixed theory.

4.3 The Mixed Theory (Membrane-Bending) of Thin Shells

Description: the cylindrical shell (pipe) of radius R, thickness h and length l is under a symmetric pressure (Fig. 4.7). The coordinates of a point is $x^1 = z$, $x^2 = R\theta$. The base is physical with the normal unit vector directed inside. So $B^2_2 = B_{22} = 1/R$. We assume the following hypotheses:

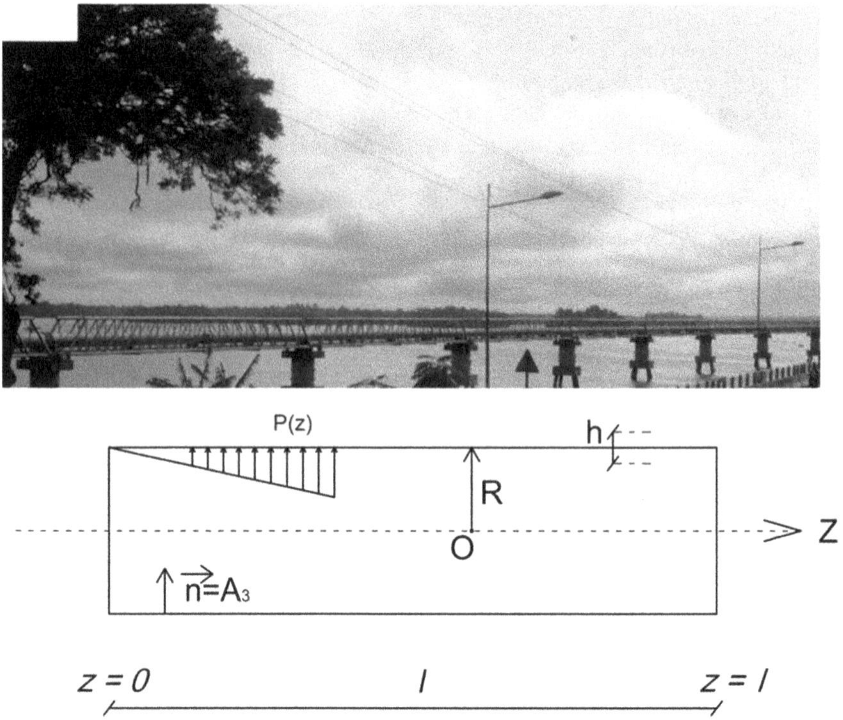

Fig. 4.7 Water pipe over the river Wouri and structural model

4.3.1 Hypotheses

- (i) There are no forces at the junction of two pipes:

$$N\overrightarrow{\nu} = N^{\alpha\beta}\nu_\beta = 0 \ at \ z = 0 \ and \ z = l \tag{4.71}$$

This is the case when flexible gaskets are inserted at joints. We deduce that

$$\begin{gathered} N_{11} = \frac{Eh}{(1-\bar{\nu}^2)}\left[(1-\bar{\nu})\,e_{11} + \bar{\nu}\,(e_{11}+e_{22})\right] = 0 \\ e_{11} + \bar{\nu}e_{22} = 0, \ \ e_{11} = -\bar{\nu}e_{22} \end{gathered} \tag{4.72}$$

- (ii) Displacement is radial

i.e. the deformation is inextensional, $u_1 = u_2 = 0, u_3 = w; e_{11} = -\nu e_{22}$ for $z\epsilon\,[0, l]$. It follows that

$$e_{22} = \nabla_2 u_2 - w B_{22} = -w/R, \quad e_{11} = \bar{\nu} w/R, \quad e_{..} = \begin{bmatrix} \bar{\nu} w/R & 0 \\ 0 & -w/R \end{bmatrix}, \tag{4.73}$$

$$K_{11} = w_{,11} - B_1^m B_{m1} w = w_{,11}, \quad K_{22} = w_{,22} - B_2^m B_{m2} w = -w/R^2,$$

$$K_{..} = \begin{bmatrix} w_{,11} & 0 \\ 0 & -w/R^2 \end{bmatrix}$$

- (iii) pressure depends only on the coordinate z, i.e. $P = P(z) = p(z)\vec{n}$. The value of the function $p(z)$ may be positive or negative. Uniform bending moments are applied on both ends of the pipe
- (iv) the radial displacement is uniform in each section, i.e. $w = w(z)$. It follows that $\bar{\theta}_\nu(0) = -\partial_{\bar{\nu}} w(0) = -\partial_{-\bar{z}} w(0) = w'(0)$ and $\bar{\theta}_\nu(l) = -\partial_{\bar{\nu}} w(l) = -\partial_{\bar{z}} w(0) = -w'(l)$.

4.3.2 Equilibrium Equation of the Pipe

We deduce from the above hypotheses that the strain energy is

$$\begin{aligned} W_{def} = & \tfrac{1}{2}\tfrac{Eh}{(1-\bar{\nu}^2)} \int_S \left[(1-\bar{\nu})\, e^2 + \bar{\nu}\,(tre)^2\right] dS \\ & + \tfrac{1}{2}\tfrac{Eh^3}{12(1-\bar{\nu}^2)} \int_S \left[(1-\bar{\nu})\, K^2 + \bar{\nu}\,(trK)^2\right] dS \\ = & \tfrac{\pi R E h^3}{12(1-\bar{\nu}^2)} \int_0^l \left[\left(w_{,11}\right)^2 + \tfrac{12(1-\bar{\nu}^2)}{R^2 h^2} w^2\right] dz \\ & + \tfrac{\pi R E h^3}{12(1-\bar{\nu}^2)} \int_0^l \left(\tfrac{w^2}{R^4} - 2\bar{\nu}\tfrac{w_{,11} w^2}{R^2}\right) dz \end{aligned} \tag{4.74}$$

We may neglect the last term. External energy reads

$$W_{ext} = +\int_0^l \int_0^{2\pi} p(z)\, w(z)\, dS + \int_{z=0} q(0)\, w(0)\, dc + \int_{z=l} q(l)\, w(l)\, dc$$
$$+ \int_{z=0} m(0)\, w'(0)\, dc - \int_{z=l} m(l)\, w'(l)\, dc \tag{4.75}$$

q is constant at the ends, along the circumference line. The total potential energy of the pipe is $I(w) = W_{def} - W_{ext}$. The solution is the stationary displacement of $I(w)$ which we rewrite

$$I(w) = D \int_0^l \left[(w'')^2 + \alpha w^2 \right] dz - 2\pi R \int_0^l p(z)\, w(z)\, d(z) - 2\pi R q(0) w(0)$$
$$-2\pi R q(l) w(l) - 2\pi R m(0)\, w'(0) + 2\pi R m(l)\, w'(l) \tag{4.76}$$

Let

$$D = \frac{1}{2} \frac{\pi R E h^3}{12\left(1 - \bar{\nu}^2\right)}, \quad \alpha = \frac{12\left(1 - \bar{\nu}^2\right)}{R^2 h^2}$$

The solution realizes the minimum because the total potential energy $I(w)$ is convex. We must have $I'(w)\,\delta w = \lim_{\varepsilon \to 0} \frac{I(w + \varepsilon \delta w) - I(w)}{\varepsilon} = 0$, i.e.

$$2D \int_0^l \left[(w'')\, \delta w'' + \alpha w \delta w \right] dz - 2\pi R \int_0^l p(z)\, \delta w(z)\, d(z)$$
$$-2\pi R q(0) \delta w(0) - 2\pi R q(l) \delta w(l) - 2\pi R m(0)\, \delta w'(0)$$
$$+2\pi R m(l)\, \delta w'(l) \qquad = 0$$

$$2D \int_0^l \left[-(w''')\, \delta w' + \alpha w \delta w \right] dz - 2\pi R \int_0^l p(z)\, \delta w(z)\, d(z)$$
$$+ \left[2D w'' \delta w' \right]_0^l - 2\pi R q(0) \delta w(0) - 2\pi R q(l) \delta w(l) \tag{4.77}$$
$$-2\pi R m(0)\, \delta w'(0) + 2\pi R m(l)\, \delta w'(l) \qquad = 0$$

$$2D \int_0^l \left[(w'''')\, \delta w + \alpha w \delta w \right] dz - 2\pi R \int_0^l p(z)\, \delta w(z)\, d(z)$$
$$+ \left[2D w'' \delta w' \right]_0^l + \left[-2D w''' \delta w \right]_0^l - 2\pi R q(0) \delta w(0)$$
$$-2\pi R q(l) \delta w(l) - 2\pi R m(0)\, \delta w'(0) + 2\pi R m(l)\, \delta w'(l) = 0$$

By taking $\delta w\,(0) = \delta w\,(l) = \delta w'\,(0) = \delta w'\,(0) = 0$ we deduce Euler's equation

$$\frac{D}{\pi R}\left(w^{(4)} + \alpha w\right) = p \tag{4.78}$$

which is a fourth order ordinary differential equation. Now letting δw and its derivatives be non-zero at the borders leads to the boundary conditions

$$Dw''\,(0) = \pi Rm\,(0)\,; \quad Dw''\,(l) = -\pi Rm\,(l) \tag{4.79}$$

$$Dw'''\,(0) = \pi Rq\,(0)\,; \quad Dw'''\,(l) = \pi Rq\,(l) \tag{4.80}$$

The four boundary conditions are sufficient to solve the equation completely. We can next calculate the strain tensor $e_{..}$, the change of curvature tensor $K_{..}$, the resultant forces and moments $N^{..}$, $M^{..}$ and at last the stresses

$$\sigma^{\alpha\beta} = \tfrac{1}{h}N^{\alpha\beta} - \tfrac{12t}{h^3}M^{\alpha\beta},\ t \in \left[-\tfrac{h}{2}, \tfrac{h}{2}\right] \tag{4.81}$$

in order to design the pipes.

4.4 Theory of Plates

A plate is a shell without curvature in its reference (or initial) configuration also called first geometry (Fig. 4.8). We therefore have $B_{\alpha\beta} = B^{\alpha}_{\beta} = 0$ and $\nabla_\alpha = D_\alpha$ in a Cartesian base with coordinates (x, y, z). The displacement vector is $U = (u,\ v,\ w)$ and

$$e_{xx} = \frac{\partial u}{\partial x};\ e_{yy} = \frac{\partial v}{\partial y};\ e_{xy} = \frac{1}{2}\left(\frac{\partial u}{\partial y} + \frac{\partial v}{\partial x}\right) \tag{4.82}$$

$$K_{xx} = \frac{\partial^2 w}{\partial x^2};\ K_{yy} = \frac{\partial^2 w}{\partial y^2};\ K_{xy} = \frac{\partial^2 w}{\partial x \partial y} \tag{4.83}$$

Plates equations obtained from the general shell equations established earlier in Chap. 2 now reduce to

$$\nabla_\alpha N^{\alpha\beta} + p^\beta = 0, \nabla_\alpha \nabla_\beta M^{\alpha\beta} - p^3 = 0 \tag{4.84}$$

In order to calculate $\nabla_\alpha \nabla_\beta M^{\gamma\delta}$ we denote $U^{\gamma\delta}_{\beta} = \nabla_\beta M^{\gamma\delta} = M^{\gamma\delta}_{,\beta} + \Gamma^{\gamma}_{\beta\rho} M^{\rho\delta} + \Gamma^{\delta}_{\beta\rho} M^{\gamma\rho}$ and

$$\nabla_\alpha \nabla_\beta M^{\gamma\delta} = \nabla_\alpha U^{\gamma\delta}_{\beta} = U^{\gamma\delta}_{\beta,\alpha} - \Gamma^{\rho}_{\beta\alpha} U^{\gamma\delta}_{\rho} + \Gamma^{\gamma}_{\alpha\rho} U^{\rho\delta}_{\beta} + \Gamma^{\delta}_{\alpha\rho} U^{\gamma\rho}_{\beta} \tag{4.85}$$

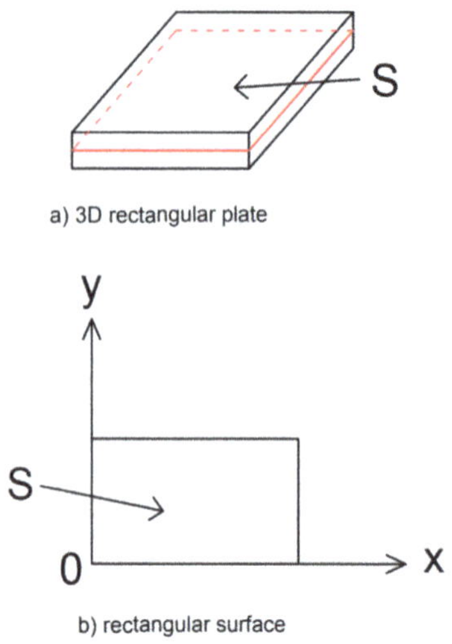

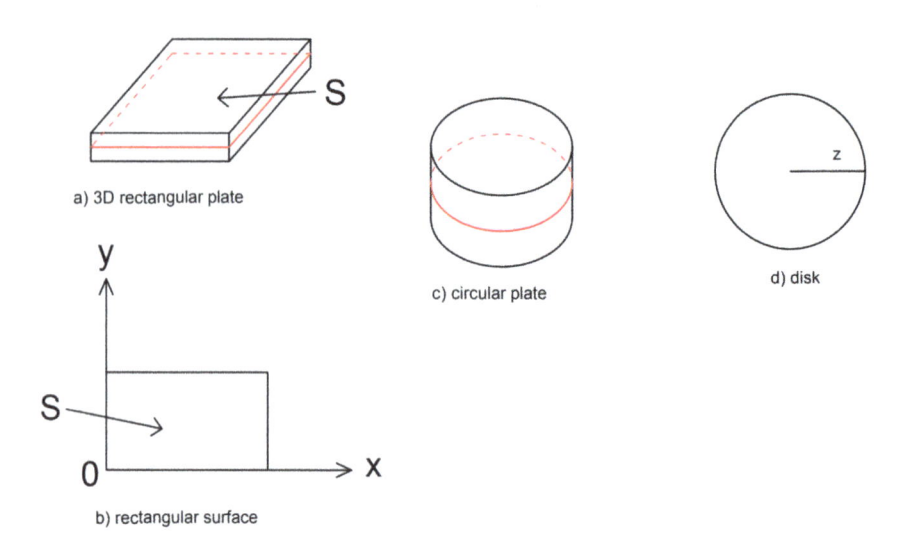

Fig. 4.8 Rectangular and circular plates

$$
\begin{aligned}
U^{\gamma\delta}_{\beta,\alpha} = &\quad M^{\gamma\delta}_{,\beta\alpha} + \Gamma^{\gamma}_{\beta\rho} M^{\rho\delta}_{,\alpha} + \Gamma^{\delta}_{\beta\rho} M^{\gamma\rho}_{,\alpha} + \Gamma^{\gamma}_{\beta\rho,\alpha} M^{\rho\delta} + \Gamma^{\delta}_{\beta\rho,\alpha} M^{\gamma\rho} \\
&= D_\alpha D_\beta M^{\gamma\delta} + \Gamma^{\gamma}_{\beta\rho} M^{\rho\delta}_{,\alpha} + \Gamma^{\delta}_{\beta\rho} M^{\gamma\rho}_{,\alpha} + \Gamma^{\gamma}_{\beta\rho,\alpha} M^{\rho\delta} + \Gamma^{\delta}_{\beta\rho,\alpha} M^{\gamma\rho}
\end{aligned} \tag{4.86}
$$

and

$$
\begin{aligned}
\nabla_\alpha \nabla_\beta M^{\gamma\delta} = D_\alpha D_\beta M^{\gamma\delta} + \Gamma^{\gamma}_{\beta\rho} M^{\rho\delta}_{,\alpha} + \Gamma^{\delta}_{\beta\rho} M^{\gamma\rho}_{,\alpha} + \Gamma^{\gamma}_{\beta\rho,\alpha} M^{\rho\delta} + \Gamma^{\delta}_{\beta\rho,\alpha} M^{\gamma\rho} \\
-\Gamma^{\rho}_{\beta\alpha} \nabla_\beta M^{\gamma\delta} + \Gamma^{\gamma}_{\alpha\rho} \nabla_\beta M^{\rho\delta} + \Gamma^{\delta}_{\alpha\rho} \nabla_\beta M^{\gamma\delta}
\end{aligned} \tag{4.87}
$$

In a plane surface Christoffel symbols are equal to zero, and in the Cartesian coordinates (x, y, z) the equations read

$$
N,^{xx}_{x} + N,^{xy}_{y} + p^x = 0, \; \mathrm{N},^{xy}_{x} + \mathrm{N},^{yy}_{y} + p^y = 0 \tag{4.88}
$$

$$
\mathrm{M},^{xx}_{xx} + 2\mathrm{M},^{xy}_{xy} + \mathrm{M},^{yy}_{yy} - p^3 = 0 \tag{4.89}
$$

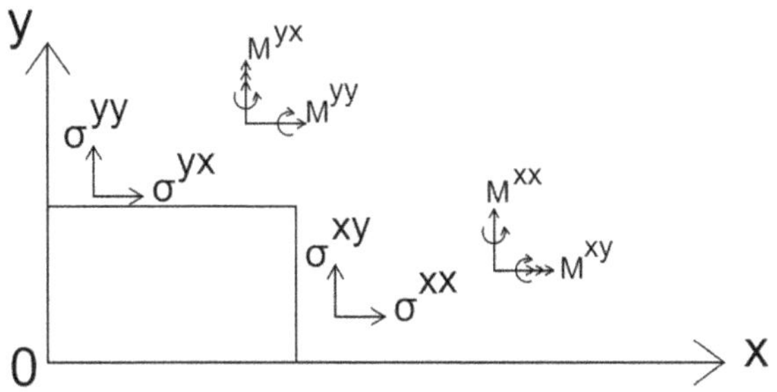

Fig. 4.9 Moments on a plate

$$M^{xx} = -\int_{\frac{-h}{2}}^{\frac{h}{2}} z\sigma^{xx} dz, \; M^{yy} = \int_{-\frac{h}{2}}^{\frac{h}{2}} -z\sigma^{yy} dz, \; M^{xy} = -\int_{-\frac{h}{2}}^{\frac{h}{2}} z\sigma^{xy} dz = M^{yx} \tag{4.90}$$

are respectively a moment of axis y which bends in the x direction, a moment of axis x which bends in the y direction, and a torque around the x axis of the planes xz and finally a torque around the y axis of the plane yz (Fig. 4.9).

We have in the Cartesian coordinates (x, y, z)

$$\begin{aligned} N_{xx} = N^{xx} &= \frac{Eh}{(1-\bar{\nu}^2)}\left[(1-\bar{\nu})\, e_{xx} + \bar{\nu}\left(e_{xx} + e_{yy}\right)\right] \\ &= \frac{Eh}{(1-\nu^2)}\left(u,_x + \bar{\nu} v,_y\right) \end{aligned} \tag{4.91}$$

$$N^{xy} = N^{yx} = \frac{Eh}{\left(1-\nu^2\right)}\left(v,_x + u,_y\right), \quad N_{yy} = \mathrm{N}^{yy} = \frac{Eh}{\left(1-\bar{\nu}^2\right)}\left(v,_y + \bar{\nu} u,_x\right) \tag{4.92}$$

$$M_{xx} = M^{xx} = \frac{Eh^3}{12\left(1-\bar{\nu}^2\right)}\left(w,_{xx} + \bar{\nu} w,_{yy}\right) = D\left(w,_{xx} + \bar{\nu} w,_{yy}\right) \tag{4.93}$$

$$M_{yy} = M^{yy} = D\left(\bar{\nu} w,_{xx} + w,_{yy}\right), \; M_{xy} = M_{yx} = Dw,_{xy} \tag{4.94}$$

These equations are completed with boundary conditions deduced from the general boundary conditions established in Chap. 2 as follows:

- (i) on the free border ($\overrightarrow{\nu}$ unit outer normal vector)

$$N^{\nu\nu} = q^{\nu}; \; N^{\nu t} = q^{t}; \; Q^{\nu} - \partial_{\overrightarrow{t}} M^{\nu t} = q^{3}; \; \left(Q^{\nu} = -\nabla_{\alpha} M^{\alpha\nu} = \nabla_{\alpha} M^{\alpha\beta}\nu_{\beta}\right); \; M^{\nu\nu} + m^{t} = 0 \tag{4.95}$$

In a rectangular plate, $\nabla_{\alpha} = D_{\alpha}$, $\overrightarrow{\nu} = \overrightarrow{i}$ or $\overrightarrow{j}$ and the same for $\overrightarrow{t}$.

- (ii) on the clamped border

$$u = v = w = 0,\ \partial_{\vec{\nu}} u = \partial_{\vec{\nu}} v = \partial_{\vec{\nu}} w = 0 \tag{4.96}$$

- (iii) simple supported border

$$u = v = w = 0;\ \ M^{\nu\nu} + m = 0 \tag{4.97}$$

From the general existence theorem established above, we eventually deduce the existence of a unique solution of the variational equation in the space of admissible displacements.

4.4.1 Theory of Pure Bending Plates

We have $e_{\alpha\beta} = \frac{1}{2}\left(\nabla_\alpha u_\beta + \nabla_\beta u_\alpha - 2b_{\alpha\beta} w\right)$. By neglecting Gauss curvature and in-plane displacements together with their derivatives, in a Cartesian coordinate system, we have

$$\begin{aligned} K_{\alpha\beta} &= \nabla_\alpha \left(\partial_\beta w + B^\rho_\beta u_\rho\right) + B^\rho_\alpha \left(\nabla_\beta u_\rho - B_{\beta\rho} w\right) \\ &= \quad \nabla_\alpha \nabla_\beta w = w,_{\alpha\beta} - w,_\rho \Gamma^\rho_{\alpha\beta} = w,_{\alpha\beta} \end{aligned} \tag{4.98}$$

$$e_{22} = \frac{1}{2}\left(\partial_y v + \partial_y v - 2\frac{w}{R}\right) = \partial_y v - w/R \simeq -w/R \tag{4.99}$$

$$K_{22} = b_{22} - B_{22} = b_{22} = 1/R \tag{4.100}$$

On a uniform loaded desk, maximum deflection is obtained at the centre and $e_{22} \simeq -\frac{f}{R}$ where f and R are respectively the deflection and radius of curvature at the centre of the desk (see Fig. 4.10). It follows that the membrane energy density is proportional to Ehe^2, i.e. $\mathcal{F}_m \propto Ehe^2 = Ehf^2/R^2$. Similarly the bending energy density is proportional to Eh^3K^2, i.e. $\mathcal{F}_f \propto Eh^3K^2 = Eh^3/R^2$ (Fig. 4.11).

Now $R^2 = (R - f)^2 + l^2/4$. We can write $l^2 \simeq 8Rf$ and $1/R \propto f/l^2$. Therefore

$$\kappa = \frac{\mathcal{F}_m}{\mathcal{F}_f} = \frac{Eh\frac{f^4}{l^4}}{Eh^3\frac{f^2}{l^4}} \approx \frac{f^2}{h^2} \tag{4.101}$$

It can be remarked that if $f = 0.2h$ then $\kappa = 0.04$, which shows that in this case membrane energy is negligible and only the bending energy should be considered. If the maximum admissible deflection is a percentage of the span, for example $f \leq l(cm)/250$, then one will have to calculate κ in order to choose one theory or the

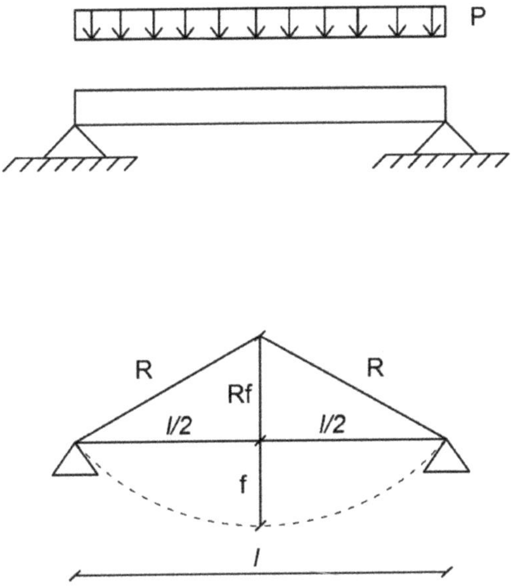

Fig. 4.10 Evaluation of the maximum deflection of a simple support plate

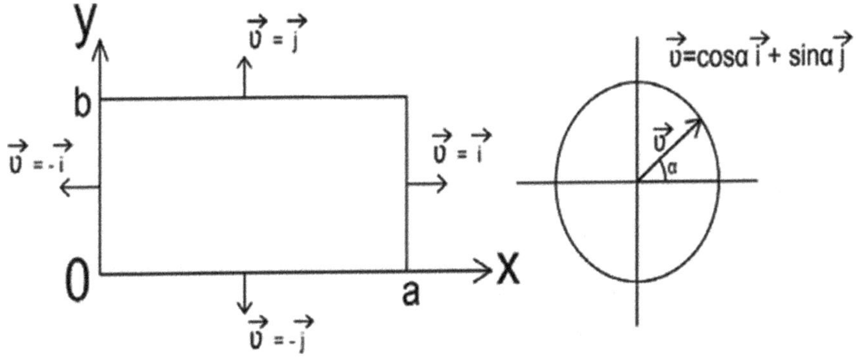

Fig. 4.11 Simple support slab

other. Any simple model of evaluating the maximum deflection can be used in order to compare energy densities. Example: $l = 10\,\text{m}, h = 0.20\,\text{m}, \kappa = 0.04$. In this case a pure bending theory is relevant. In a Cartesian coordinate system, for a homogeneous isotropic material, we have

$$\begin{cases} D_\alpha D_\beta M^{\alpha\beta} - P & = 0 \\ \\ boundary\ conditions & \end{cases} \tag{4.102}$$

$$M_{..} = D\left[(1-\bar{\nu})K + \bar{\nu}trK\right],\ K_{..} = \begin{bmatrix} w_{,xx} & w_{,xy} \\ w_{,xy} & w_{,yy} \end{bmatrix},\ D = \frac{Eh^3}{12\left(1-\bar{\nu}^2\right)} \tag{4.103}$$

equivalent to

$$\begin{cases} D\left(\frac{\partial^4 w}{\partial x^4} + 2\frac{\partial^4 w}{\partial x^2 \partial y^2} + \frac{\partial^4 w}{\partial y^4}\right) = P \\ \text{boundary conditions} \end{cases} \tag{4.104}$$

or

$$\begin{cases} D\Delta\Delta w = P \\ \text{boundary conditions} \end{cases} \tag{4.105}$$

Under simple supported border, we have $w = 0$ and $m^s = 0$ on C; on clamped border condition, $w = 0$ and $\partial_{\vec{\nu}} w = 0$ on C. The variational equation reads

$$\begin{aligned} \int_S M^{\alpha\beta} D_\alpha D_\beta v dS + \int_C D_\beta M^{\alpha\beta} \nu_\alpha v dC \\ - \int_C M^{\alpha\beta} \nu_\beta D_\alpha v dC \quad &= \int_S P v dS = \prec P, v \succ \end{aligned} \tag{4.106}$$

$w,\ v \in H^2(S) \cap H_0^1(S)$ for the simple supported border and $w,\ v \in H_0^2(S)$ for the clamped border. But $M^{\alpha\beta}\nu_\beta = M^{\alpha\nu} = M^{..}\vec{\nu} = M^{\nu\nu}\vec{\nu} + M^{\nu t}\vec{t}$. Therefore $M^{\alpha\beta}\nu_\beta D_\alpha v = (M^{\nu\nu}\vec{\nu} + M^{\nu t}\vec{t}) \cdot \nabla v = M^{\nu\nu}\partial_{\vec{\nu}} v + M^{\nu t}\partial_{\vec{t}} v$. For the simple supported border, $\partial_{\vec{t}} v = 0$ and $M^{\nu\nu} = 0$ since $m^t = 0$. For the clamped border $\partial_{\vec{\nu}} v = 0$ and $\partial_{\vec{t}} v = 0$. Therefore, in both situations $M^{\alpha\beta}\nu_\beta D_\alpha v = 0$ and $D_\beta M^{\alpha\beta}\nu_\alpha v = 0$ on the border C. The variational formulation now reads

$$\int_S M^{\alpha\beta} D_\alpha D_\beta v dS = \int_S P v dS = \prec P, v \succ \tag{4.107}$$

$$\begin{aligned} \int_S D\left((1-\bar{\nu})K^{\alpha\beta} v_{,\alpha\beta} + \bar{\nu} K^\rho_\rho v_{,\gamma\gamma}\right) dS &= \int_S D((1-\bar{\nu}) w_{,\alpha\beta} v_{,\alpha\beta} \\ &\quad + \bar{\nu} w_{,\rho\rho} v_{,\gamma\gamma}) dS \\ &= \quad \prec P, v \succ \end{aligned} \tag{4.108}$$

or

$$\int_S D\Delta w \Delta v dS = \prec P, v \succ \tag{4.109}$$

It should be noted that this variational equation is valid in a Cartesian coordinate system regardless of the shape of the plate and the loads (distributed or concentrated). In the curvilinear coordinate system, the Cartesian derivation $D_\alpha D_\beta M^{\alpha\beta}$ is replaced by $\nabla_\alpha \nabla_\beta M^{\alpha\beta}$. In the polar coordinate system, we have

$$A_r = \begin{pmatrix} cos\theta \\ sin\theta \end{pmatrix}, \; A_\theta = r \begin{pmatrix} -sin\theta \\ cos\theta \end{pmatrix}, \; A_{..} = \begin{bmatrix} 1 & 0 \\ 0 & r^2 \end{bmatrix}, \tag{4.110}$$

$$A^{..} = \begin{bmatrix} 1 & 0 \\ 0 & 1/r^2 \end{bmatrix}, \; \sqrt{A} = r,$$

$$\Gamma^r_{\alpha\beta} = \begin{bmatrix} 0 & 0 \\ 0 & -r \end{bmatrix}, \; \Gamma^\theta_{\alpha\beta} = \begin{bmatrix} 0 & 1/r \\ 1/r & 0 \end{bmatrix}, \tag{4.111}$$

$$K_{..} = \begin{bmatrix} w'' & w_{,r\theta} - \frac{1}{r} w_{,\theta} \\ w_{,r\theta} - \frac{1}{r} w_{,\theta} & w_{,\theta} + r w' \end{bmatrix}$$

and in a physical base $K_{..} = b_{..} - B_{..} = b_{..}$, i.e.

$$(K_{\alpha\beta}) = \begin{bmatrix} w'' & \frac{1}{r} w_{,r\theta} - \frac{1}{r^2} w_{,\theta} \\ \frac{1}{r} w_{,r\theta} - \frac{1}{r^2} w_{,\theta} & \frac{1}{r^2} w_{,\theta} + \frac{1}{r} w' \end{bmatrix}$$

Whatever the shape of the plate is, we deduce from $div V = \nabla_\alpha V^\alpha = \frac{1}{\sqrt{A}} (\sqrt{A} V^\alpha)_{,\alpha}$ and $V = \nabla w = w_{,\beta} A^\beta = w_{,\beta} A^{\beta\alpha} A_\alpha = V^\alpha A_\alpha$ that

$$\Delta w = div \nabla w = \frac{1}{\sqrt{A}} \left(\sqrt{A} A^{\alpha\beta} w_{,\beta} \right)_{,\alpha} \tag{4.112}$$

and

$$\Delta w = \frac{1}{r} \left(r w_{,r} \right)_{,r} + \frac{1}{r^2} w_{,\theta\theta} \tag{4.113}$$

$$\Delta\Delta w = \frac{1}{r}\left(r\left[\frac{1}{r}(rw_{,r})_{,r} + \frac{1}{r^2}w_{,\theta\theta}\right]_{,r}\right)_{,r} + \frac{1}{r^2}\left(\frac{1}{r}\left(rw_{,r}\right)_{,r} + \frac{1}{r^2}w_{,\theta\theta}\right)_{,\theta\theta} \tag{4.114}$$

Example 1: rectangular plate, $S = [0, a] \times [0, b]$ simply supported on its border C (the four sides)

$$\begin{cases} D\Delta\Delta w = P & in\ S \\ w = 0 & on\ C \\ C = \{x = 0\} \cup \{x = a\} & \cup \{y = 0\} \cup \{y = b\} \\ M^{\nu\nu} + m^t = 0 \end{cases} \tag{4.115}$$

$$\begin{cases} D\Delta\Delta w = P \text{ in } S \\ w = 0 \text{ on } C \\ M^{\nu\nu} = \bar{\nu}\Delta w + (1+\bar{\nu})\left(\frac{\partial^2 w}{\partial x^2}cos^2\alpha + 2\frac{\partial^2 w}{\partial x \partial y}sin\alpha cos\alpha + sin^2\alpha\frac{\partial^2 w}{\partial y^2}\right) = 0 \end{cases} \tag{4.116}$$

$$\begin{cases} on\quad \{x = 0\} \cup \{x = a\}, \ \frac{\partial w}{\partial y} = \frac{\partial^2 w}{\partial y^2} = 0 \\ on\quad \{y = 0\} \cup \{y = b\}, \ \frac{\partial w}{\partial x} = \frac{\partial^2 w}{\partial x^2} = 0 \end{cases} \tag{4.117}$$

We can find w in the form $w = w_{mn} sin\left(\frac{m\pi x}{a}\right) sin\left(\frac{n\pi y}{b}\right) = w_{mn}\bar{w}_{mn}(x, y)$; w satisfies the boundary conditions above. We assume the load $P(x, y)$ can also be decomposed in a Fourier functions base $\bar{w}_{mn}(x, y)$ by $P(x, y) = P_{mn}\bar{w}_{mn}(x, y)$. Then, we have (without summation on indexes)

$$\Delta\Delta\bar{w}_{mn}(x, y) = \left(\frac{m^2\pi^2}{a^2} + \frac{n^2\pi^2}{b^2}\right)^2 \bar{w}_{mn}(x, y) = \pi^4\left(\frac{m^2}{a^2} + \frac{n^2}{b^2}\right)^2 \bar{w}_{mn}(x, y) \tag{4.118}$$

which shows that the functions $\bar{w}_{mn}$ are eigenfunctions of the operator $\Delta\Delta$. So $D\Delta\Delta w = P$ turns down to looking for constants w_{mn} such that (without summation on m *and* n)

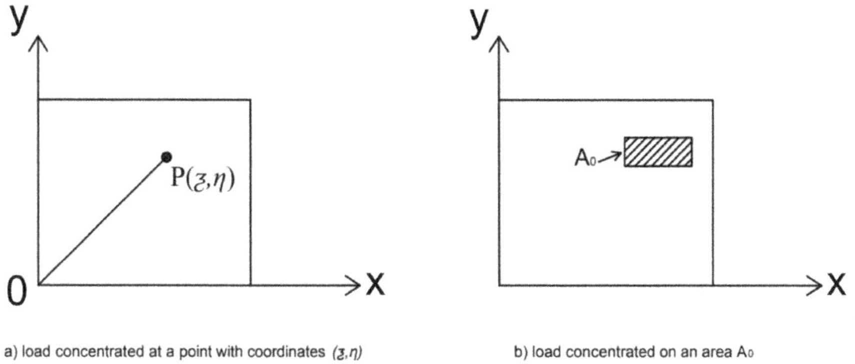

Fig. 4.12 Slab under concentrated loads

$$\pi^4 D\left(\frac{m^2}{a^2}+\frac{n^2}{b^2}\right)^2 w_{mn} = P_{mn} \tag{4.119}$$

If P is a concentrated load, then for odd numbers m and n, $P_{mn} = \frac{16P}{\pi^2 mn}$ or 0 otherwise. We obtain

$$\bar{w}_{11}(x, y) = \frac{16P}{\pi^6 D}\frac{1}{\left(\frac{1}{a^2}+\frac{1}{b^2}\right)^2} sin\left(\frac{\pi x}{a}\right) sin\left(\frac{\pi y}{b}\right) \tag{4.120}$$

Let a be the length of a square plate, then, $w\left(\frac{a}{2}, \frac{a}{2}\right) \simeq \bar{w}\left(\frac{a}{2}, \frac{a}{2}\right) = \frac{4Pa^4}{\pi^6 D}$ to one per cent. Because of the regularity of the function $w(x, y)$ ($w \in H^2(S) \subset C^0(S)$), Euler's equations are also valid for a concentrated load at a point (τ, η) on the plate (Fig. 4.12).

A concentrated load $P(\tau, \eta)$ can also be decomposed as

$$P(\tau, \eta) = P_{pq}\bar{w}_{pq}(\tau, \eta),\ P_{pq} = \frac{4P(\tau,\eta)}{ab} sin\left(\frac{p\pi\tau}{a}\right) sin\left(\frac{q\pi\eta}{b}\right) \tag{4.121}$$

and in the same manner, the solution reads

$$\begin{aligned} w(x, y, \tau, \eta) = &\frac{4P(\tau,\eta)}{\pi^4 Dab}\sum_{m,n}\frac{1}{mn\left(\frac{m^2}{a^2}+\frac{n^2}{b^2}\right)^2}\times \\ &sin\left(\frac{p\pi\tau}{a}\right) sin\left(\frac{q\pi\eta}{b}\right) sin\left[\frac{m\pi x}{a}\right] sin\left[\frac{n\pi y}{b}\right] \end{aligned} \tag{4.122}$$

Using the function $w(x, y, \tau, \eta)$, we calculate the solution for a concentrated load on a portion A_0 of the plate as follows:

$$w(x,y) = \int_{A_0} w(x,y,\tau,\eta)\,d\tau d\eta \tag{4.123}$$

In the above example, at each point, we can calculate from w, the different tensors and deduce from the constitutive law the moments. The stress components at each point now read

$$\sigma^{\alpha\beta} = -\frac{12z}{h^3} M^{\alpha\beta} \tag{4.124}$$

Under constant volume loads f^α, f^3 and the equilibrium equations

$$\sigma^{\alpha\beta}_{,\beta} + \sigma^{\alpha z}_{,z} = -f^\alpha \tag{4.125}$$

we deduce the in-plane shear stress through the thickness from the equations

$$\sigma^{\alpha z}_{,z} = -f^\alpha - \sigma^{\alpha\beta}_{,\beta} = \frac{12z}{h^3} M^{\alpha\beta}_{,\beta} - f^\alpha \ and\ \sigma^{\alpha z} = \frac{6z^2}{h^3} M^{\alpha\beta}_{,\beta} - zf^\alpha + c(x,y) \tag{4.126}$$

The constant c is determined from boundary conditions $\sigma^{\alpha z}$ at $z = -h/2$, $z = h/2$ and by assuming $\sigma^{\alpha h/2} = \sigma^{\alpha - h/2} = 0$, we obtain $c(x,y) = -\frac{6}{4h} M^{\alpha\beta}_{,\beta}$ even if $f^\alpha = 0$. We also deduce the pinch stress distribution through the thickness from the equation

$$\sigma^{\alpha z}_{,\alpha} + \sigma^{zz}_{,z} = -f^3 \tag{4.127}$$

which is equivalent to

$$\sigma^{zz}_{,z} = -f^3 - \sigma^{\alpha z}_{,\alpha} = -f^3 - \frac{6z^2}{h^3} M^{\alpha\beta}_{,\alpha\beta} - c_{,x}(x,y) - c_{,y}(x,y) \tag{4.128}$$

and finally

$$\sigma^{zz} = -f^3 z - \frac{2z^3}{h^3} M^{\alpha\beta}_{,\alpha\beta} - zc_{,x}(x,y) - zc_{,y}(x,y) + d \tag{4.129}$$

The constant d is also determined by using the boundary condition σ^{zz} at $z = h/2$ and letting $\sigma^{zh/2} = P - f^3$. The rotation of a cross section at a border is calculated by $\theta_{support} = \partial_{\vec{\nu}} w$, $\vec{\nu} = \vec{i}$ at $x = a$ or $-\vec{i}$ at $x = -a$; $\vec{\nu} = \vec{j}$ at $y = b$ or $-\vec{j}$ at $y = -b$. The reactions R_a at supports are calculated by

$$Q^\nu - \partial_s M^{\nu t} = R_a;\ \left(Q^\nu = -\nabla_\alpha M^{\alpha\beta}\nu_\beta\right),\ \partial_s = \partial_y\ or\ \partial_x \tag{4.130}$$

Example2: clamped disc of radius R (Fig. 4.13)

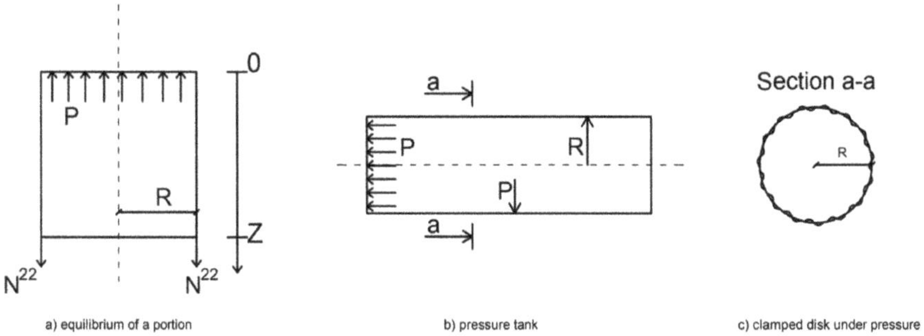

Fig. 4.13 Pressurized pipe with end discs

The variational equation remains unchanged. Now, if the load is sufficiently smooth, we can still consider Euler's equation

$$
\begin{aligned}
D\Delta\Delta w &= \frac{D}{r}(r[\frac{1}{r}(rw_{,r})_{,r} + \frac{1}{r^2}w_{,\theta\theta}]_{,r})_{,r} + \frac{1}{r^2}(\frac{1}{r}(rw_{,r})_{,r} + \frac{1}{r^2}w_{,\theta\theta})_{,\theta\theta} \\
&= P \text{ on S}
\end{aligned}
\tag{4.131}
$$

$$
\text{w} = 0 \text{ and } \partial w/\partial r = 0 \text{ on } C \tag{4.132}
$$

If the load P is uniformly constant or axisymmetric, then the displacement w does not depend on θ and the above equation becomes

$$
\Delta\Delta w = \frac{1}{r}\left(r\left[\frac{1}{r}(rw_{,r})_{,r}\right]_{,r}\right)_{,r} = P/D \tag{4.133}
$$

By integration we obtain

$$
w(r) = \frac{P}{64D}r^4 + a(\frac{r^4}{4}Logr - \frac{r^4}{4}) + b\frac{r^2}{4} + cLogr + d \tag{4.134}
$$

Displacement at the centre ($r = 0$) of the disc is finite. So, the constants a and c must be equal to zero. From the clamped edge equations $w(R) = 0$ and $w'(R) = 0$, it follows that $w(r) = \frac{PR^4}{64D}(\frac{r^4}{R^4} - \frac{2}{R^2}r^2 + 1)$. Reaction moment and force at the clamped edge are

$$\begin{aligned} M^{\nu\nu} = M^{rr} = M_{rr} &= D(K_{11} + \bar{\nu} K_{22}) \\ &= \quad D(w''(R) + \bar{\nu}\frac{w'(R)}{R}) \\ &= \quad Dw''(R) = \frac{PR^2}{8} \end{aligned} \tag{4.135}$$

$$R_a = Q^{\nu} - \partial_{R\theta} M^{12} = Q^r = -M^{rr}_{,r}(R) + \frac{1}{R}(M^{rr}(R) - M^{22}(R)) = \frac{PR}{2} \tag{4.136}$$

Other boundary conditions such as simple support with or without applied bending moment can be completely solved. Once the constants are all determined, moments and stress can be calculated for design purpose.

It may happen that the deformation is such that $RN^{\alpha\beta} \gg M^{\alpha\beta}$ (R is the radius of curvature). We are then in the second geometry and some terms in the third shell general equilibrium equation become significant. It is recommended to consider the deformed state as the new reference configuration which is no longer plane. Such an analysis is relevant especially if in-plane loads do exist. Let TS be the transformation of the initial mid-surface S of the plate in the initial reference configuration (first geometry) of the plate and let $(u(x, y),\ v(x, y),\ w(x, y))$ be the displacement field, then

$$TS = \{(x + u(, x, y),\ y + v(x, y),\ w(x, y));\ (x, y) \in S\}$$

the covariant base, the metric tensor and curvature tensor in both configurations are as follows:

$$A_x = \vec{i},\ A_y = \vec{j},\ A_{\alpha\beta} = \delta_{\alpha\beta},\ B_{\alpha\beta} = 0;$$

$$a_x = \begin{pmatrix} 1 + u_{,x} \\ v_{,x} \\ w_{,x} \end{pmatrix},\ a_y = \begin{pmatrix} u_{,y} \\ 1 + v_{,y} \\ w_{,y} \end{pmatrix},$$

$$a_x \times a_y = \begin{pmatrix} v_{,x} w_{,y} - w_{,x} - w_{,x} w_{,y} \\ w_{,x} w_{,y} - w_{,y} - w_{,x} w_{,y} \\ 1 + w_{,y} + u_{,x} + u_{,x} w_{,y} \end{pmatrix} \simeq \begin{pmatrix} -w_{,x} \\ -w_{,y} \\ 1 \end{pmatrix},$$

$$a_3 \simeq \frac{1}{\sqrt{1 + w_{,x}^2 + w_{,y}^2}} \begin{pmatrix} -w_{,x} \\ -w_{,y} \\ 1 \end{pmatrix}$$

$$a_{x,x} = \begin{pmatrix} u_{,xx} \\ v_{,xx} \\ w_{,xx} \end{pmatrix}, \quad a_{y,y} = \begin{pmatrix} u_{,yy} \\ v_{,yy} \\ w_{,yy} \end{pmatrix}, \quad a_{x,y} = \begin{pmatrix} u_{,xy} \\ v_{,xy} \\ w_{,xy} \end{pmatrix}$$

By neglecting some small terms, we obtain $a_{xx} = a_x \cdot a_x = 1 + 2u_{,x} + w_{,x}^2$, $a_{yy} = a_y \cdot a_y = 1 + 2v_{,y} + w_{,y}^2$, $a_{xy} = a_x \cdot a_y = u_{,y} + v_{,x} + w_{,x}w_{,y} = a_{yx}$ In the same manner we also obtain

$$e_{xx} = \frac{1}{2}(a_{xx} - A_{xx}) = u_{,x} + \frac{1}{2}w_{,x}^2 \quad e_{yy} = \frac{1}{2}(a_{yy} - A_{yy}) = v_{,y} + \frac{1}{2}w_{,y}^2,$$

$$e_{xy} = \frac{1}{2}(a_{xy} - A_{xy}) = \frac{1}{2}(u_{,y} + v_{,x} + w_{,x}w_{,y});$$

$$K_{xx} = b_{xx} - B_{xx} = b_{xx} = a_{x,x} \cdot a_3 \simeq w_{,xx},$$

$$K_{yy} = b_{yy} - B_{yy} = b_{yy} = a_{y,y} \cdot a_3 \simeq w_{,yy}$$

$$K_{xy} = b_{xy} - B_{xy} = b_{xy} = a_{x,y} \cdot a_3 \simeq w_{,xy} = K_{yx}$$

We here summarize all the important fields of the deformed plate TS:

$$e_{..} = \begin{bmatrix} u_{,x} + \frac{1}{2}w_{,x}^2 & \frac{1}{2}(u_{,y} + v_{,x}) + \frac{1}{2}w_{,x}w_{,y} \\ \frac{1}{2}(u_{,y} + v_{,x}) + \frac{1}{2}w_{,x}w_{,y} & v_{,y} + \frac{1}{2}w_{,y}^2 \end{bmatrix}; \tag{4.137}$$

$$K_{..} = \begin{bmatrix} w_{,xx} & w_{,xy} \\ w_{,xy} & w_{,yy} \end{bmatrix}$$

The equations now read:

$$N^{xx} = N_{xx} = \frac{Eh}{1 - \bar{\nu}^2}\left(e_{xx} + \bar{\nu}e_{yy}\right),$$

$$N^{yy} = N_{yy} = \frac{Eh}{1 - \bar{\nu}^2}\left(e_{yy} + \bar{\nu}e_{xx}\right),$$

$$N^{xy} = N_{xy} = \frac{Eh}{1 - \bar{\nu}^2}\left(e_{xy}\right) = N^{yx}$$

$$M^{xx} = M_{xx} = \frac{Eh^3}{1-\bar{\nu}^2}\left(K_{xx} + \bar{\nu}K_{yy}\right),$$

$$M^{yy} = M_{yy} = \frac{Eh^3}{1-\bar{\nu}^2}\left(K_{yy} + \bar{\nu}K_{xx}\right)$$

$$M^{xy} = M_{xy} = \frac{Eh^3}{1-\bar{\nu}^2}\left(K_{xy}\right) = M^{yx}$$

$$D_\alpha N^{\alpha\beta} + p^\beta = 0 \tag{4.138}$$

$$D_\alpha D_\beta M^{\alpha\beta} - b_{\alpha\beta}N^{\alpha\beta} - p^3 = 0 \tag{4.139}$$

Under simple support boundary conditions without applied bending moment, or under clamped edge, the variational equation reads:

$$\begin{cases} \text{Find (u,v,w) such that} \\ \int_S M^{\alpha\beta}\partial_\alpha\partial_\beta\bar{w}_3 dS - \int_S N^{\alpha\beta}b_{\alpha\beta}\bar{w}_3 dS + \int_S N^{\alpha\beta}\partial_\alpha\bar{w}_\beta dS = \int_S p^i\bar{w}_i dS \end{cases} \tag{4.140}$$

$\bar{w}_3 \in H^2(S) \cap H_0^1(S)$ for simple support and $\bar{w}_3 \in H_0^2(S)$ for clamped edge, $u, v, \bar{w}_\alpha \in H_0^1(S)$. These equations are nonlinear in both variational and Euler's forms because of the constitutive law. They therefore need a particular more elaborated analysis for their resolution. In-plane stresses may be known as stiffening or none stiffening prestresses, oriented in particular directions. They can be written in the form $N^{\alpha\beta}$ by using tensor algebra formulas.

4.4.2 The Von Karman Equations

In practice, it is often to deal with plates without membrane (or in-plane) loads p^β. The stresses $N^{\alpha\beta}$ may be due to curvature created by the deformation or prestresses applied on the edges (Fig. 4.14).

By introducing Airy's stress functions on $N^{\alpha\beta} = N_{\alpha\beta}$, we obtain new equations. From

$$e_{xx,yy} + e_{yy,xx} - 2e_{xy,xy} = (w_{,xy})^2 - w_{,xx}w_{,yy} \tag{4.141}$$

we deduce that

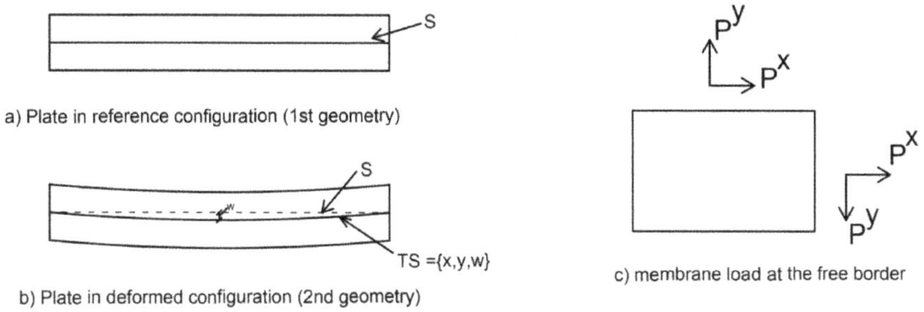

Fig. 4.14 Bent slab under additional membrane loads

$$(N_{xx} - \bar{\nu}N_{yy})_{,yy} + (N_{yy} - \bar{\nu}N_{xx})_{,xx} - 2(1+\bar{\nu})(N_{xy})_{,xy} = Eh(w_{,xy}^2 - w_{,xx}w_{,yy}) \tag{4.142}$$

Let Φ denote Airy stress function. Then $N_{xx} = \Phi_{,yy}$; $N_{yy} = \Phi_{,xx}$; $N_{xy} = -\Phi_{,xy}$. By replacing $N_{\alpha\beta}$ by these relations in the above equation and the third equilibrium equation of the plate, we obtain the equations

$$\Delta\Delta\Phi = Eh(w_{,xy}^2 - w_{,xx}w_{,yy}) \tag{4.143}$$

$$D\Delta\Delta w - \Phi_{,yy}w_{,xx} - \Phi_{,xx}w_{,yy} + 2\Phi_{,xy}w_{,xy} = p^3 \tag{4.144}$$

which arc the Von Karman equations. For some particular boundary conditions, under some simplifying hypotheses on Φ and w, an analytic approximate solution of the Von Karman equations can be developed in series.
Exercise: On a rectangular plate we assume that $N_{xy} = N_{yy} = 0$, $N_{xx} = p^x$ is a constant prestress applied on the edge $x = 0$, $x = l$ and $w(0, y) = w(l, y) = 0$. Find $w(x, y)$ in the form of a series and describe the relation between the applied loads (the prestress and the transverse load p^3). The load p^3 will also be decomposed in the same series.

4.5 Theory of Orthotropic Plates

4.5.1 The Huber Equation

A plate is orthotropic (unlike isotropic) if its mechanical properties (moduli) differ according to directions, for example, a concrete slab unevenly reinforced in two perpendicular directions, a stiffened isotropic plate, etc. The constitutive law is then

$$\sigma_{xx} = E_x\varepsilon_{xx} + E'\varepsilon_{yy};\;\; \sigma_{yy} = E'\varepsilon_{xx} + E_y\varepsilon_{yy};\;\; \tau_{xy} = \bar{\mu}\gamma_{xy};\;\; \bar{\mu} = E/2(1+\bar{\nu}) \tag{4.145}$$

$$\varepsilon_{xx} = -zw_{,xx};\ \ \varepsilon_{yy} = -zw_{,yy};\ \ \gamma_{xy} = -zw_{,xy} \tag{4.146}$$

$$M_{xx} = -\int_{\frac{-h}{2}}^{\frac{h}{2}} z\left(\sigma_{xx}\right)dz;\ M_{xy} = -\int_{\frac{-h}{2}}^{\frac{h}{2}} z\left(\sigma_{xy}\right)dz;\ M_{yy} = -\int_{\frac{-h}{2}}^{\frac{h}{2}} z\left(\sigma_{yy}\right)dy \tag{4.147}$$

It follows that

$$M_{xx} = D_x w_{,xx} + D_1 w_{,yy};\ M_{yy} = D_y w_{,yy} + D_1 w_{,xx};\ M_{xy} = M_{yx} = D_{xy} w_{,xy}$$

$$D_x = \frac{E_x h^3}{12};\ \ D_y = \frac{E_y h^3}{12};\ \ D_1 = \frac{Eh^3}{12};\ \ D_{xy} = \frac{\bar{\mu} h^3}{12} \tag{4.148}$$

The variational equations are still the same.

Find w such that

$$\begin{aligned} \int_S M^{\alpha\beta}\partial_\alpha\partial_\beta v dS = & \int_S (M^{xx} v_{,xx} + 2M^{xy} v_{,xy} + M^{yy} v_{,yy}) dS \\ & = \int_S ((D_x w_{,xx} + D_1 w_{,yy}) v_{,xx} + 2D_{xy} w_{,xy} v_{,xy} \\ & \quad + (D_y w_{,yy} + D_1 w_{,xx}) v_{,yy}) dS \\ = & \quad \int_S P v dS \end{aligned} \tag{4.149}$$

$w,\ v \in H^2(S) \cap H_0^1(S)$ for simple support and $w,\ v \in H_0^2(S)$ for clamped edge, the load P being distributed, concentrated on a portion or on a point, and the stiffness D_x, D_y, D_1 and D_{xy} constant or variable. Let us assume they are constant and let $T = -divM$ denote the shear force, then

$$\begin{aligned} T_x &= -M_{xx,x} - M_{xy,y} = -D_x w_{,xxx} - \left(D_1 + D_{xy}\right) w_{,xyy}; \\ T_y &= -D_y w_{,yyy} - \left(D_1 + D_{xy}\right) w_{,xxy} \end{aligned} \tag{4.150}$$

and $divdivM = P$ becomes

$$D_x w_{,x^4} + 2\left(D_1 + 2D_{xy}\right) w_{,x^2y^2} + D_y w_{,y^4} = P \tag{4.151}$$

which is arranged in the form

$$D_x w_{,x^4} + 2H w_{,x^2y^2} + D_y w_{,y^4} = P \tag{4.152}$$

It is Huber's equation. It must be completed with boundary conditions. For a simple support on the edges $x = a$ and $x = -a$, $w = 0$ and the reaction on the support is

$$R_x = T_x - M_{xy,y} = -D_x w_{,xxx} - \left(2D_{xy} + D_1\right) w_{,xyy} \tag{4.153}$$

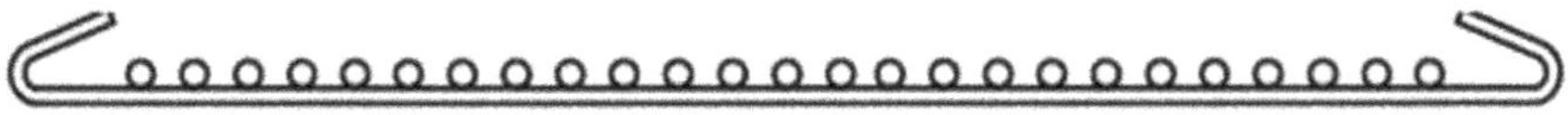

Fig. 4.15 Unevenly reinforced slab

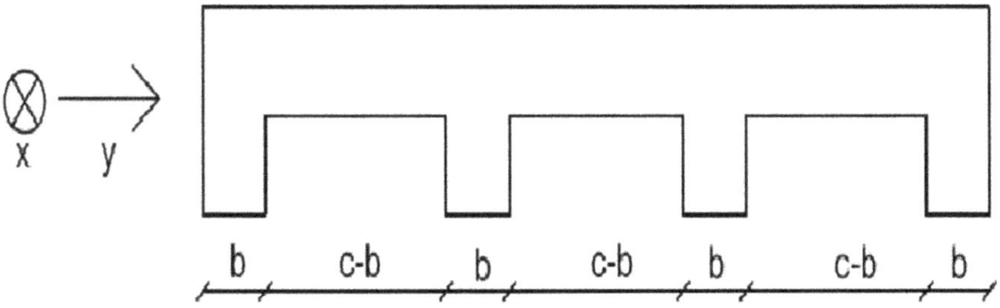

Fig. 4.16 x-direction ribbed slab

Similarly on $y = b$ and $y = -b$, $w = 0$ and the reaction on the support is

$$R_y = T_y - M_{xy,x} = -D_y w_{,yyy} - \left(2D_{xy} + D_1\right) w_{,xxy} \tag{4.154}$$

In an isotropic plate, $E_x = E_y = \frac{E}{1-\bar{\nu}^2}$; $E' = \frac{\nu E}{1-\bar{\nu}^2}$; $D_x = D_y = H = \frac{Eh^3}{12\left(1-\bar{\nu}^2\right)}$. It appears that an isotropic plate is a particular orthotropic structure. The stiffness most often vary in practice. They are replaced with constant stiffness obtained through homogenization (Figs. 4.15 and 4.16).

4.5.2 Examples

(1) Unevenly reinforced slab

$H = \sqrt{D_x D_y}$

(2) Ribbed slab in the x direction with beam drop

$2H = 2C_p + \frac{C_R}{c}$

$D_x = \frac{EI_x}{c}$; $D_1 = 0$; $\frac{c}{D_y} = \frac{c-b}{D_y^p} + \frac{b}{D_y^1}$

$2C_p \equiv$ torsional stiffness without rib

$C_R \equiv$ torsional stiffness of an isolated rib

$D_y^p \equiv$ bending stiffness of the slab without rib

$D_y^1 \equiv$ bending stiffness of the slab with the additional height of the rib.

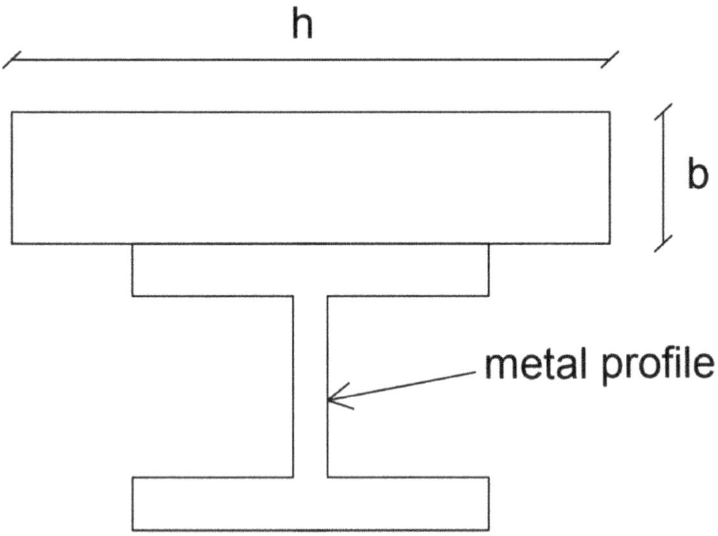

Fig. 4.17 Composite profile

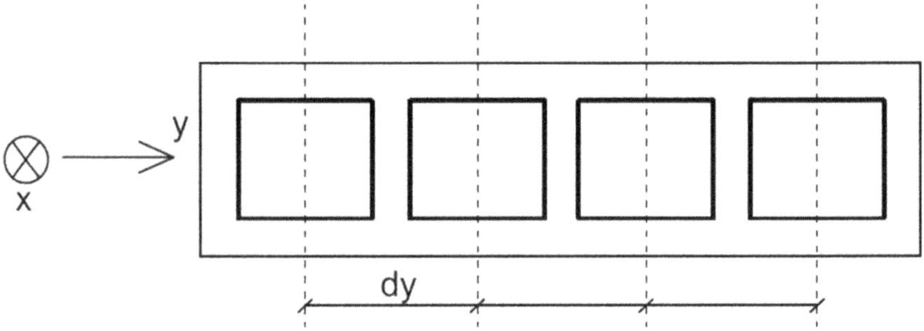

Fig. 4.18 Prestressed floor

Torsional stiffness is equal to zero if it is insignificant, i.e. $C_p = C_R = 0$, which yields $H = 0$.

(3) Mixed structure metal profile—concrete (Fig. 4.17)

$2H = C = C_x^{metal} + \frac{1}{2}\frac{\bar{\mu}_b h b^3}{3}$, $\bar{\mu}_b = E/2(1+\bar{\nu})$, C_x or C_y, D_x *or* D_y are calculated.

(4) Prestressed concrete slab (Fig. 4.18)

$$H = \sqrt{D_x D_y}, \quad B = EI, \quad D_x = \frac{B_x}{d_y}, \quad D_y = \frac{B_y}{d_x}.$$

(5) Grid beams Fig. 4.19)

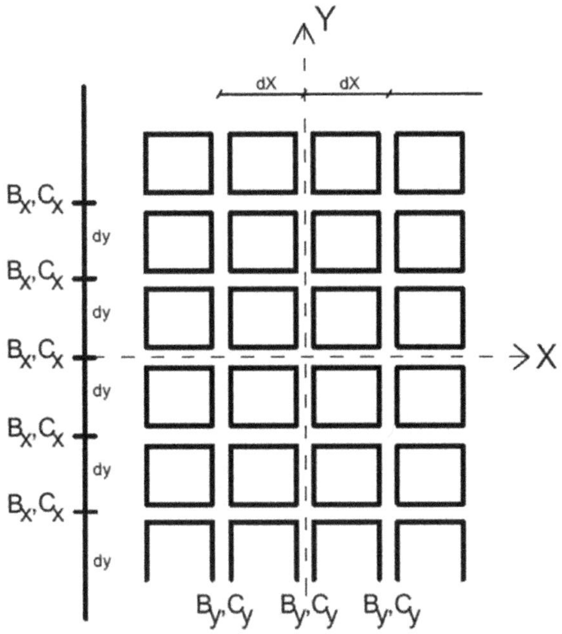

Fig. 4.19 Grid beams

$$2H = \frac{C_x}{d_y} + \frac{C_y}{d_x}, \quad D_x = \frac{B_x}{d_y}, \quad D_y = \frac{B_y}{d_x}.$$

The torsional stiffness of a rectangular section ab, a being the smaller side, is calculated with the formula $C = \lambda b a^3$ where $\lambda = \lambda(b/a)$ is given in Table 4.1.

Table 4.1 Built-up section

b/a	1	1.2	1.5	2	3	4	5	10	∞
λ	0.141	0.166	0.196	0.229	0.23	0.281	0.291	0.312	1/3

Chapter 5
Numerical Methods

Shell equations are so complex that analytical solutions can only exist in a few rare exceptions obtained under sometimes very restrictive assumptions. Numerical methods are essential especially when the problem to be solved is real and such simplified assumptions are unrealistic. The Finite Element Method (FEM) has been very successful over the Finite Difference Method (FDM). Finite Volume Method (FVM) is recent and has not yet sufficiently impacted numerical methods as the FEM. The main objective of this chapter is to briefly present the FEM and to show some specificities in calculating shell structures.

5.1 Generalities of the 2D FEM

5.1.1 Description

The Finite Element Method (FEM) consists in subdividing the domain or the structure in small elements in which the solution of the problem can be approximated by some elementary functions that depend on a finite number of unknowns. The variational equation of the whole structure is obtained by summation over each element, of the restriction of the variational equation. Similarly the total potential energy of the structure is obtained. This procedure leads to a discrete formulation of the initial continuous problem, i.e. a variational problem which depends on a finite number of unknowns. A mesh consists of elements which are close convex sets with non empty interior, which are such that the intersection of two elements is either an empty set, a summit-node or a complete edge of both elements. A mesh is characterized by the maximum edge dimension of all the constitutive elements that is denoted by h_e. So a mesh made of triangular elements, whose maximum edge dimension is h_e, will be denoted by T_{h_e}.

An element is characterized by its

R. Nzengwa, *Plate and Shell Models*,
https://doi.org/10.1007/978-981-97-2780-3_5

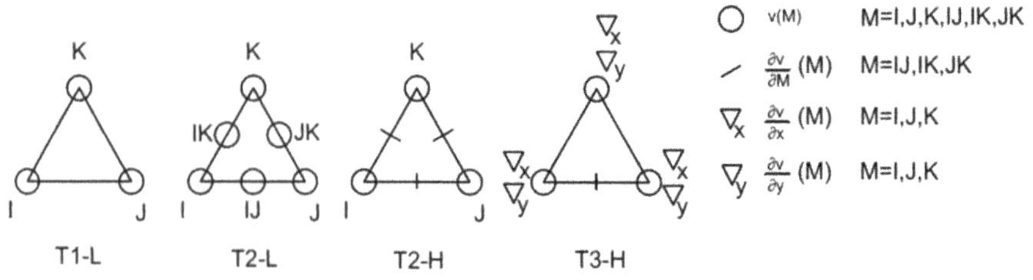

Fig. 5.1 Lagrange and Hermite elements

- geometry (a convex close set)
- parametrization (values of functions or of their derivatives); example: $w,\ \partial_x w,\ \partial_y w,\ \partial_{xx} w,\ \partial_{xy} w,\ \partial_{yy} w$,...)
- shape functions (or interpolation functions are functions which approach better the restriction of the solution in an element). They constitute a finite dimension vector space F_e
- set I_e of points with a finite cardinal (points where the parameters are taken to completely define the shape function. They are usually nodes, midpoints of edges, etc.).

For plates or shells, which are 2D structures, an element is a triangle with straight line or curved line edges. Triangles are close convex sets which are completely defined from their summit-nodes or vertices. Interpolation functions must reproduce per element a constant virtual power or work. In other words, if the variational formulation contains derivatives of order m of a function, the m-order derivative of a shape function should not be equal to zero. Also, shape functions should be able to reproduce rigid motions and ensure continuity at common edges up to a certain order less or equal to m of their derivatives (Fig. 5.1).

5.1.2 Element Stiffness Matrix

Let E_e be an element and $\varphi \in F_e$. Let U_e denote the nodal vector whose components are values of the parameters taken at the points in I_e. It is called the local nodal vector of degrees of freedom (dof). A base N of F_e can be chosen such that, for each function $\varphi \in F_e$, we have

$$\varphi(x, y) = NU_e \tag{5.1}$$

Similarly, for a vector field $u = (u_1, u_2, u_3)$, we define $\{F_e\} = (F_e^1, F_e^2, F_e^3)$, $\{U_e\} = (U_e^1, U_e^2, U_e^3)$, the diagonal matrix bloc $[N] = (N^1, N^2, N^3)$ and write

$$u(x, y) = [N]\{U_e\};\ \ v(x, y) = [N]\{V_e\} \tag{5.2}$$

Strain and stress tensors in an element are also written as

$$\epsilon(u) = [B]\{U_e\},\ \sigma = [D]\epsilon(u) = [C]\{U_e\},\ \epsilon(v) = [B]\{V_e\}, \tag{5.3}$$

and the virtual internal work reads

$$\delta W^e_{def} = \int\limits_{E_e} \left\{{}^t\epsilon(v)\right\}\{\sigma(u)\}\,dE_e = \{V_e\}\,K_e\,\{U_e\} \tag{5.4}$$

where the matrix K_e is the element stiffness matrix.

5.1.3 Element Nodal Force Vector

In an element, the virtual work of external forces is

$$\delta W^e_{ext} = \int\limits_{E_e} p v\,dE_e + \int\limits_{\partial E_e \cap \gamma_1} q v\,dE_e + \int\limits_{\partial E_e \cap \gamma_1} m \cdot \bar{\theta}(v)\,d\gamma = \{V_e\} \cdot \{f_e\} \tag{5.5}$$

The nodal vector $\{f_e\}$ is the local nodal force vector. Virtual internal and external works being determined locally, the global virtual works are obtained by summation over all the elements of the mesh.

Let Q denote the global nodal vector of the different dof $\{U_e\}$. Let V denote the global nodal vector of the local dof $\{V_e\}$. Let $[R_e]$ denote the position matrix such that

$$\{U_e\} = [R_e]\,Q \text{ and } \{V_e\} = [R_e]\,\mathrm{V} \tag{5.6}$$

Then

$$\delta W^e_{ext} = V \cdot {}^t[R_e]\{f_e\} = V \cdot \left\{f_e^G\right\}, \tag{5.7}$$

$$\delta W^e_{def} = V \cdot {}^t[R_e]\,K_e\,[R_e]\,Q = V \cdot K_e^G Q \tag{5.8}$$

$$\delta W_{ext} = \sum \delta W^e_{ext} = V \cdot F,\ \delta W_{def} = \sum \delta W^e_{def} = V \cdot K Q; \tag{5.9}$$

$$F = \sum \left\{f_e^G\right\};\ K = \sum K_e^G \tag{5.10}$$

5.1.4 Numerical Resolution

The equivalent discrete problem is the linear system

$$KQ = F \tag{5.11}$$

The efficiency of the resolution depends on the property of the global stiffness matrix K. Though K is symmetric and positive definite, it can be a sparse and unstable matrix depending on the arrangement (or numbering) of the dof. The nodal vector Q is organized in two parts: $Q = (Q_1,\ Q_2)$ where Q_2 groups all dof related to boundary conditions. Also, the nodal force vector is arranged as $F = (F_1,\ F_2)$ where $F_2 = R$ is the nodal reaction vector or some relations on $Q_2 \in \mathbb{R}^m$. The matrix K is consequently rearranged as follows:

$$K = \begin{bmatrix} K_{11} & K_{12} \\ K_{21} & K_{22} \end{bmatrix} \tag{5.12}$$

and the linear system becomes

$$K_{11}Q_1 = F_1 - K_{12}Q_2,\ K_{21}Q_1 + K_{22}Q_2 = R \tag{5.13}$$

The boundary conditions may lead to an equation of the form $CQ_2 = 0$ where C is a $l \times m$ matrix. A Lagrangian multiplier $\lambda = (\lambda_1,\ \lambda_2,\ ....,\ \lambda_l)$ is introduced and the solution consists in finding the extremum of the function

$$J(Q_1,\ Q_2,\ \lambda) = \frac{1}{2}KQ \cdot Q - F_1 \cdot Q_1 - \lambda \cdot CQ_2 \tag{5.14}$$

which is also equivalent to solve the linear system

$$\begin{bmatrix} K_{11} & K_{12} & 0 \\ K_{21} & K_{22} & -{}^tC \\ 0 & -C & 0 \end{bmatrix} \begin{pmatrix} Q_1 \\ Q_2 \\ \lambda \end{pmatrix} = \begin{pmatrix} F_1 \\ 0 \\ 0 \end{pmatrix} \tag{5.15}$$

This matrix is also symmetric and reactions are calculated as above. The different linear systems can be solved by all the different numerical methods developed for this purpose, especially the pre-conditioned conjugate gradient method. The efficiency of the resolution scheme is intimately related to the compact (non-sparse) form of the matrix K. In order to obtain a tridiagonal or tridiagonal bloc matrix, it is necessary to renumber the dof. The following algorithm may be used.

Let NN denote the total number of nodes of the mesh

For $i = 1,\ NN$, $P_i = \dfrac{1}{n_i}\sum_{k=1}^{n_i} M_k$;

n_i is the number of elements to which belongs the node i, M_k is the number of each node of the element related to i. Nodes are renumbered according to their weight P_i

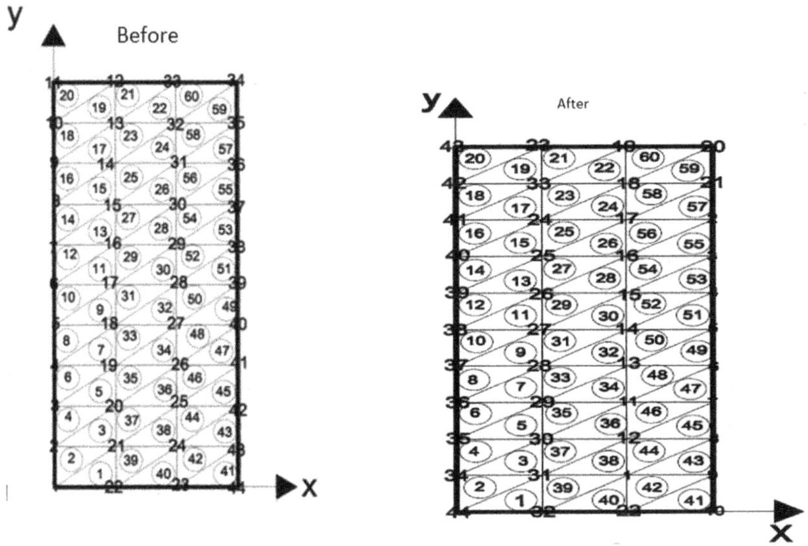

Fig. 5.2 Example of renumbering

in an ascending or descending order. The process is applied separately on free dof and dof related to boundary conditions. Recall that all the dof of a node may not be related to boundary conditions (Fig. 5.2).

In order for the numerical solution to converge, as the characteristic dimension of the mesh $h_e \to 0$, to the continuous solution, the mesh should not contain flat elements, i.e. elements in which the ratio h_e/r tends to infinity, r being the radius of the largest circle inscribed in the element. The numerical solution is also realistic if the mesh is refined in areas of concentrated loads or heat, etc. where the solution may vary very rapidly.

5.2 C^0 Finite Elements

5.2.1 Finite Element Spaces

Recall that, in static analysis, the kinematics of the models N-T and N are respectively

$$U_\alpha(x, z) = u_\alpha(x) - z(u_{3,\alpha}(x) + 2B^\rho_\alpha u_\rho(x)) + z^2\left(B^\tau_\alpha B^\nu_\tau u_\nu(x) + B^\tau_\alpha u_{3,\tau}(x)\right); \quad U_3(x, z) = u_3(x) \tag{5.16}$$

and

$$
\begin{aligned}
U_\alpha(x,z) &= u_\alpha(x) - z(u_{3,\alpha}(x) + 2B_\alpha^\rho u_\rho(x)) + z^2\left(B_\alpha^\tau B_\tau^\nu u_\nu(x) + B_\alpha^\tau u_{3,\tau}(x)\right) \\
U_3(x,z) &= u_3(x) + w(z)\bar{q}(x)
\end{aligned}
\tag{5.17}
$$

In the Reissner-Mindlin form, they are rewritten respectively as

$$U_\alpha(x,z) = u_\alpha(x) - z\vartheta_\alpha(x) + z^2\varphi(x);\; U_3(x,z) = u_3(x); \tag{5.18}$$

$$U_\alpha(x,z) = u_\alpha(x) - z\vartheta_\alpha(x) + z^2\varphi(x);\; U_3(x,z) = u_3(x) + w(z)\bar{q}(x) \tag{5.19}$$

These kinematics suggest to implement finite elements of class C^0, i.e. the solution is globally continuous. Variational solutions are found in the space $U_{ad} \subset \left(H^1(S)\right)^2 \times H^2(S)$. Generally in the literature, the transverse component of the solution should be of class at least C^1 in order that summation over elements of local virtual works be exactly equal to the global virtual work defined in the space of admissible displacements. Very many different finite elements have been defined in order to satisfy this regularity requirement on the transverse displacement. Consider a triangulation mesh T_{h_e} of the shell's mid-surface S, and let N_{h_e} denote the number of elements. Consider U_{ad}, the space of admissible displacement and let $\mho_{h_e}$ denote the finite element space defined by

$$\mho_{h_e} = \left\{ v^{h_e} \in U_{ad},\; v^{h_e}_{\alpha|T} \in P_1(T),\; v^{h_e}_{3|T} \in P_2(T) \text{ for any triangle } T \in T_{h_e} \right\}$$

where $P_k(T)$ is the space of polynomials of degree k in $x,\ y$; in-plane displacements are $v^{h_e}_\alpha$ $\alpha = 1,\ 2$ while the transverse displacement is $v^{h_e}_3$. Recall that if a vector field v is such that v_α, v_3 are respectively C^0 and C^1, then $v \in \left(H^1(S)\right)^2 \times H^2(S)$.

Let $T1$ and $T2$ be two adjacent triangles of the mesh T_{h_e} with respective summit-nodes $I,\ J,\ K$ and $K,\ J,\ L$. Consider $v^{h_e} \in \mho_{h_e}$ such that ${}^tV1 = \left(v^{h_e}_\alpha(I),\ v^{h_e}_\alpha(J),\ v^{h_e}_\alpha(K)\right)$, ${}^tV2 = \left(v^{h_e}_\alpha(K),\ v^{h_e}_\alpha(J),\ v^{h_e}_\alpha(L)\right)$, N_1^1 and N_2^1 the base of shape functions such that

$$
\begin{gathered}
v^{h_e}_{\alpha|T1} = N_1^1 V1,\; v^{h_e}_{\alpha|T2} = N_2^1 V2,\; \partial_\beta v^{h_e}_{\alpha|T1} = N_1^{1\prime} V1,\; \partial_\beta v^{h_e}_{\alpha|T2} = N_2^{1\prime} V2 \\
N_\alpha^{1\prime} = \partial_\beta N_\alpha^1,\; \partial_\beta = \partial_x \text{ or } \partial_y
\end{gathered}
\tag{5.20}
$$

It can be remarked that the gradients $\nabla v^{h_e}_\alpha$ on common nodes are not necessary equal if v_α is globally of class C^0. Consequently, their values at common nodes depend on the element considered and the class of the displacement v_α. It is also the same on the transverse displacement $v^{h_e}_3$ which is defined using its value on summit-nodes and mid-edge nodes, i.e. six values. Even though v_3 is C^1, second derivatives at common nodes are not necessarily equal. We conclude that the finite element space

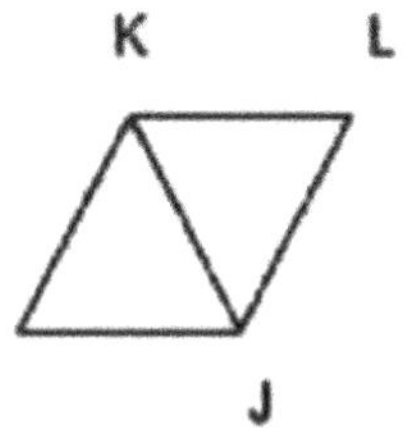

Fig. 5.3 Adjacent triangles

defined by $\mho_{h_e}$ does not necessarily ensure continuity of gradients of shape functions at common nodes or edges. Consequently strains and stresses may be discontinuous at common edges (Fig. 5.3).

In order to avoid this discontinuity at common edges, some sophisticated finite elements have been defined with higher order interpolation functions, thereby increasing the number of dof. However, calculating time, cost and even precision are not necessarily improved. A usual approach on the transverse displacement consists in defining at each node the value of the function and its derivatives up to the second order $\{v_3,\ v_{3,x},\ v_{3,y},\ v_{3,xx},\ v_{3,xy},\ v_{3,yy}\}$, i.e. 6 dof per node and 18 dof per element; or the value of the function and its first derivative per node $\{v_3,\ v_{3,x},\ v_{3,y}\}$. Shape functions in each element are therefore of the form

$$\begin{aligned} v_3\,(x,y) = \;& \alpha_1 + \alpha_2 x + \alpha_3 y + \alpha_4 x^2 + \alpha_5 xy + \alpha_6 y^2 + \alpha_7 x^3 \\ & + \alpha_8 x^2 y + \alpha_9 xy^2 + \alpha_{10} y^3 + \alpha_{12} x^4 + \alpha_{13} y^4 + \alpha_{14} x^5 \\ & + \alpha_{15}(x^3y^2 + x^2y^3) + \alpha_{16} x^4 y + \alpha_{17} xy^4 + \alpha_{18} y^5 \end{aligned} \tag{5.21}$$

for the first approach, or

$$v_3\,(x,y) = \alpha_1 + \alpha_2 x + \alpha_3 y + \alpha_4 x^2 + \alpha_5 xy + \alpha_6 y^2 + \alpha_7 x^3 + \alpha_8 (xy^2 + x^2 y) + \alpha_9 y^3 \tag{5.22}$$

for the second. The polynomials are of degree 5 and 3 respectively. But they can also be defined differently.

Let a function $v \in C^0(\bar{S})$ be such that $v_{|T} \in P_1$ in any triangle T. As shown above, $\partial_\beta v(J)_{|T1} = N_1^{1\prime} V1$ is different from $\partial_\beta v(J)_{|T2} = N_2^{1\prime} V2$. So ∇v is not defined at the node J. How can a better unique approximation of ∇v be defined at the node J, independently of the elements to which it belongs? The Gradient Polynomial Preserving Recovery method has provided a satisfactory solution.

5.2.2 Gradient Recovery Method (GR)

Several researchers (see references in Feumo et al. [1]) have proposed simple average, weighted average, L^2-projection or the discrete least square methods on elements having the node I as common summit-node. NAGA and ZHANG [7] proposed a polynomial recovery method which is super-convergent. The method consists in defining a local operator G_{h_e} such that $G_{h_e} v_{h_e}(I)$ *is unique independently of the path used for the calculation* and that the gap $\mid G_{h_e} v_{h_e} - \nabla v \mid$ is better than $\mid \nabla v_{h_e} - \nabla v \mid$.

Let Z_i denote the node where the gradient ∇v is to be determined. Consider all the nodes Z_i^j of all the elements T_j having Z_i common. Assume that $v_{h_e|T_j} \in P_{k+1}$, the space of polynomials of degree $k+1$, then the gradient recovery consists in finding the polynomial $p \in P_{k+1}$ such that

$$\sum_{j=1}^{n}(p - v_{h_e})^2(Z_i^j) = \min_{q \in P_{k+1}} \sum_{j=1}^{n} \left(q - v_{h_e}\right)^2 (Z_i^j), \tag{5.23}$$

where n is the total number of nodes including Z_i and define $G_{h_e} v_{h_e}(Z_i) = \nabla p(Z_i)$.

In a 2D space, NAGA and ZHANG [7] showed that if $n \geqq m = (k+2)(k+3)/2$ and the sum of two adjacent angles in the mesh is less or equal to π, then $G_{h_e} v_{h_e}(Z_i)$ is unique for any Z_i. These conditions on angles and the number of nodes to be considered will be satisfied henceforth in any mesh. We shall proceed as follows to select nodes.

Let (x_i, y_i) be the coordinates of a node Z_i of a triangle element and let h_i be the longest edge of the element. We denote $B(Z_i, h_i)$ as the disc centred in Z_i with radius h_i. Then we consider the number of nodes n contained in the successive circles $B(Z_i, h_i)$, $B(Z_i, 2h_i)$, $B(Z_i, 3h_i)$, $\ldots$, $B(Z_i, Nh_i)$ until the condition $n \geqq m$ is satisfied. Then the polynomial $p(x, y) = \left(1\, x\, y\, x^2 ... x^{k+1}\, x^k y ... y^{k+1}\right) a$ where ${}^t a = (a_1, a_2, a_3, .., a_m)$ satisfies the equation $Aa = b$ where $p(Z_i^j) = v_h(Z_i^j) = b_j \ for \ j = 1, \ldots, n$. A is a $n \times m$ matrix, ${}^t b = (b_1, b_2, .., b_n)$. This equation is also equivalent to

$${}^t A A a = {}^t A b \tag{5.24}$$

The matrix ${}^t AA$ is of order m and is also invertible. It is then deduced that

$$\begin{aligned} p(x, y) = & \left(1\, x\, y\, x^2 ... x^{k+1}\, x^k y ... y^{k+1}\right) a \\ = & \left(1\, x\, y\, x^2 ... x^{k+1}\, x^k y ... y^{k+1}\right) \left({}^t AA\right)^{-1} tAb \\ = & P_m X b \end{aligned} \tag{5.25}$$

The gradient is then

$$G_{h_e} u_{h_e}(Z_i) = \nabla p(x_i, y_i) \simeq \nabla v(x_i, y_i) \tag{5.26}$$

So by letting $B = {}^t AA$

$$\begin{aligned} \partial_x v(x_i, y_i) = \partial_x p(x_i, y_i) &= \left(0\,1\,0\,2x_i \ldots (k+1)x_i^k\, kx_i^{k-1} y_i \ldots 0\right) B^{-1t} Ab \\ &= P_m X^1 b \\ \partial_y v(x_i, y_i) = \partial_y p(x_i, y_i) &= \left(0\,0\,1\,0 \ldots 0\, x_i^k \ldots (k+1) y_i^k\right) B^{-1t} Ab \\ &= P_m X^2 b \end{aligned} \tag{5.27}$$

For degree 1 polynomials (k = 0) we have

$$\begin{aligned} \partial_x v(x_i, y_i) &= \partial_x p(x_i, y_i) = (0\,1\,0)\, B^{-1t} Ab = P_1 X^1 b \\ \partial_y v(x_i, y_i) &= \partial_y p(x_i, y_i) = (0\,0\,1)\, B^{-1t} Ab = P_1 X^2 b \end{aligned} \tag{5.28}$$

and degree 2 (k = 1)

$$\begin{aligned} \partial_x v(x_i, y_i) &= \partial_x p(x_i, y_i) = (0\,1\,0\,2x_i\; y_i\; 0)\, B^{-1t} Ab = P_2 X^1 b \\ \partial_y v(x_i, y_i) &= \partial_y p(x_i, y_i) = (0\,0\,0\,0\,x_i\; 2y_i)\, B^{-1t} Ab = P_2 X^2 b \end{aligned} \tag{5.29}$$

$$\begin{aligned} \partial_{xx} v(x_i, y_i) &= \partial_{xx} p(x_i, y_i) = (0\,0\,0\,2\,0\,0)\, B^{-1t} Ab = P_2 X^{11} b, \\ \partial_{yy} v(x_i, y_i) &= \partial_{yy} p(x_i, y_i) = (0\,0\,0\,0\,0\,2)\, B^{-1t} Ab = P_2 X^{22} b, \\ \partial_{xy} v(x_i, y_i) &= \partial_{xy} p(x_i, y_i) = (0\,0\,0\,0\,1\,0)\, B^{-1t} Ab = P_2 X^{12} b \end{aligned} \tag{5.30}$$

Membrane displacements v_α and transverse displacement v_3 restricted in each element are polynomials of order 1 and 2 respectively. Their derivatives at the node Z_i shall be denoted respectively by $P_1 X_\alpha^1 b$, $P_1 X_\alpha^2 b$, $P_2 X_3^1 b$, $P_2 X_3^2 b$, $P_2 X_3^{11} b$, $P_2 X_3^{22} b$ and $P_2 X_3^{12} b$. Henceforth, every mesh will consist of isosceles triangles with two perpendicular edges of the same length.

In order to determine the number of related nodes necessary to recover the gradient at a node Z_i, we distinguish eight different situations that we can describe by using three groups of notations as follows:

(1) The node Z_i is inside the domain: we use the following notations: $P_1 I X_\alpha^1 b$, $P_1 I X_\alpha^2 b$, $P_2 I X_3^1 b$, $P_2 I X_3^2 b$, $P_2 I X_3^{11} b$, $P_2 I X_3^{22} b$, $P_2 I X_3^{12} b$.
(2) The node Z_i is on the border, at the corner of the domain: $P_1 C X_\alpha^1 b$, $P_1 C X_\alpha^2 b$, $P_2 C X_3^1 b$, $P_2 C X_3^2 b$, $P_2 C X_3^{11} b$, $P_2 C X_3^{22} b$, $P_2 C X_3^{12} b$.
(3) The node Z_i is on the border, but not at the corner: $P_1 M X_\alpha^1 b$, $P_1 M X_\alpha^2 b$, $P_2 M X_3^1 b$, $P_2 M X_3^2 b$, $P_2 M X_3^{11} b$, $P_2 M X_3^{22} b$, $P_2 M X_3^{12} b$.

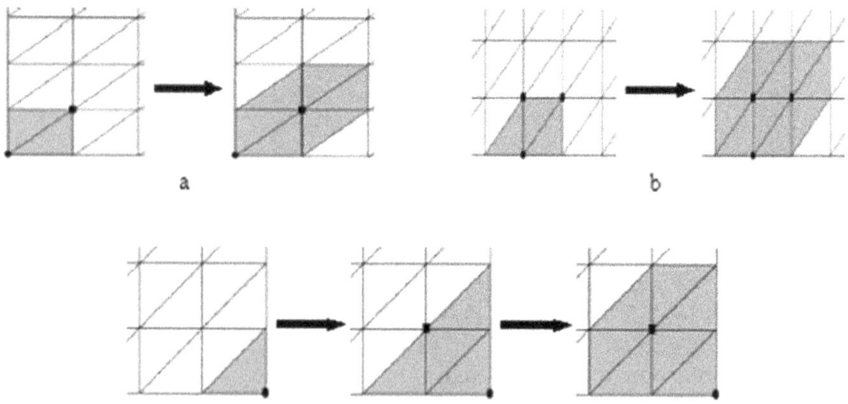

Fig. 5.4 Designation of nodes on a triangle

Let T_j be an element of the triangulation mesh. The summit-nodes will be denoted by a_1, a_2, a_3, starting from the node at the right angle, turning clockwise or anticlockwise. Nodes on mid-sides will be denoted by a_4, a_5, a_6 for the sides a_1a_2, a_2a_3 and a_3a_1 respectively. Related nodes necessary to recover the gradient at Z_i are numbered, turning in the same way as on the summit-nodes of the triangular element (Fig. 5.4).

Let T_j be a triangle whose summit-nodes have the coordinates $a_1 = (x_1, y_1), a_2 = (x_2, y_2)$ *and* $a_3 = (x_3, y_3)$. Let P_1 and P_2 denote spaces of polynomials of degree 1 and 2 respectively, generated by the bases $\{1, x, y\}$ and $\{1, x, y, x^2, y^2, xy\}$ then the barycentric functions $\{\lambda_1, \lambda_2, \lambda_3\}$ and $\{\lambda_1, \lambda_2, \lambda_3, 4\lambda_1\lambda_2, 4\lambda_1\lambda_3, 4\lambda_2\lambda_3\}$ also constitute bases of P_1 and P_2 respectively. They are defined as follows:

$$\begin{cases} \lambda_1(x, y) = \frac{1}{2\Delta}[(y_3 - y_2)(x_2 - x) - (x_3 - x_2)(y_2 - y)] \\ \lambda_2(x, y) = \frac{1}{2\Delta}[(y_1 - y_3)(x_3 - x) - (x_1 - x_3)(y_3 - y)] \\ \lambda_3(x, y) = \frac{1}{2\Delta}[(y_2 - y_1)(x_1 - x) - (x_2 - x_1)(y_1 - y)] \end{cases} \tag{5.31}$$

$$2\,\Delta = 2\int_{T_j} dxdy = 2det \begin{bmatrix} 1 & 1 & 1 \\ x_1 & x_2 & x_3 \\ y_1 & y_2 & y_3 \end{bmatrix} = (x_3 - x_2)(y_1 - y_2) - (x_1 - x_2)(y_3 - y_2) \tag{5.32}$$

where Δ is the area of the triangle T_j. Consider $v_\alpha,\ v_3 \in C^0(\bar{S})$ such that $v^{h_e}_{\alpha|T_j} \in P_1$ and $v^{h_e}_{3|T_j} \in P_2$ then

Fig. 5.5 Node positioning

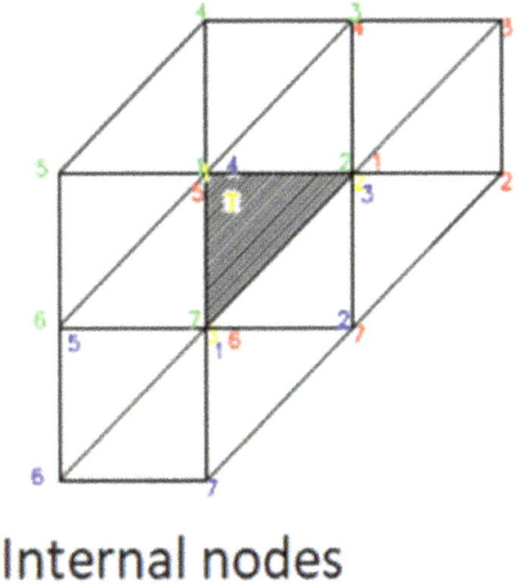

$$\begin{cases} v_\alpha^{h_e}(x,y)\mid_{T_j} = & \sum_{k=1}^{3} v_\alpha^{h_e}(a_k)\lambda_k(x,y) = \sum_{k=1}^{3} \bar{v}_\alpha^k \lambda_k(x,y) \\ v_3^{h_e}(x,y)\mid_{T_j} = & \sum_{k=1}^{3} v_3^{h_e}(a_k)\lambda_k(x,y) + \sum_{k=4}^{6} v_3^{h_e}(a_k)\lambda_{k-3}(x,y)\lambda_{k-2}(x,y) \\ = & \sum_{k=1}^{3} \bar{v}_3^k \lambda_k(x,y) + \sum_{k=4}^{6} \bar{v}_3^k \lambda_{k-3}(x,y)\lambda_{k-2}(x,y) \\ \partial_\alpha v_l^{h_e}(x,y)\mid_{T_j} = & \sum_{k=1}^{3} \partial_\alpha v_l^{h_e}(a_k)\lambda_k(x,y),\ \partial_\alpha v_l^{h_e}(a_k) = G_{h_e}(v_l^{h_e})(a_k) \\ \partial_{\alpha\beta}^2 v_3^{h_e}(x,y)\mid_{T_j} = & \sum_{k=1}^{3} \partial_{\alpha\beta}^2 v_3^{h_e}(a_k)\lambda_k(x,y),\ \partial_{\alpha\beta}^2 v_3^{h_e}(a_k) = G_{h_e}(\partial_\alpha v_3^{h_e})(a_k) \end{cases} \tag{5.33}$$

where gradients restricted to the element T_j are calculated as follows:

(i) the three nodes of the element are inside the domain.

The related nodes to recover the gradient are arranged in a vector form ${}^t(a_1, a_2, 0, 0, 0, 0, a_3)$, ${}^t(a_2, 0, 0, 0, a_1, a_3, 0)$, ${}^t(a_3, 0, a_2, a_1, 0, 0, 0)$, for the nodes a_1, a_2, a_3 respectively. We deduce the form of the vector b and the matrix $[A_i]$ which are position matrices (Fig. 5.5):

$$\begin{cases} \partial_\alpha v_i^{h_e}(a_1) & = [P_1 I X_i^\alpha]^t [\overline{v}_i^1, \overline{v}_i^2, 0, 0, 0, 0, \overline{v}_i^3] \\ & = [P_1 I X_i^\alpha]^t [A_1]^t [\overline{v}_i^1, \overline{v}_i^2, \overline{v}_i^3] \quad i = 1, 2 \\ \partial_\alpha v_3^{h_e}(a_1) & = [P_2 I X_3^\alpha]^t [\overline{v}_3^1, \overline{v}_3^2, 0, 0, 0, 0, \overline{v}_3^3] \\ & = [P_2 I X_3^\alpha]^t [A_1]^t [\overline{v}_3^1, \overline{v}_3^2, \overline{v}_3^3] \\ \partial^2_{\alpha\beta} v_3^{h_e}(a_1) & = [P_2 I X_3^{\alpha\beta}]^t [\overline{v}_3^1, \overline{v}_3^2, 0, 0, 0, 0, \overline{v}_3^3] \\ & = [P_2 I X_3^{\alpha\beta}]^t [A_1]^t [\overline{v}_3^1, \overline{v}_3^2, \overline{v}_3^3] \end{cases} \tag{5.34}$$

$$\begin{cases} \partial_\alpha v_i^{h_e}(a_2) & = [P_1 I X_i^\alpha]^t [\overline{v}_i^2, 0, 0, 0, \overline{v}_i^1, \overline{v}_i^3, 0] \\ & = [P_1 I X_i^\alpha]^t [A_2]^t [\overline{v}_i^1, \overline{v}_i^2, \overline{v}_i^3] \quad i = 1, 2 \\ \partial_\alpha v_3^{h_e}(a_2) & = [P_2 I X_3^\alpha]^t [\overline{v}_3^2, 0, 0, 0, \overline{v}_3^1, \overline{v}_3^3, 0] \\ & = [P_2 I X_3^\alpha]^t [A_2]^t [\overline{v}_3^1, \overline{v}_3^2, \overline{v}_3^3] \\ \partial^2_{\alpha\beta} v_3^{h_e}(a_2) & = [P_2 I X_3^{\alpha\beta}]^t [\overline{v}_3^2, 0, 0, 0, \overline{v}_3^1, \overline{v}_3^3, 0] \\ & = [P_2 I X_3^{\alpha\beta}]^t [A_2]^t [\overline{v}_3^1, \overline{v}_3^2, \overline{v}_3^3] \end{cases} \tag{5.35}$$

$$\begin{cases} \partial_\alpha v_i^{h_e}(a_3) & = [P_1 I X_i^\alpha]^t [\overline{v}_i^3, 0, \overline{v}_i^2, \overline{v}_i^1, 0, 0, 0] \\ & = [P_1 I X_i^\alpha][A_3]^t [\overline{v}_i^1, \overline{v}_i^2, \overline{v}_i^3] \quad i = 1, 2 \\ \partial_\alpha v_3^{h_e}(a_3) & = [P_2 I X_3^\alpha]^t [\overline{v}_3^3, 0, \overline{v}_3^2, \overline{v}_3^1, 0, 0, 0] \\ & = [[P_2 I X_3^\alpha][A_3]^t [\overline{v}_3^1, \overline{v}_3^2, \overline{v}_3^3] \\ \partial^2_{\alpha\beta} v_3^{h_e}(a_3) & = [P_2 I X_3^{\alpha\beta}]^t [\overline{v}_3^3, 0, \overline{v}_3^2, \overline{v}_3^1, 0, 0, 0] \\ & = [[P_2 I X_3^{\alpha\beta}]^t [A_3]^t [\overline{v}_3^1, \overline{v}_3^2, \overline{v}_3^3] \end{cases} \tag{5.36}$$

Fig. 5.6 Position of node, Case 2

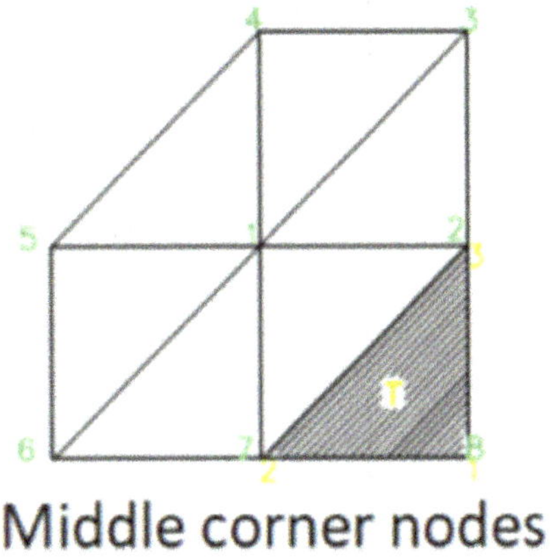

$$\begin{cases} [A_1] = \begin{bmatrix} 1\,0\,0\,0\,0\,0\,0 \\ 0\,1\,0\,0\,0\,0\,0 \\ 0\,0\,0\,0\,0\,0\,1 \end{bmatrix} \\ [A_2] = \begin{bmatrix} 0\,0\,0\,0\,1\,0\,0 \\ 1\,0\,0\,0\,0\,0\,0 \\ 0\,0\,0\,0\,0\,1\,0 \end{bmatrix} \\ [A_3] = \begin{bmatrix} 0\,0\,0\,1\,0\,0\,0 \\ 0\,0\,1\,0\,0\,0\,0 \\ 1\,0\,0\,0\,0\,0\,0 \end{bmatrix} \end{cases}$$

(ii) Among the three nodes of the element, the third is inside, and the two others on the border, the second being at the corner (Fig. 5.6).

Proceeding in the same way, we have

$$\begin{cases} \partial_\alpha v_i^{h_e}(a_1) & = [P_1 M X_i^\alpha]^t[\overline{v}_i^3, 0, 0, 0, 0, \overline{v}_i^2, \overline{v}_i^1, 0, 0, 0] \\ & = [P_1 M X_1^\alpha]^t[C_1]^t[\overline{v}_i^1, \overline{v}_i^2, \overline{v}_i^3] \quad i = 1, 2 \\ \partial_\alpha v_3^{h_e}(a_1) & = [P_2 M X_3^\alpha]^t[\overline{v}_3^3, 0, 0, 0, 0, \overline{v}_3^2, \overline{v}_3^1, 0, 0, 0] \\ & = [P_2 M X_3^\alpha]^t[C_1]^t[\overline{v}_3^1, \overline{v}_3^2, \overline{v}_3^3] \\ \partial^2_{\alpha\beta} v_3^{h_e}(a_1) & = [P_2 M X_3^{\alpha\beta}]^t[\overline{v}_3^3, 0, 0, 0, 0, \overline{v}_3^2, \overline{v}_3^1, 0, 0, 0] \\ & = [P_2 M X_3^{\alpha\beta}]^t[C_1]^t[\overline{v}_3^{j1}, \overline{v}_3^{j2}, \overline{v}_3^{j3}] \end{cases}$$

$$\begin{cases} \partial_\alpha v_3^{h_e}(a_2) & = [P_1 C X_i^\alpha]^t[\overline{v}_3^3, 0, 0, 0, 0, \overline{v}_3^2, \overline{v}_3^1] \\ & = [P_2 C X_3^\alpha]^t[C_2]^t[\overline{v}_3^1, \overline{v}_3^2, \overline{v}_3^3] \\ \partial_\alpha v_3^{h_e}(a_2) & = [P_2 C X_3^\alpha]^t[\overline{v}_3^3, 0, 0, 0, 0, \overline{v}_3^2, \overline{v}_3^1] \\ & = [P_2 C X_3^\alpha]^t[C_2]^t[\overline{v}_3^1, \overline{v}_3^2, \overline{v}_3^3] \\ \partial^2_{\alpha\beta} v_3^{h_e}(a_2) & = [P_2 C X_3^{\alpha\beta}]^t[\overline{v}_3^3, 0, 0, 0, 0, \overline{v}_3^2, \overline{v}_3^1] \\ & = [P_2 C X_3^{\alpha\beta}]^t[C_2]^t[\overline{v}_3^1, \overline{v}_3^2, \overline{v}_3^3] \end{cases}$$

$$\begin{cases} \partial_\alpha v_i^{h_e}(a_3) & = [P_1 I X_i^\alpha]^t[\overline{v}_i^3, 0, 0, 0, 0, \overline{v}_i^2, \overline{v}_i^1] \\ & = [P_1 I X_i^\alpha][C_3]^t[\overline{v}_i^1, \overline{v}_i^2, \overline{v}_i^3] \quad i = 1, 2 \\ \partial_\alpha v_3^{h_e}(a_3) & = [P_2 I X_3^\alpha]^t[\overline{v}_3^3, 0, 0, 0, 0, \overline{v}_3^2, \overline{v}_3^1] \\ & = [[P_2 I_3^\alpha]^t[C_3]^t[\overline{v}_3^1, \overline{v}_3^2, \overline{v}_3^3] \\ \partial^2_{\alpha\beta} v_3^{h_e}(a_3) & = [P_2 I X_3^{\alpha\beta}]^t[\overline{v}_3^3, 0, 0, 0, 0, \overline{v}_3^2, \overline{v}_3^1] \\ & = [[P_2 I X_3^{\alpha\beta}]^t[C_3]^t[\overline{v}_3^1, \overline{v}_3^2, \overline{v}_3^3] \end{cases}$$

$$\begin{cases} [C_1] = \begin{bmatrix} 0\,0\,0\,0\,0\,0\,1\,0\,0\,0 \\ 0\,0\,0\,0\,0\,1\,0\,0\,0\,0 \\ 1\,0\,0\,0\,0\,0\,0\,0\,0\,0 \end{bmatrix} \\ [C_2] = \begin{bmatrix} 0\,0\,0\,0\,0\,0\,1 \\ 0\,0\,0\,0\,0\,1\,0 \\ 1\,0\,0\,0\,0\,0\,0 \end{bmatrix} \\ [C_3] = \begin{bmatrix} 0\,0\,0\,0\,0\,0\,1 \\ 0\,0\,0\,0\,0\,1\,0 \\ 1\,0\,0\,0\,0\,0\,0 \end{bmatrix} \end{cases} \tag{5.37}$$

We proceed as above for (see Fig. 5.7)

(iii) two nodes are inside and one is on the border

(iv) all three nodes are on the borders, one being at the corner

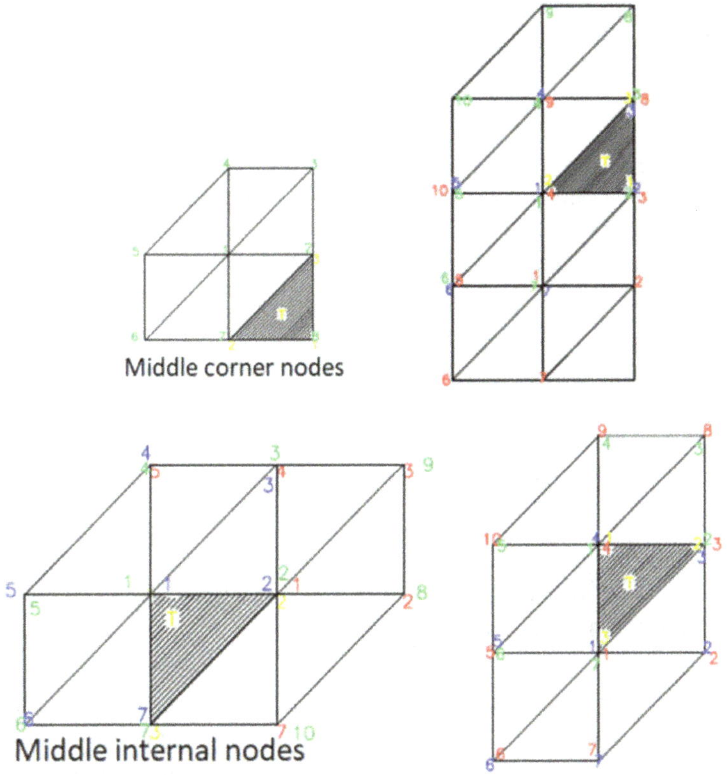

Fig. 5.7 Different positions of nodes

(v) one node is inside and the two others are on the border

(vi) one node is on the border while the two others are inside; etc.

This method enables the calculation of the different gradients at each node of the mesh, using only values of a function even if it is globally only of class C^0 (see Feumo et al. [1]). This suggests to define the following C^0 finite element spaces. Let $\mathcal{T}_{h_e}$ be a regular triangulation of the mid-surface S, characterized by h_e and comprising n_{h_e} elements. Let us denote

$$X^1_{h_e} = \{v_{h_e} \in C^0\left(\bar{S}\right), \quad v_{h_e|T_j} \in P_1\left(T_j\right), \quad \forall T_j,\ j = 1, ...n_{h_e}, \\ \partial_\alpha v_{h_e} = G_{h_e} v_{h_e},\ v_{h_e} = 0\ on\ \gamma_0\} \tag{5.38}$$

the subspace of dimension N_{h_e} (the number of nodes) and

$$X^2_{h_e} = \{v_{h_e} \in C^0\left(\bar{S}\right), v_{h_e|T_j} \in P_2\left(T_j\right), \forall T_j,\ j = 1, ..n_{h_e}\ and \\ \partial_\alpha v_{h_e} = G_{h_e} v_{h_e}, \partial_{\alpha\beta} v_{h_e} = G_{h_e} \partial_\rho v_{h_e}, \\ v_{h_e} = \partial_{\vec{\nu}} v_{h_e} = 0 \text{ on } \gamma_0\} \tag{5.39}$$

the subspace of dimension M_{h_e} (the number of summit-nodes and mid-segment nodes). The subspace of admissible displacements $U_{ad} = (H^1_{\gamma_0}(S))^2 \times H^2_{\gamma_0}(S)$ is approximated by the C^0 finite element space $\mho^0_{h_e} = X^1_{h_e} \times X^1_{h_e} \times X^2_{h_e}$ whose dimension is $M = 2N_{h_e} + M_{h_e}$.

The gradient recovery operator assures continuity (no jump) of strains and consequently stresses as a by-product, across common edges for globally continuous functions by using only their values at the required nodes. This means that the Lagrange finite element thus described has 3dof (values at each summit-node) per membrane displacement (or in-plane displacement) and 6 dof (values at each summit-node and each mid-point of an edge) for the transverse displacement, i.e. a total of 12 dof per element. The element stiffness matrix and nodal force vector are therefore of order 12×12 and 12×1 respectively:

$$v_1 = \alpha_1 + \alpha_2 x + \alpha_3 y, \\ v_2 = \alpha_4 + \alpha_5 x + \alpha_6 y, \\ v_3 = \alpha_7 + \alpha_8 x + \alpha_9 y + \alpha_{10} x^2 + \alpha_{11} xy + \alpha_{12} y^2 \tag{5.40}$$

5.3 Curved Triangular Elements and Assumed Strain Approach for Shells

This method is based on the determination of a shape function as a sum of the homogeneous and particular solutions of the rigid motion equation of shells. It is a question of solving the equation $\epsilon_{ij}(U) = 0$ and deducing the form of the solution in each element. Recall that the displacements read

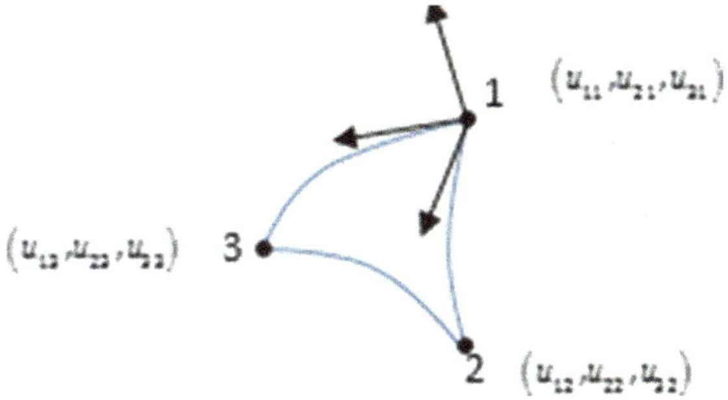

Fig. 5.8 Triangles with 18 dof

$$
\begin{aligned}
U_\alpha(x, y, z) = u_\alpha(x, y) - z\left(\partial_\alpha u_3(x, y) + 2B_\alpha^\tau u_\tau(x, y)\right) \\
+z^2\left(B_\alpha^\tau B_\tau^\nu u_\nu(x, y) + B_\alpha^\tau u_{3,\tau}(x, y)\right) \qquad (5.41) \\
U_3(x, y) \quad = \quad u_3(x, y) + w(z)\bar{q}(x, y)
\end{aligned}
$$

The rigid motion equations are equivalent to $\epsilon_{\alpha\beta}(U) = 0$ and $\epsilon_{ij}(U) = 0$ for the N-T and N models respectively. We must solve for the N-T model ($\bar{q} \equiv 0$) the equations

$$e_{\alpha\beta}(u_1, u_2, u_3) = 0, \;\; K_{\alpha\beta}(u_1, u_2, u_3) = 0, \;\; Q_{\alpha\beta}(u_1, u_2, u_3) = 0; \qquad (5.42)$$

and for the N model

$$e_{\alpha\beta}(u_1, u_2, u_3) = 0, \quad K_{\alpha\beta}(u_1, u_2, u_3) = 0, \quad Q_{\alpha\beta}(u_1, u_2, u_3) = 0, \; \bar{q} = 0 \qquad (5.43)$$

These equations are linear and solvable for cylindrical and spherical shells as follows (Fig. 5.8).

5.3.1 *Curved Triangle Element for Cylindrical Shells*

We consider the coordinates $(x, \; y) = (x, \; \theta)$ where x is the coordinate on the straight generic line while θ is the angle coordinate on the circle of radius R. A point on the mid-surface is defined by the position vector $\overrightarrow{OM} = (x, \;\; R\cos\theta, \;\; R\sin\theta)$. We denote $u_1(x, \theta) = u_1$, $u_2(x, \theta) = u_2$, $u_3(x, \theta) = u_3$ and $U = U(u)$. Then

$$e_{\alpha\beta}(u) = \frac{1}{2}(\nabla_\alpha u_\beta + \nabla_\beta u_\alpha - 2u_3 B_{\alpha\beta}) = 0 \qquad (5.44)$$

is equivalent to

$$e_{xx} = u_{1,x} = 0;\ \ 2e_{x\theta} = \frac{1}{R}u_{1,\theta} + u_{2,x} = 0,\ \ e_{\theta\theta} = \frac{1}{R}(u_{2,\theta} + u_3) = 0 \tag{5.45}$$

$$K_{\alpha\beta}(u) = \nabla_\alpha(\nabla_\beta u_3 + B^\rho_\beta u_\rho) + B^\rho_\alpha(\nabla_\beta u_\rho - u_3 B_{\beta\rho}) = 0 \tag{5.46}$$

is equivalent to

$$K_{xx} = u_{3,xx} = 0;\ \ K_{x\theta} = \frac{1}{R}(u_{3,x\theta} - u_{2,x}) = 0,\ \ K_{\theta\theta} = -\frac{1}{R^2}(2u_{2,\theta} + u_3 - u_{3,\theta\theta}) = 0 \tag{5.47}$$

$$Q_{\alpha\beta}(u) = \frac{1}{2}(B^\nu_\alpha \nabla_\beta(\nabla_\nu u_3 + B^\rho_\nu u_\rho) + B^\nu_\beta \nabla_\alpha(\nabla u_3 + B^\rho_\nu u_\rho)) = 0 \tag{5.48}$$

is equivalent to

$$Q_{xx} = 0;\ \ Q_{x\theta} = \frac{1}{R^2}(u_{2,x} - u_{3,x\theta}) = 0,\ \ Q_{\theta\theta} = -\frac{2}{R^3}(u_{2,\theta} - u_{3,\theta\theta}) = 0 \tag{5.49}$$

The solution of these equations on a cylindrical shell with an open angle $2\theta_0$ is $u_0 = (u_{10},\ u_{20},\ u_{30})$ for the rigid motion, to be completed by a particular solution $u_p = (u_{1p},\ u_{2p},\ u_{3p})$ as follows:

$$\begin{pmatrix} u_{10} \\ u_{20} \\ u_{30} \end{pmatrix} = \begin{bmatrix} 1 & R(cos\theta - cos\theta_0) & -Rsin\theta & 0 & 0 & 0 \\ 0 & xsin\theta & xcos\theta & -Rsin^2\theta & -sin\theta & cos\theta \\ 0 & -xcos\theta & xsin\theta & Rsin\theta cos\theta & cos\theta & sin\theta \end{bmatrix} \begin{pmatrix} a_1 \\ a_2 \\ a_3 \\ a_4 \\ a_5 \\ a_6 \end{pmatrix} \tag{5.50}$$

$$\begin{pmatrix} u_{1p} \\ u_{2p} \\ u_{3p} \end{pmatrix} = \begin{bmatrix} Rx & R\theta & Rx\theta & 0 & 0 & 0 & 0 & 0 & 0 & 0 & 0 & 0 \\ 0 & 0 & 0 & \theta & x\theta & 0 & 0 & 0 & 0 & 0 & 0 & 0 \\ 0 & 0 & 0 & 0 & 0 & x^2 & x\theta & \theta^2 & x^3 & x^2\theta & x\theta^2 & \theta^3 \end{bmatrix} \begin{pmatrix} a_1 \\ a_2 \\ a_3 \\ a_4 \\ a_5 \\ a_6 \\ a_7 \\ a_8 \\ a_9 \\ a_{10} \\ a_{11} \\ a_{12} \end{pmatrix} \tag{5.51}$$

such that $u = u_0 + u_p$. So 18 unknowns or dof are needed to characterize the shape function in an element. Let $\varphi_\alpha = -(\nabla_\alpha u_3 + B_\alpha^\rho u_\rho), \varphi_3 = \frac{1}{2}(\nabla_1 u_2 - \nabla_2 u_1) - \nabla_{12} u_3$ which are two rotations around the axes A_α and an adjusted third component of the rotational around the normal axis A_3. On each node of the triangle, there are 3 displacements and 3 rotations, i.e. 6 dof $(u_i,\ \varphi_i)$ per node and 18 dof per element. This is also a Lagrange finite element which assures no jump on strains across common edges.

5.3.2 *Shifted Lagrange Curved Finite Element (sh-L)*

It is a three-node element with shifted Lagrange interpolation functions requiring only 3 dof per function and defined as follows: consider two non-zero real numbers α and β, x, y curved coordinates of the mid-surface (Fig. 5.9). We define another coordinate by $X(x) = \alpha x + \beta$, $Y(y) = \alpha y + \beta$ and consider the space of polynomials generated by monomials of the form $X^i Y^j$ i,= 1, 2, 3, ...p, j,= 1, 2, 3, ...m. It is the space of shifted Lagrange functions. These functions must reproduce rigid motion as required above. Let us denote $\epsilon(U) = \epsilon(u)$. Then we must have $u = u_0 + u_p$, $\epsilon_{\alpha\beta}(u_0) = 0$. We deduce from $\epsilon_{\alpha\beta}(U) = e_{\alpha\beta}(u) - zK_{\alpha\beta}(u) + z^2 Q_{\alpha\beta}(u)$ (polynomial of degree 2 in z) that $\epsilon_{\alpha\beta}(u_0) = e_{\alpha\beta}(u_0) - zK_{\alpha\beta}(u_0) + z^2 Q_{\alpha\beta}(u_0) = 0$ if and only if $e_{\alpha\beta}(u_0) = 0$, $K_{\alpha\beta}(u_0) = 0$, $Q_{\alpha\beta}(u_0) = 0$. A trivial solution for any shell is $u_0 = 0$. So a particular solution u_p must no longer reproduce a rigid motion. In other words the displacement calculated from the particular solution u_p must not be zero. Such a displacement constructed over sh-L functions must satisfy some additional conditions called compatibility conditions obtained as follows.

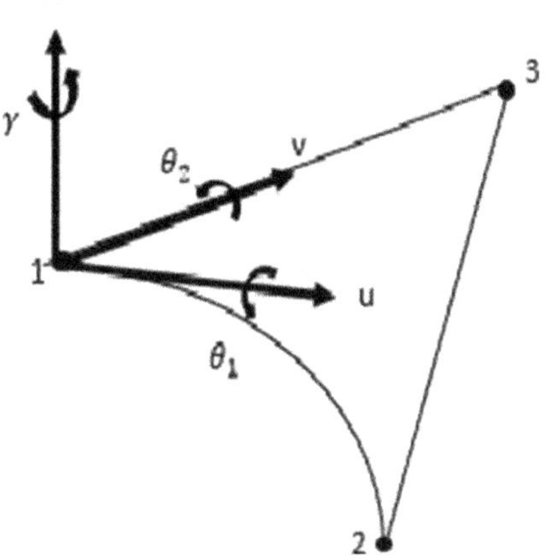

Fig. 5.9 Triangles with 9 degrees of freedom (dof)

Let us denote $u_p = u$ and consider the 9 equation terms $e_{\alpha\beta}$, $K_{\alpha\beta}$ and $Q_{\alpha\beta}$, assuming that they are independent

$$\begin{aligned} e_{\alpha\beta}(u) &= \tfrac{1}{2}\left(\nabla_\alpha u_\beta + \nabla_\beta u_\alpha - 2u_3 B_{\alpha\beta}\right) \\ K_{\alpha\beta}(u) &= \nabla_\alpha B^\rho_\beta u_\rho + B^\rho_\alpha \nabla_\beta u_\rho + B^\rho_\beta \nabla_\alpha u_\rho + \nabla_\alpha \nabla_\beta u_3 - B^\rho_\alpha B_{\rho\beta} u_3 \\ Q_{\alpha\beta}(u) &= \tfrac{1}{2}\left(B^\delta_\alpha \nabla_\beta(\nabla_\delta u_3 + B^\rho_\delta u_\rho) + B^\delta_\beta \nabla_\alpha(\nabla_\delta u_3 + B^\rho_\delta u_\rho)\right) \end{aligned} \tag{5.52}$$

In a diagonal base, in order to eliminate u in these equations we need 3 additional equations

$$\frac{1}{2}\left(\frac{1}{R_\alpha} + \frac{1}{R_\beta}\right)e_{\alpha\beta} - K_{\alpha\beta} + \frac{1}{2}(R_\alpha + R_\beta)Q_{\alpha\beta} = 0 \tag{5.53}$$

R_α and R_β are radii of principal curvature. These 3 equations constitute *the compatibility conditions*. So a displacement field is eligible as a shape function in the assumed strain approach for a shell with finite radii, if the tensors $e_{\alpha\beta}$, $K_{\alpha\beta}$ and $Q_{\alpha\beta}$ constructed satisfy the compatibility conditions. Now, concerning the stretch function $q = w(z)\bar{q}(x, y)$, the condition $Rot\nabla q = 0$ is always satisfied for any function. So no extra condition is required on the choice of $\bar{q}(x, y)$.

Henceforth we choose $p = 1, m = 3$ for u_α and $p = 2, m = 6$ for u_3. Shape functions are defined by $u_1 = a_1 + a_2 X + a_3 Y$, $u_2 = a_4 + a_5 X + a_6 Y$, $u_3 = a_7 X^2 + a_8 XY + a_9 Y^2$; i.e. 9dof for the N-T model or $\bar{q} = a_{10} + a_{11} X + a_{12} Y$ yielding 12dof for the model N. The shape functions thus defined satisfy the compatibility conditions and have proven to converge very rapidly to the theoretical solution with very few elements for both models with $\alpha = 1$.

5.3.3 *Stiffness Matrix and Nodal Force Vector per Element in the GR Method*

The variational formulation of the best first order approximation of the model N-T, in the finite element space, consists of

$$\begin{cases} \text{find } u_{h_e} \in \mho^0_{h_e} = X^1_{h_e} \times X^1_{h_e} \times X^2_{h_e} \text{ such that} \\ A_1(u_{h_e}, v_{h_e}) = L(v_{h_e}) \ \forall v_{h_e} \in \mho^0_{h_e} \end{cases} \tag{5.54}$$

$$
\begin{aligned}
A_1(u_{h_e}, v_{h_e}) = & \sum_{j=1}^{n_{he}}\Big(\tfrac{Eh}{1-\bar{\nu}^2}\int_{T_j} \left(\bar{\nu} e^{\rho}_{\rho}\left(u_{h_e}\right) A^{\alpha\beta} + (1-\bar{\nu})\, e^{\alpha\beta}\left(u_v\right)\right) e_{\alpha\beta}\left(v_{h_e}\right) dS \\
& + \tfrac{Eh^3}{12(1-\bar{\nu}^2)}\int_{T_j} \left(\bar{\nu} K^{\rho}_{\rho}\left(u_{h_e}\right) A^{\alpha\beta} + (1-\bar{\nu})\, K^{\alpha\beta}\left(u_{h_e}\right)\right) K_{\alpha\beta}\left(v_{h_e}\right) dS \\
& + \tfrac{Eh^3}{12(1-\bar{\nu}^2)}\int_{T_j} \left(\bar{\nu} e^{\rho}_{\rho}\left(u_{h_e}\right) A^{\alpha\beta} + (1-\bar{\nu})\, e^{\alpha\beta}\left(u_{h_e}\right)\right) Q_{\alpha\beta}\left(v_{h_e}\right) dS \\
& + \tfrac{Eh^3}{12(1-\bar{\nu}^2)}\int_{T_j} \left(\bar{\nu} Q^{\rho}_{\rho}\left(u_{h_e}\right) A^{\alpha\beta} + (1-\bar{\nu})\, Q^{\alpha\beta}\left(u_{h_e}\right)\right) e_{\alpha\beta}\left(v_{h_e}\right) dS \\
& + \tfrac{Eh^5}{80(1-\bar{\nu}^2)}\int_{T_j} \left(\bar{\nu} Q^{\rho}_{\rho}\left(u_{h_e}\right) A^{\alpha\beta} + (1-\bar{\nu})\, Q^{\alpha\beta}\left(u_{h_e}\right)\right) Q_{\alpha\beta}\left(v_{h_e}\right) dS\Big)
\end{aligned}
\tag{5.55}
$$

$$
L(v_{h_e}) = \sum_{j=1}^{n_{he}}\Big(\int_{T_j} p \cdot v_{h_e} dx dy + \int_{\partial T_j \cap \gamma_1} q \cdot v_{h_e} d\gamma + \int_{\partial T_j \cap \gamma_1} m \cdot \bar{\theta}(v_{h_e}) d\gamma\Big) \tag{5.56}
$$

$S \subseteq \cup T_j,\ j = 1, \ldots . n_{h_e}$. For any element T_j with summit-nodes a_1, a_2, a_3, let

$$
\begin{aligned}
{}^t U_j &= (u_1^1,\ u_1^2,\ u_1^3,\ u_2^1,\ u_2^2,\ u_2^3,\ u_3^1,\ u_3^2,\ u_3^3,\ u_3^4,\ u_3^5,\ u_3^6); \\
{}^t V_j &= (v_1^1,\ v_1^2,\ v_1^3,\ v_2^1,\ v_2^2,\ v_2^3,\ v_3^1,\ v_3^2,\ v_3^3,\ v_3^4,\ v_3^5,\ v_3^6)
\end{aligned}
\tag{5.57}
$$

be respectively the 12dof nodal displacement and virtual displacement vectors of the element; the 1×19 deformation source nodal vectors E_u^j, E_v^j are defined as follows:

$$
\begin{aligned}
{}^t E_u^j = \ & [u_1; \partial_1 u_1; \partial_2 u_1; u_2; \partial_1 u_2; \partial_2 u_2; u_3; -2B_1^1 u_1 - 2B_1^2 u_2 - \partial_1 u_3; \\
& -2B_1^1 \partial_1 u_1 - 2B_1^2 \partial_1 u_2 - 2B_1^1 \partial_2 u_1 - 2B_1^2 \partial_2 u_2; -2b_2^1 u_1 - 2B_2^2 u_2 \\
& -\partial_2 u_3; -2B_2^1 \partial_1 u_1 - 2B_2^2 \partial_1 u_2; -2B_2^1 \partial_2 u_1 - 2B_2^2 \partial_2 u_2; \overline{B}^1 u_1 + \\
& \overline{\overline{B}}^1 u_2 + B_1^1 \partial_1 u_3 + B_1^2 \partial_2 u_3; \overline{B}^1 \partial_1 u_1 + \overline{\overline{B}}^1 \partial_1 u_2; \overline{B}^1 \partial_2 u_1 + \overline{\overline{B}}^1 \partial_2 u_2; \\
& \overline{B}^2 u_1 + \overline{\overline{B}}^2 u_2 + B_2^1 \partial_1 u_3 + B_2^2 \partial_2 u_3; \overline{B}^2 u_1 + \overline{\overline{B}}^2 u_2 + B_2^1 \partial_1 u_3 \\
& + B_2^2 \partial_2 u_3; \overline{B}^2 \partial_1 u_1 + \overline{\overline{B}}^2 \partial_1 u_2; \overline{B}^2 \partial_2 u_1 + \overline{\overline{B}}^2 \partial_2 u_2]
\end{aligned}
\tag{5.58}
$$

$$
\overline{B}^1 = B_1^1 B_1^1 + B_2^1 B_1^2,\ \overline{\overline{B}}^1 = B_1^2 B_1^1 + B_2^2 B_1^2, \tag{5.59}
$$

$$\overline{B}^2 = B_1^1 B_2^1 + B_2^1 B_2^2,\ \overline{B}^2 = B_1^2 B_2^1 + B_2^2 B_2^2$$

and the same for E_v^j; let

$$\begin{aligned}
\alpha_i &= \lambda_i(x, y) \quad i = 1, 2, 3 \\
\rho_k &= 4\lambda_{k-3}(x, y)\lambda_{k-2}(x, y),\ k = 4, 5, 6 \\
\beta_i^j &= [P_1 I X_i^j][A_1](:, i) + [P_1 I X_i^j][A_2](:, i) + [P_1 I X_i^j][A_3](:, i) \\
\gamma_i^\alpha &= [P_2 I X_3^\alpha][A_1](:, i) + [P_2 I X_3^\alpha][A_2](:, i) + [P_2 I X_3^\alpha][A_3](:, i) \\
\tau_i^{\alpha\beta} &= [P_2 I X_3^{\alpha\beta}][A_1](:, i) + [P_2 I X_3^{\alpha\beta}][A_2](:, i) + [P_2 I X_3^{\alpha\beta}][A_3](:, i)
\end{aligned} \tag{5.60}$$

where I, $[A_n]$ are replaced by C or M according to the position of the node in the mesh as treated earlier; the 19×12 matrix B^j is defined as follows:

$$
\left[B^{T_j}\right]^t = \begin{bmatrix}
\alpha_1 & \alpha_2 & \alpha_3 & 0 & 0 & 0 & 0 & 0 & 0 & 0 & 0 & 0 \\
\beta_1^1 & \beta_2^1 & \beta_3^1 & 0 & 0 & 0 & 0 & 0 & 0 & 0 & 0 & 0 \\
\beta_1^2 & \beta_1^2 & \beta_1^2 & 0 & 0 & 0 & 0 & 0 & 0 & 0 & 0 & 0 \\
0 & 0 & 0 & \alpha_1 & \alpha_2 & \alpha_3 & 0 & 0 & 0 & 0 & 0 & 0 \\
0 & 0 & 0 & \beta_1^1 & \beta_2^1 & \beta_3^1 & 0 & 0 & 0 & 0 & 0 & 0 \\
0 & 0 & 0 & \beta_1^2 & \beta_2^2 & \beta_3^2 & 0 & 0 & 0 & 0 & 0 & 0 \\
0 & 0 & 0 & 0 & 0 & 0 & \alpha_1 & \alpha_2 & \alpha_3 & \rho_4 & \rho_5 & \rho_6 \\
-2b_1^1\alpha_1 & -2b_1^1\alpha_2 & -2b_1^1\alpha_3 & -2b_1^2\alpha_1 & -2b_1^2\alpha_2 & -2b_1^2\alpha_3 & -\gamma_1^1 & -\gamma_2^1 & -\gamma_3^1 & 0 & 0 & 0 \\
-2b_1^1\beta_1^1 & -2b_1^1\beta_2^1 & -2b_1^1\beta_3^1 & -2b_1^2\beta_1^1 & -2b_1^2\beta_2^1 & -2b_1^2\beta_3^1 & -\tau_1^{11} & -\tau_2^{11} & -\tau_3^{11} & 0 & 0 & 0 \\
-2b_1^1\beta_1^2 & -2b_1^1\beta_2^2 & -2b_1^1\beta_3^2 & -2b_1^2\beta_1^2 & -2b_1^2\beta_2^2 & -2b_1^2\beta_3^2 & -\tau_1^{12} & -\tau_2^{12} & -\tau_3^{12} & 0 & 0 & 0 \\
-2b_2^1\alpha_1 & -2b_2^1\alpha_2 & -2b_2^1\alpha_3 & -2b_2^2\alpha_1 & -2b_2^2\alpha_2 & -2b_2^2\alpha_3 & -\gamma_1^2 & -\gamma_2^2 & -\gamma_3^2 & 0 & 0 & 0 \\
-2b_2^1\beta_1^1 & -2b_2^1\beta_2^1 & -2b_2^1\beta_3^1 & -2b_2^2\beta_1^1 & -2b_2^2\beta_2^1 & -2b_2^2\beta_3^1 & -\tau_1^{12} & -\tau_2^{12} & -\tau_3^{12} & 0 & 0 & 0 \\
-2b_2^1\beta_1^2 & -2b_2^1\beta_2^2 & -2b_2^1\beta_3^2 & -2b_2^2\beta_1^2 & -2b_2^2\beta_2^2 & -2b_2^2\beta_3^2 & -\tau_1^{22} & -\tau_2^{22} & -\tau_3^{22} & 0 & 0 & 0 \\
\bar{b}^1\alpha_1 & \bar{b}^1\alpha_2 & \bar{b}^1\alpha_3 & \bar{\bar{b}}^1\alpha_1 & \bar{\bar{b}}^1\alpha_2 & \bar{\bar{b}}^1\alpha_3 & b_1^1\gamma_1^1 + b_1^2\gamma_1^2 & b_1^1\gamma_2^1 + b_1^2\gamma_2^2 & b_1^1\gamma_3^1 + b_1^2\gamma_3^2 & 0 & 0 & 0 \\
\bar{b}^1\beta_1^1 & \bar{b}^1\beta_2^1 & \bar{b}^1\beta_2^1 & \bar{\bar{b}}^1\beta_1^1 & \bar{\bar{b}}^1\beta_2^1 & \bar{\bar{b}}^1\beta_3^1 & b_1^1\tau_1^{11} + b_1^2\tau_1^{12} & b_1^1\tau_2^{11} + b_1^2\tau_2^{12} & b_1^1\tau_3^{11} + b_1^2\tau_3^{12} & 0 & 0 & 0 \\
\bar{b}^1\beta_1^2 & \bar{b}^1\beta_2^2 & \bar{b}^1\beta_2^2 & \bar{\bar{b}}^1\beta_1^2 & \bar{\bar{b}}^1\beta_2^2 & \bar{\bar{b}}^1\beta_3^2 & b_1^1\tau_1^{12} + b_1^2\tau_1^{22} & b_1^1\tau_2^{12} + b_1^2\tau_2^{22} & b_1^1\tau_3^{12} + b_1^2\tau_3^{22} & 0 & 0 & 0 \\
\bar{b}^2\alpha_1 & \bar{b}^2\alpha_2 & \bar{b}^2\alpha_3 & \bar{\bar{b}}^2\alpha_1 & \bar{\bar{b}}^2\alpha_2 & \bar{\bar{b}}^2\alpha_3 & b_2^1\gamma_1^1 + b_2^2\gamma_1^2 & b_2^1\gamma_2^1 + b_2^2\gamma_2^2 & b_2^1\gamma_3^1 + b_2^2\gamma_3^2 & 0 & 0 & 0 \\
\bar{b}^2\beta_1^1 & \bar{b}^2\beta_2^1 & \bar{b}^2\beta_3^1 & \bar{\bar{b}}^2\beta_1^1 & \bar{\bar{b}}^2\beta_2^1 & \bar{\bar{b}}^2\beta_3^1 & b_2^1\tau_1^{11} + b_2^2\tau_1^{12} & b_2^1\tau_2^{11} + b_2^2\tau_2^{12} & b_2^1\tau_3^{11} + b_2^2\tau_3^{12} & 0 & 0 & 0 \\
\bar{b}^2\beta_1^2 & \bar{b}^2\beta_2^2 & \bar{b}^2\beta_3^2 & \bar{\bar{b}}^2\beta_1^2 & \bar{\bar{b}}^2\beta_2^2 & \bar{\bar{b}}^2\beta_3^2 & b_2^1\tau_1^{12} + b_2^2\tau_1^{22} & b_2^1\tau_2^{12} + b_2^2\tau_2^{22} & b_2^1\tau_3^{12} + b_2^2\tau_3^{22} & 0 & 0 & 0
\end{bmatrix}
$$

we have (without summation on j)

$$E_u^j = B^j U_j, \ E_v^j = B^j V_j \tag{5.61}$$

Let $(x, y) = (x^1, x^2)$ be the coordinates on the mid-surface S and $(\lambda_1, \lambda_2, \lambda_3)$ the barycentric coordinates of a point in the triangle. Consider a function $f(x, y) = f(x(\lambda), y(\lambda))$ $\lambda = (\lambda_1, \lambda_2, \lambda_3)$, $1 - \lambda_2 - \lambda_3 = \lambda_1$, by denoting $\lambda_2 = \xi$, $\lambda_3 = \eta$; we write

$$\begin{aligned}\begin{pmatrix}\frac{\partial f}{\partial \xi}\\ \frac{\partial f}{\partial \eta}\end{pmatrix} = \begin{bmatrix} x_{,\xi} & y_{,\xi}\\ x_{,\eta} & y_{,\eta}\end{bmatrix}\begin{pmatrix}\frac{\partial f}{\partial x}\\ \frac{\partial f}{\partial y}\end{pmatrix} &= \begin{bmatrix} x_2 - x_1 & y_2 - y_1\\ x_3 - x_1 & y_3 - y_1\end{bmatrix}\begin{pmatrix}\frac{\partial f}{\partial x}\\ \frac{\partial f}{\partial y}\end{pmatrix}\\ &= [J]\begin{pmatrix}\frac{\partial f}{\partial x}\\ \frac{\partial f}{\partial y}\end{pmatrix}\end{aligned} \tag{5.62}$$

$$\begin{aligned}\int_T f(x, y)dxdy &= 2\triangle \int_{\bar{T}} f(\lambda_1, \lambda_2, \lambda_3)d\xi d\eta\\ &\approx \tfrac{2\triangle mes(\bar{T})}{3}(f(a_4) + f(a_5) + f(a_6))\\ &\approx \tfrac{\triangle}{3}(f(a_4) + f(a_5) + f(a_6))\\ &\approx \tfrac{\triangle}{3}(f(\tfrac{2}{3}, \tfrac{1}{6}, \tfrac{1}{6}) + f(\tfrac{1}{6}, \tfrac{2}{3}, \tfrac{1}{6}) + f(\tfrac{1}{6}, \tfrac{1}{6}, \tfrac{2}{3}))\end{aligned} \tag{5.63}$$

where $\bar{T}$ is the reference triangle defined by $\lambda_2 = \xi$, $\lambda_3 = \eta$ and whose $mes(\bar{T}) = 1/2$. This is the Gauss integration formula. Any other integration formula may be used. In any element T_j the membrane strain tensor can be expressed using its mixed component as follows:

$$\begin{aligned}e_\beta^\alpha(u) = A^{\alpha\lambda}e_{\beta\lambda}(u) &= \tfrac{1}{2}A^{\alpha\lambda}\left(u_{\beta,\lambda} + u_{\lambda,\beta}\right) - A^{\alpha\lambda}\bar{\Gamma}_{\lambda\beta}^{\rho}u_\rho - A^{\alpha\lambda}B_{\lambda\beta}u_3\\ &= \left[D_e^j\right]_\beta^\alpha E_u^j = \left[D_e^j\right]_\beta^\alpha B^j U_j\end{aligned} \tag{5.64}$$

$$\begin{aligned}(\left[D_e^j\right]_\beta^\alpha) = [&-A^{\alpha\nu}\bar{\Gamma}_{\beta\nu}^1;\ A^{\alpha 1}I_{\beta 1};\ \tfrac{1}{2}A^{\alpha\lambda}J_{\lambda\beta};\ -A^{\alpha\lambda}\bar{\Gamma}_{\beta\lambda}^2;\ \tfrac{1}{2}A^{\alpha\lambda}J_{\lambda\beta};\ A^{\alpha 2}J_{\beta 2};\\ &-B_\beta^\alpha;\ 0;\ 0;\ 0;\ 0;\ 0;\ 0;\ 0;\ 0;\ 0;\ 0;\ 0;\ 0]\end{aligned} \tag{5.65}$$

$$I = \begin{bmatrix}1 & 0\\ 0 & 1\end{bmatrix}; \quad J = \begin{bmatrix}0 & 1\\ 1 & 0\end{bmatrix}$$

Similarly

$$K^{\alpha}_{\beta}(u) = \left[D^{j}_{K}\right]^{\alpha}_{\beta} E^{j}_{u} = \left[D^{j}_{K}\right]^{\alpha}_{\beta} B^{j} U_{j} \tag{5.66}$$

$$\begin{aligned}
(\left[D^{j}_{K}\right]^{\alpha}_{\beta}) = \quad & [\frac{1}{2}A^{\alpha\mu}(B^{1}_{\mu,\beta} + \bar{\Gamma}^{1}_{\lambda\beta}B^{\lambda}_{\mu} - \bar{\Gamma}^{\lambda}_{\mu\beta}B^{1}_{\lambda}); 0; 0; \frac{1}{2}A^{\alpha\mu}(B^{2}_{\mu,\beta} \\
& +\bar{\Gamma}^{2}_{\lambda\beta}B^{\lambda}_{\mu} - \bar{\Gamma}^{\lambda}_{\mu\beta}B^{2}_{\lambda}); 0; 0; B^{\alpha\mu}B_{\mu\beta}; -A^{\alpha\mu}\bar{\Gamma}^{1}_{\mu\beta}; -I_{\alpha 1}A^{\alpha 1}; \\
& \frac{1}{2}J_{\alpha\beta}A^{\alpha\mu}; -A^{\alpha\mu}\bar{\Gamma}^{2}_{\mu\beta}; \frac{1}{2}J_{\mu\beta}A^{\alpha\mu}; I_{\alpha 2}A^{\alpha 2}; 0; 0; 0; 0; 0; 0]
\end{aligned} \tag{5.67}$$

$$Q^{\alpha}_{\beta}(u) = \left[D^{j}_{Q}\right]^{\alpha}_{\beta} E^{j}_{u} = \left[D^{j}_{Q}\right]^{\alpha}_{\beta} B^{j} U_{j} \tag{5.68}$$

and from

$$\begin{aligned}
\left[D^{j}_{Q}\right]^{\alpha}_{\beta} = & [A^{\alpha\mu}B^{1}_{\nu}(B^{\nu}_{\beta,\mu} + \bar{\Gamma}^{\nu}_{\gamma\beta}B^{\gamma}_{\mu} - \bar{\Gamma}^{\gamma}_{\mu\beta}B^{\nu}_{\gamma}); 0; 0; A^{\alpha\mu}B^{2}_{\nu}(B^{\nu}_{\beta,\mu} + \bar{\Gamma}^{\nu}_{\gamma\beta}B^{\gamma}_{\mu} - \bar{\Gamma}^{\gamma}_{\mu\beta}B^{\nu}_{\gamma}); \\
& 0; 0; 0; (B^{1}_{\beta,\mu} + \bar{\Gamma}^{1}_{\gamma\beta}B^{\gamma}_{\mu} - \bar{\Gamma}^{\gamma}_{\mu\beta}B^{1}_{\gamma}); 0; 0; (B^{2}_{\beta,\mu} + \bar{\Gamma}^{2}_{\gamma\beta}B^{\gamma}_{\mu} - \bar{\Gamma}^{\gamma}_{\mu\beta}B^{2}_{\gamma}); 0; 0; \\
& -A^{\alpha\mu}\bar{\Gamma}^{1}_{\mu\beta}; I_{\alpha 1}A^{\alpha 1}; \tfrac{1}{2}J_{\alpha\beta}A^{\alpha\mu}; -A^{\alpha\mu}\bar{\Gamma}^{2}_{\mu\beta}; \tfrac{1}{2}J_{\beta\mu}A^{\alpha\mu}; I_{\alpha 2}A^{\alpha 2}]
\end{aligned} \tag{5.69}$$

$$\begin{aligned}
A_1(u_{h_e}, v_{h_e}) = \quad & \sum_{j=1}^{n_{h_e}}(\frac{Eh}{1-\bar{\nu}^2}\int_{T_j}\left(\bar{\nu}e^{\rho}_{\rho}(u_{h_e})\,e^{\beta}_{\beta}(v_{h_e}) + (1-\bar{\nu})\,e^{\alpha}_{\beta}(u_{h_e})\,e^{\beta}_{\alpha}(v_{h_e})\right)dS \\
& +\frac{Eh^3}{12(1-\bar{\nu}^2)}\int_{T_j}\left(\bar{\nu}K^{\rho}_{\rho}(u_{h_e})\,K^{\beta}_{\beta}(v_{h_e}) + (1-\bar{\nu})\,K^{\alpha}_{\beta}(u_{h_e})\,K^{\beta}_{\alpha}(v_{h_e})\right)dS \\
& +\frac{Eh^3}{12(1-\bar{\nu}^2)}\int_{T_j}\left(\bar{\nu}e^{\rho}_{\rho}(u_{h_e})\,Q^{\beta}_{\beta}(v_{h_e}) + (1-\bar{\nu})\,e^{\alpha}_{\beta}(u_{h_e})\,Q^{\beta}_{\alpha}(v_{h_e})\right)dS \\
& +\frac{Eh^3}{12(1-\bar{\nu}^2)}\int_{T_j}\left(\bar{\nu}Q^{\rho}_{\rho}(u_{h_e})\,e^{\beta}_{\beta}(v_{h_e}) + (1-\bar{\nu})\,Q^{\alpha}_{\beta}(u_{h_e})\,e^{\beta}_{\alpha}(v_{h_e})\right)dS \\
& +\frac{Eh^5}{80(1-\bar{\nu}^2)}\int_{T_j}\left(\bar{\nu}Q^{\rho}_{\rho}(u_{h_e})\,Q^{\beta}_{\beta}(v_{h_e}) + (1-\bar{\nu})\,Q^{\alpha}_{\beta}(u_{h_e})\,Q^{\beta}_{\alpha}(v_{h_e})\right)dS) \\
= \quad & \sum_{j=1}^{n_{h_e}}({}^{t}V_j\int_{T_j}{}^{t}\left[B^{j}\right]C^{1}[B^{j}]dxdy)\,U_j) \\
= \quad & \sum_{j=1}^{n_{h_e}}({}^{t}V_j\int_{T_j}[F_{..}(x,y)]\,dxdy)\,U_j) \\
= \quad & \sum_{j=1}^{n_{h_e}}({}^{t}V_j.K^{j}_{e}U_j)
\end{aligned} \tag{5.70}$$

$$
\begin{aligned}
[C^1] = \; & \frac{Eh}{1-\bar{\nu}^2}(1-\bar{\nu})\left({}^t\,[D_e]^{\alpha}_{\beta}\right)\left([D_e]^{\beta}_{\alpha}\right) + \tfrac{Eh}{1-\bar{\nu}^2}\bar{\nu}\left({}^t\,[D_e]^{\alpha}_{\alpha}\right)\left([D_e]^{\lambda}_{\lambda}\right) \\
& + \frac{Eh^3}{12(1-\bar{\nu}^2)}(1-\bar{\nu})\left({}^t\,[D_K]^{\alpha}_{\beta}\right)\left([D_K]^{\beta}_{\alpha}\right) + \tfrac{Eh^3}{12(1-\bar{\nu}^2)}\bar{\nu}\left({}^t\,[D_K]^{\alpha}_{\alpha}\right)\left([D_K]^{\lambda}_{\lambda}\right) \\
& + \frac{Eh^3}{12(1-\bar{\nu}^2)}(1-\bar{\nu})\left({}^t\,[D_Q]^{\alpha}_{\beta}\right)\left([D_e]^{\beta}_{\alpha}\right) + \tfrac{Eh^3}{12(1-\bar{\nu}^2)}\bar{\nu}\left({}^t\,[D_Q]^{\alpha}_{\alpha}\right)\left([D_k]^{\lambda}_{\lambda}\right) \\
& + \frac{Eh^3}{12(1-\bar{\nu}^2)}(1-\bar{\nu})\left({}^t\,[D_e]^{\alpha}_{\beta}\right)\left([D_Q]^{\beta}_{\alpha}\right) + \tfrac{Eh^3}{12(1-\bar{\nu}^2)}\bar{\nu}\left({}^t\,[D_e]^{\alpha}_{\alpha}\right)\left([D_Q]^{\lambda}_{\lambda}\right) \\
& + \frac{Eh^5}{80(1-\bar{\nu}^2)}(1-\bar{\nu})\left({}^t\,[D_Q]^{\alpha}_{\beta}\right)\left([D_Q]^{\beta}_{\alpha}\right) + \tfrac{Eh^5}{80(1-\bar{\nu}^2)}\bar{\nu}\left({}^t\,[D_Q]^{\alpha}_{\alpha}\right)\left([D_Q]^{\lambda}_{\lambda}\right)
\end{aligned}
\tag{5.71}
$$

$$
\begin{aligned}
[K_e^j] = \; & \int_{T_j} F_{..}(x, y)dxdy \\
= \; & \frac{\triangle}{3}F_{..}\left(\tfrac{2}{3}x_1 + \tfrac{1}{6}x_2 + \tfrac{1}{6}x_3, \tfrac{2}{3}y_1 + \tfrac{1}{6}y_2 + \tfrac{1}{6}y_3\right) \\
& + \frac{\triangle}{3}F_{..}\left(\tfrac{1}{6}x_1 + \tfrac{2}{3}x_2 + \tfrac{1}{6}x_3, \tfrac{1}{6}y_1 + \tfrac{2}{3}y_2 + \tfrac{1}{6}y_3\right) \\
& + \frac{\triangle}{3}F_{..}\left(\tfrac{1}{6}x_1 + \tfrac{1}{6}x_2 + \tfrac{2}{3}x_3, \tfrac{1}{6}y_1 + \tfrac{1}{6}y_2 + \tfrac{2}{3}y_3\right)
\end{aligned}
\tag{5.72}
$$

is the 12×12 element stiffness matrix calculated by using Gauss integration formula. The nodal force vector is calculated as follows:

$$
\begin{aligned}
L(v_{h_e}) = \; & \sum_{j=1}^{n_{h_e}} \int_{T_j} {}^tV_j^t[B^j][P]dS + \oint_{\varpi_1}^{*\varpi_2*} {}^tV_j[B^j][G] \mid J_* \mid d\zeta \\
= \; & \sum_{j=1}^{n_{h_e}} {}^tV_j\left[f^j\right]
\end{aligned}
\tag{5.73}
$$

$$
\left[f^j\right] = {}^t\left[F_1^1, F_1^2, F_1^3, F_2^1, F_2^2, F_2^3, F_3^1, F_3^2, F_3^3, F_3^4, F_3^5, F_3^6\right]
$$

where force vectors are [P] , [G] while $\mid J_* \mid d\zeta$ is a line element on part of the border γ_1. Nodal concentrated loads are added on the global nodal force vector (Fig. 5.10).
Remarks: The first two terms in $A_1(u_{h_e}, v_{h_e})$ are found in the K-L thin shell theory and lead to the stiffness matrix of thin shells. The number of dof of the element displacement nodal vector U_j is 12. In the assumed strain approach, U_j has 18 dof, while the number is reduced to 9 when shifted Lagrange polynomials are used (Nkongho, Nzengwa et al. [56]). In this assumed strain approach the element stiffness matrices are respectively 18×18 and 9×9, while nodal force vectors have 18 and 9

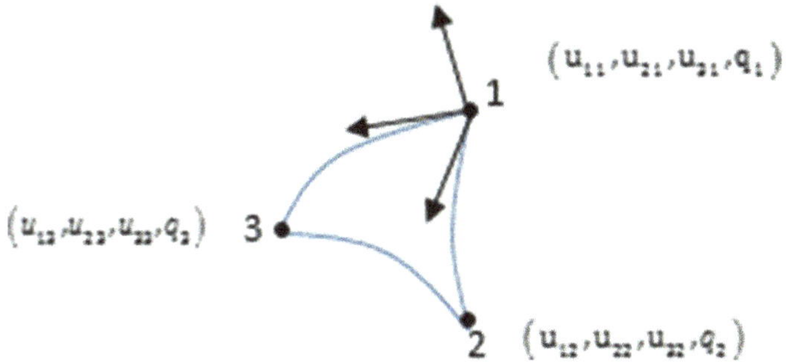

Fig. 5.10 Lagrange element

dof respectively. In order to calculate the stiffness matrix per element, it is necessary to define, using the shape functions, the matrix $[B^j]$ such that $E_u^j = [B^j]U_j$. The bilinear form $A_1(u_{h_e}, v_{h_e})$ and linear form $L(v_{h_e})$ are approximated by numerical integration, and the mesh does not recover the domain exactly. However, it is proved (see CIARLET [29]), at the cost of certain properties of the different forms, that the numerical solutions converge to the real solution as the characteristic parameter of the mesh tends to zero. Now, for the static analysis of the best first order of the model N, by letting $q(x, z) = w(z)\bar{q}(x)$, the variational equation of the continuous problem reads

$$\begin{cases} \text{find } (u, \bar{q}) \in U_{ad} = H^1_{\gamma_o}(S) \times H^1_{\gamma_o}(S) \times H^2_{\gamma_o}(S) \times H^1_{\gamma_o}(S) \text{ such that} \\ \\ A_1^w\left((u, \bar{q}); (v, \bar{y})\right) = L^w(v, \bar{y}); \end{cases} \tag{5.74}$$

$$\begin{aligned} A_1^w\left((u, \bar{q}); (v, \bar{y})\right) = & \int_S \int_{\frac{-h}{2}}^{\frac{h}{2}} \sigma^{ij}(u, q)\epsilon_{ij}(v, w\bar{y})\, dz dS \\ = & \int_S \int_{\frac{-h}{2}}^{\frac{h}{2}} [\bar{\lambda}\left(\epsilon_T^T(u) + \epsilon_l^l(q)\right) A^{\alpha\beta} + 2\bar{\mu}(\epsilon^{\alpha\beta}(u) \\ & +\epsilon^{\alpha\beta}(q))](e_{\alpha\beta}(v) - zK_{\alpha\beta}(v) + z^2 Q_{\alpha\beta}(v) + \\ & w\bar{y}\Upsilon_{\alpha\beta})dzdS + \int_S \int_{\frac{-h}{2}}^{\frac{h}{2}} [\bar{\mu}A^{\alpha\beta}w^2\partial_\beta\bar{q}\partial_\alpha\bar{y} \\ & +\bar{\lambda}(A^{\alpha\beta}\epsilon_{\alpha\beta}(u) + w\bar{q}A^{\alpha\beta}\Upsilon_{\alpha\beta})w'\bar{y} \\ & +\left(\bar{\lambda} + 2\bar{\mu}\right)\bar{q}\left(w'\right)^2 \bar{y}]dzdS \end{aligned} \tag{5.75}$$

$$L^w(v,\bar{y}) = \int\limits_S \left(\bar{p}^i v_i + \bar{p}^4 \bar{y}\right) dS + \int\limits_{\gamma_1} \left(q^i v_i + q^4 \bar{y}\right) d\gamma + \int\limits_{\gamma_1} m^\alpha \bar{\theta}_\alpha(v) d\gamma$$

$$\bar{p}^4 = \int_{\frac{-h}{2}}^{\frac{h}{2}} f^3 w dz + w(h/2) p_+^3 - w(-h/2) p_-^3, \; q^4 = \int_{\frac{-h}{2}}^{\frac{h}{2}} p^3 w dz; \tag{5.76}$$

$$\begin{cases} \epsilon_{\alpha\beta}(u) = e_{\alpha\beta}(u) - zK_{\alpha\beta}(u) + z^2 Q_{\alpha\beta}(u) \\ \Upsilon_{\alpha\beta} = -\left(\mu_\alpha^\rho B_{\rho\beta} + \mu_\beta^\rho B_{\rho\alpha}\right)/2 \\ \epsilon_{\alpha\beta}(w\bar{q}) = w\bar{q}\Upsilon_{\alpha\beta}, \; \epsilon_{\alpha 3}(w\bar{q}) = w\partial_\alpha \bar{q}/2, \; \epsilon_{33}(w\bar{q}) = w'\bar{q} \end{cases} \tag{5.77}$$

$\bar{p}^i$, q^i and m^α have been defined above. The different terms $\int_{\frac{-h}{2}}^{\frac{h}{2}}(..)dz$ are explicit when the function $w(z)$ is chosen. The discrete equations are obtained similarly.

5.4 Applications

5.4.1 *Cylindrical Shell with Gradient Recovery (GR) Method*

We construct a finite element called "Cylindrical Shell Finite Element" (CSFE), based on the GR method and the N-T model for linearly elastic thick shells. Results are compared with those of some well-known benchmarks such as the pinched cylinder. A cylinder subject to two diametrically opposite concentrated loads and based on two rigid plane diaphragms at the edges is shown in Fig. 5.11a. Through variation of the thickness, the characteristic parameter ranges from thin to thick shell values. Complex membrane deformations coupled with bending without surface expansion around load points are observed. Moreover, these tests have revealed the performance and the aptitude of the finite element CSFE to better simulate cylindrical structures. Geometrical, mechanical properties and results are presented below in Fig. 5.11b, c.

A reference solution is presented in Feumo et al. [1]: W_C is transverse displacement (*z-direction*) at the point C under the load; $\tilde{W}_C = \frac{-EhW_C}{P} = 164.24$ and $\bar{V}_D = \frac{-EhV_D}{P} = 4.11$ is the membrane displacement V_D in the *y-direction* at the point D; $\bar{V}_D = \frac{-EhV_D}{P} = 4.11$. These reference results on symmetric loaded shells are considered as 3D exact solutions. Only an eighth portion of the pinched cylinder with diaphragms was meshed. It turns out that, for a thin shell analysis with the finite element CSFE, whatever the type (A or B) of regular mesh used, K-L and N-T models converge very well. A comparison with other finite elements such as the DKT12, DKT18 and SFE3 confirms its convergence at C and D as shown in Fig. 5.11b, c.

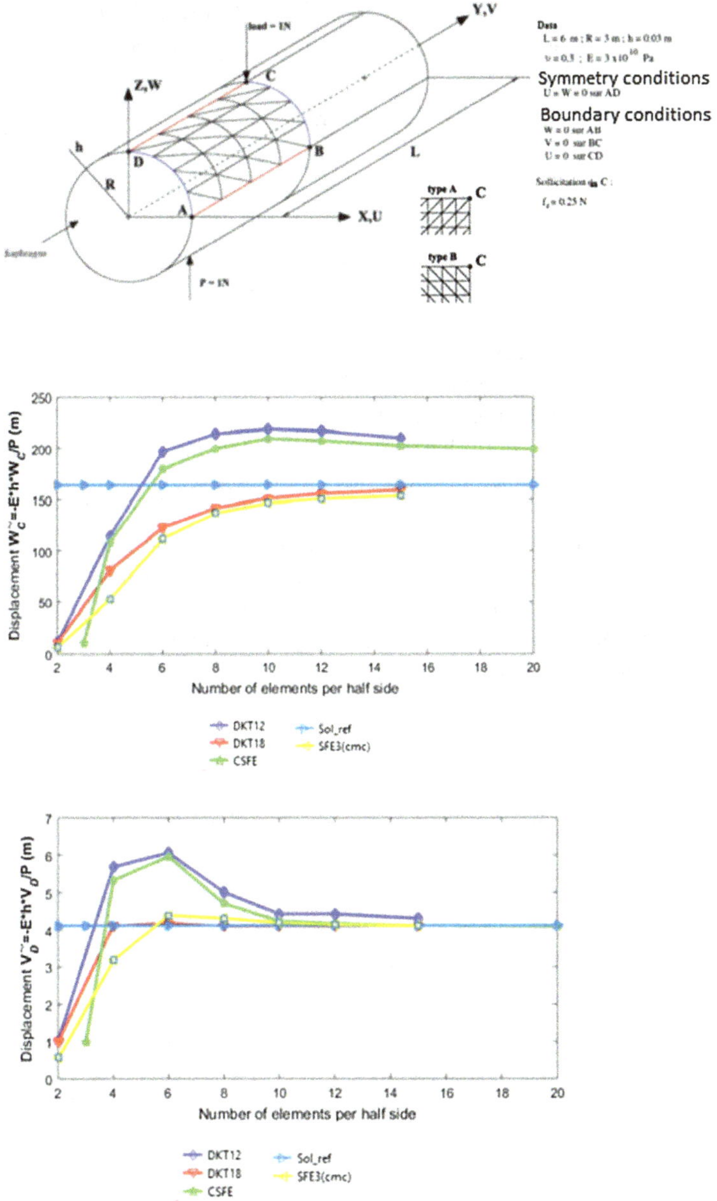

Fig. 5.11 Benchmark of a pinched cylindrical roof on rigid diaphragms. Convergence at load points C and D

5.4.2 Cylindrical N-T Shell Under the Assumed Strain Approach with Shifted Lagrange Polynomials

5.4.3 Spherical Shell with GR Method

The Spherical Shell Finite Element (SSFE) is used with the GR method to implement the K-L and N-T models. Results obtained are compared with those obtained with the finite elements DKT12, DKT18 and SFE3. Figure 5.14 below shows displacements at load point A, variations and convergence rate. Consistency and stability of numerical convergence are clearly established. Consequently, the memoryless greedy finite element SSFE, which needs only a globally continuous function, is as efficient as DKT and SFE for the displacement at the load point A (Figs. 5.12 and 5.13).

In order to evaluate the performance of the SSFE element and the behaviour of the model N-T, the following values 0.1, 0.3, 0.325, 0.4 and 0.5 of the characteristic parameter $2\chi = h/R$ of the spherical shell were considered to plot the different curves in Fig. 5.15. Here, the thickness h varies while the radius is fixed.

We conclude from the curves Fig. 5.16a–d that membrane displacements at the load point A are the same for the two models R-M and N-T when $0 < \chi < 0.099$, while they differ beyond this value; transverse displacements differ during inextensible membrane deformation. More precisely, accordingly, $2\chi \simeq \sqrt{1/10}$ roughly delimits thin and thick shells.

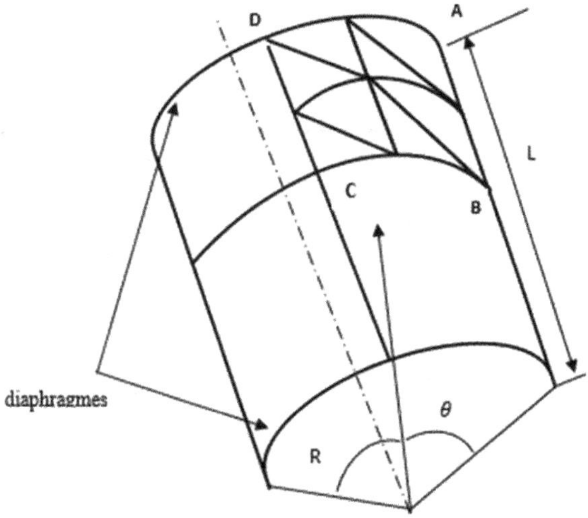

Fig. 5.12 Self-weighted cylindrical roof

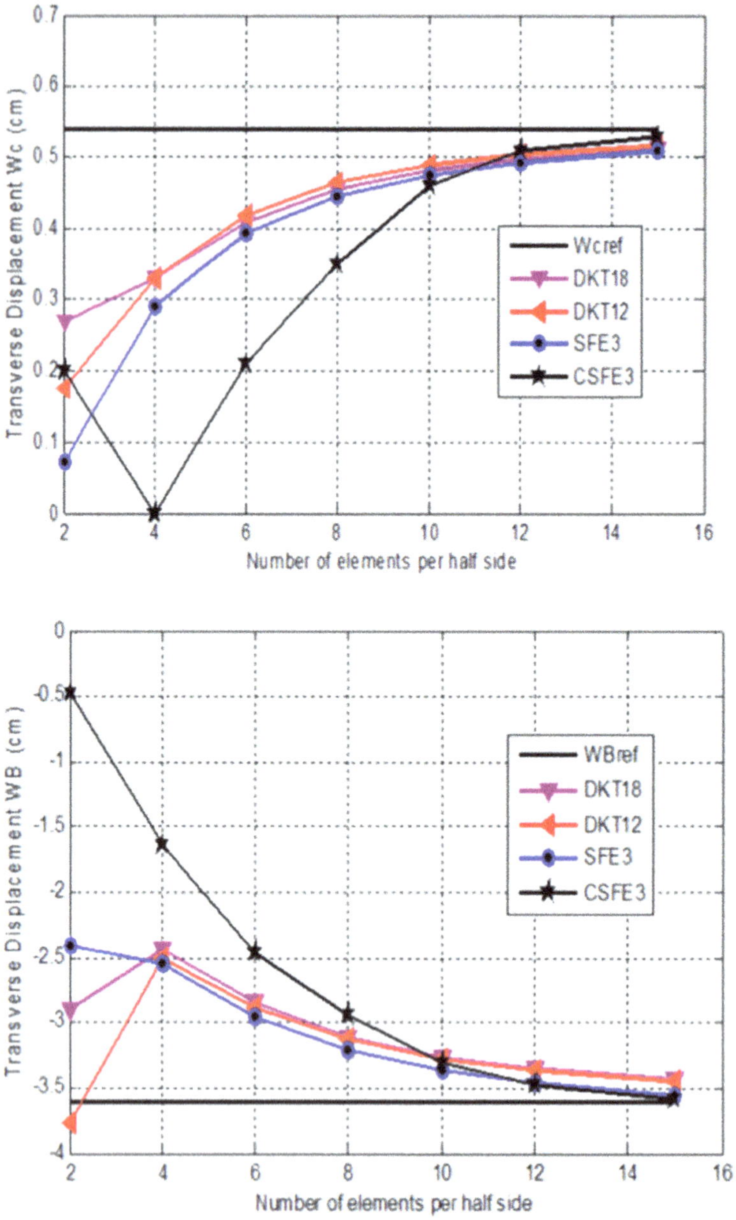

Fig. 5.13 Convergence curves at load points B and C on transverse displacement of the cylindrical roof

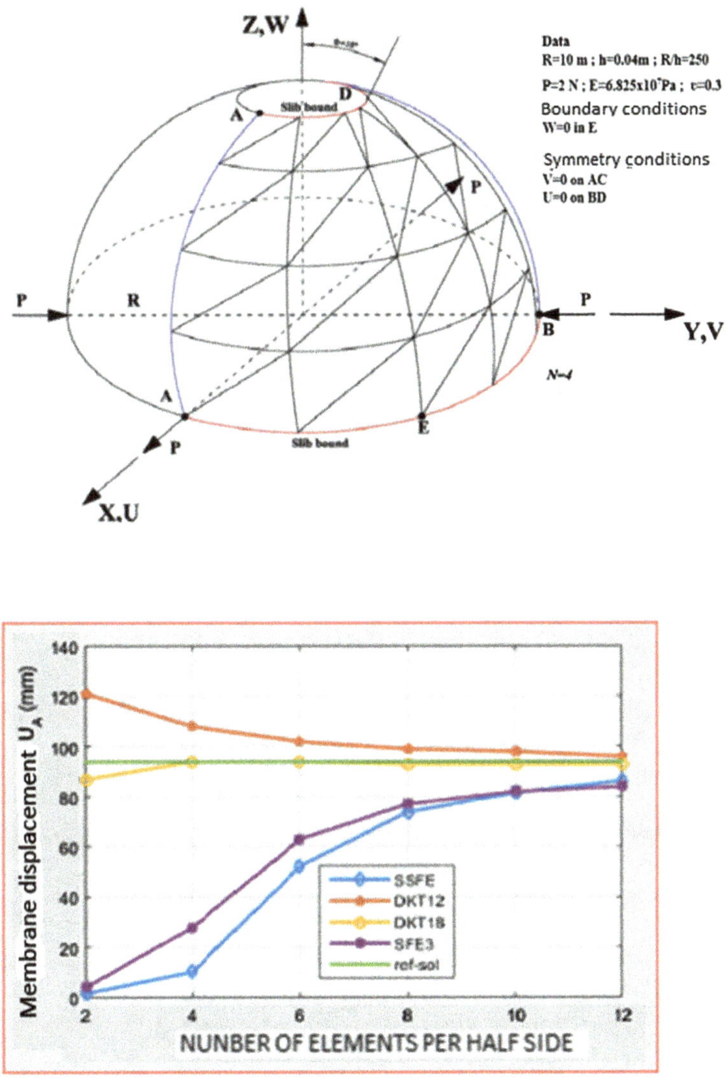

Fig. 5.14 Benchmark of a hemispherical shell; convergence at point A

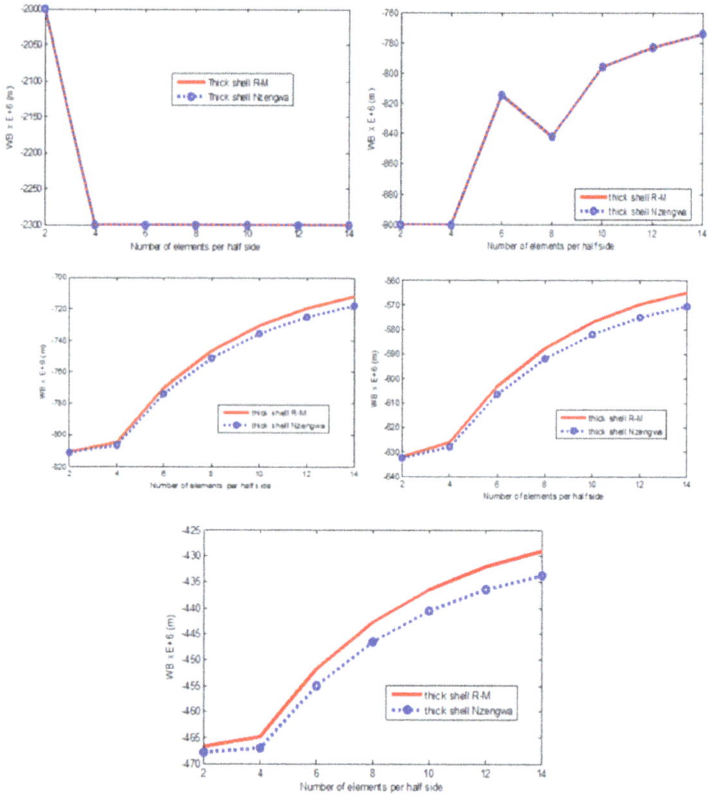

Fig. 5.15 Effect of the variation of h/R = 0.10; 0.30; 0.325; 0.40; 0.50

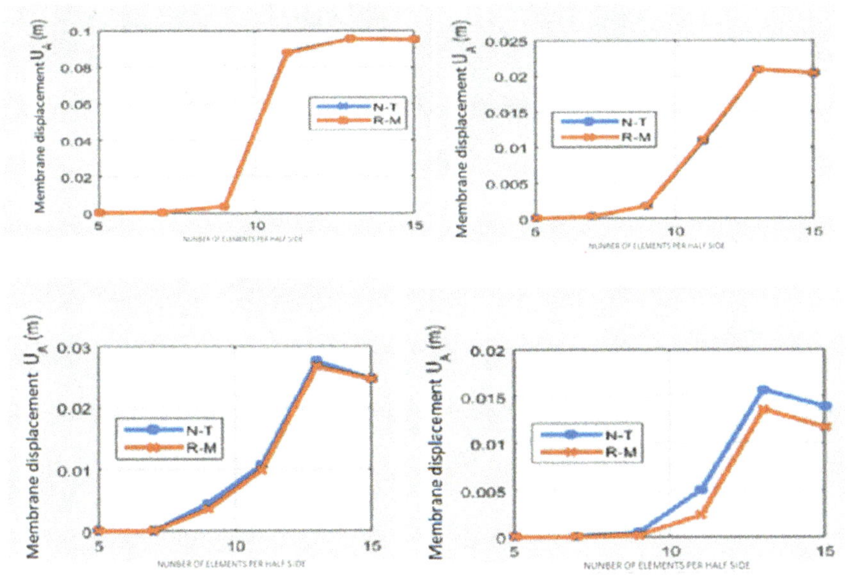

Fig. 5.16 Variation of membrane displacement U_A at A with regard to h/2R = 0.006; 0.099; 0.12; 0.15

Let us examine the gap between the two models R-M and N-T at the load point A. For the different values $\chi = 0.006,\ 0.099,\ 0.12,\ 0.15$, the curves plotted in Fig. 5.16a–d clearly show the gaps in displacement. The gaps become more significant with mesh refinement and increasing values of χ. This is due to the impact of the contribution of the Gauss deformation in the total strain energy in the model N-T.

Chapter 6
Other Models

6.1 Stiffened, Thermoelastic and Homogeneous Anisotropic Shells

6.1.1 Variational Equation of Stiffened Shells

In practice, very many shell structures are not constantly thick. They comprise on the inner surface ($z = -h/2$) called intrados or on the outer surface ($z = h/2$) called extrados, localized extra thickness (stiffeners), to the constant thickness h of the shell with mid-surface S. We can cite for example a cylindrical tank stiffened by hoops or stiffened longitudinally or with a grid network of stiffeners (Fig. 6.1). Stiffeners are constituted by transverse distributions of material which make the thickness to vary locally between $g(x, y)$ and $-h/2$ at the intrados or between $h/2$ and $f(x, y)$ at the extrados. In certain stiffened shell models, stiffeners are considered as beams (Combescure [36]). Let $e_m = max\{| f |, | g |\}$. Let $r = min \mid R \mid$, R being the radius of curvature. If $\chi = e_m/r < 1$, then the stiffened shell can be embedded in a shell with the same mid-surface and $g(x, y) \leqslant z \leqslant f(x, y)$ (Nzengwa et al. 1999). The variational formulations of the models N-T or N consist in using the corresponding kinematic and to integrate with respect to z in the intervals $[g(x, y), -h/2]$, $[-h/2, h/2]$ and $[h/2, f(x, y)]$, i.e.

$$\int_S\int_{g(x,y)}^{\frac{-h}{2}} \sigma^{ij}\epsilon_{ij}\psi dzdS + \int_S\int_{\frac{-h}{2}}^{\frac{h}{2}} \sigma^{ij}\epsilon_{ij}\psi dzdS + \int_S\int_{\frac{h}{2}}^{f(x,y)} \sigma^{ij}\epsilon_{ij}\psi dzdS = L(v) \tag{6.1}$$

R. Nzengwa, *Plate and Shell Models*,
https://doi.org/10.1007/978-981-97-2780-3_6

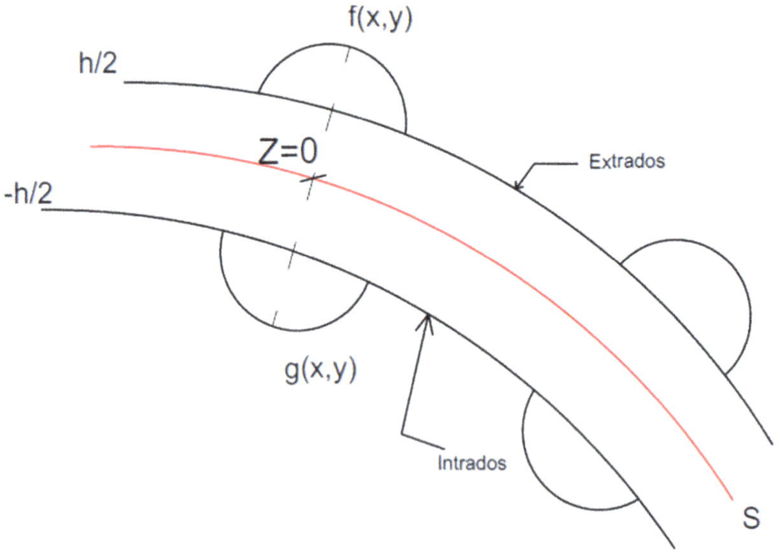

Fig. 6.1 Stiffened shells

6.1.2 First-Order Variational Equations of Stiffened Shells

We can also use the N-T and N first-order models to establish variational equations of stiffened shells. By applying the first-order N-T model we have

$$\begin{aligned} A_{1R}(u,v) = \int_S \int_{\frac{-h}{2}}^{\frac{h}{2}} \sigma^{\alpha\beta}(u)\epsilon_{\alpha\beta}(v)dzdS + \int_S \int_{g(x,y)}^{\frac{-h}{2}} \sigma^{\alpha\beta}(u)\epsilon_{\alpha\beta}(v)dzdS \\ + \int_S \int_{\frac{h}{2}}^{f(x,y)} \sigma^{\alpha\beta}(u)\epsilon_{\alpha\beta}(v)dzdS = L(v) \end{aligned} \tag{6.2}$$

$$\begin{aligned} A_{1R}(u,v) &= A_1(u,v) + \frac{E}{(1-\bar{\nu}^2)}\int_S \left(-\frac{h}{2}-g\right)((1-\bar{\nu})e^{\alpha\beta}(u)\,e_{\alpha\beta}(v) + \bar{\nu}e^{\rho}_{\rho}(u)\,e^{\rho}_{\rho}(v))dS \\ &- \frac{E}{2(1-\bar{\nu}^2)}\int_S \left((-\frac{h}{2})^2 - g^2\right)((1-\bar{\nu})e^{\alpha\beta}(u)\,K_{\alpha\beta}(v) + \bar{\nu}e^{\rho}_{\rho}(u)\,K^{\rho}_{\rho}(v))dS \\ &+ \frac{E}{3(1-\bar{\nu}^2)}\int_S \left((-\frac{h}{2})^3 - g^3\right)((1-\bar{\nu})e^{\alpha\beta}(u)\,Q_{\alpha\beta}(v) + \bar{\nu}e^{\rho}_{\rho}(u)\,Q^{\rho}_{\rho}(v))dS \\ &- \frac{E}{2(1-\bar{\nu}^2)}\int_S \left((-\frac{h}{2})^2 - g^2\right)((1-\bar{\nu})K^{\alpha\beta}(u)\,e_{\alpha\beta}(v) + \bar{\nu}K^{\rho}_{\rho}(u)\,e^{\rho}_{\rho}(v))dS \\ &+ \frac{E}{3(1-\bar{\nu}^2)}\int_S \left((-\frac{h}{2})^3 - g^3\right)((1-\bar{\nu})K^{\alpha\beta}(u)\,K_{\alpha\beta}(v) + \bar{\nu}K^{\rho}_{\rho}(u)\,K^{\rho}_{\rho}(v))dS \end{aligned} \tag{6.3}$$

$$
\begin{aligned}
&-\frac{E}{4(1-\bar{\nu}^2)}\int_S \left((-\frac{h}{2})^4 - g^4\right)((1-\bar{\nu})K^{\alpha\beta}(u)\,Q_{\alpha\beta}(v) + \bar{\nu}K^\rho_\rho(u)\,Q^\rho_\rho(v))dS \\
&+\frac{E}{3(1-\bar{\nu}^2)}\int_S \left((-\frac{h}{2})^3 - g^3\right)((1-\bar{\nu})Q^{\alpha\beta}(u)\,e_{\alpha\beta}(v) + \bar{\nu}Q^\rho_\rho(u)\,e^\rho_\rho(v))dS \\
&-\frac{E}{4(1-\bar{\nu}^2)}\int_S \left((-\frac{h}{2})^4 - g^4\right)((1-\bar{\nu})Q^{\alpha\beta}(u)\,K_{\alpha\beta}(v) + \bar{\nu}Q^\rho_\rho(u)\,K^\rho_\rho(v))dS \\
&+\frac{E}{5(1-\bar{\nu}^2)}\int_S \left((-\frac{h}{2})^5 - g^5\right)((1-\bar{\nu})Q^{\alpha\beta}(u)\,Q_{\alpha\beta}(v) + \bar{\nu}Q^\rho_\rho(u)\,Q^\rho_\rho(v))dS \\
&+\frac{E}{(1-\bar{\nu}^2)}\int_S \left(f - \frac{h}{2}\right)((1-\bar{\nu})e^{\alpha\beta}(u)\,e_{\alpha\beta}(v) + \bar{\nu}e^\rho_\rho(u)\,e^\rho_\rho(v))dS \\
&-\frac{E}{2(1-\bar{\nu}^2)}\int_S \left(f^2 - (\frac{h}{2})^2\right)((1-\bar{\nu})e^{\alpha\beta}(u)\,K_{\alpha\beta}(v) + \bar{\nu}e^\rho_\rho(u)\,K^\rho_\rho(v))dS \\
&+\frac{E}{3(1-\bar{\nu}^2)}\int_S \left(f^3 - (\frac{h}{2})^3\right)((1-\bar{\nu})e^{\alpha\beta}(u)\,Q_{\alpha\beta}(v) + \bar{\nu}e^\rho_\rho(u)\,Q^\rho_\rho(v))dS \\
&-\frac{E}{2(1-\bar{\nu}^2)}\int_S \left(f^2 - (\frac{h}{2})^2\right)((1-\bar{\nu})K^{\alpha\beta}(u)\,e_{\alpha\beta}(v) + \bar{\nu}K^\rho_\rho(u)\,e^\rho_\rho(v))dS \\
&+\frac{E}{3(1-\bar{\nu}^2)}\int_S \left(f^3 - (\frac{h}{2})^3\right)((1-\bar{\nu})K^{\alpha\beta}(u)\,K_{\alpha\beta}(v) + \bar{\nu}K^\rho_\rho(u)\,K^\rho_\rho(v))dS \\
&-\frac{E}{4(1-\bar{\nu}^2)}\int_S \left(f^4 - (\frac{h}{2})^4\right)((1-\bar{\nu})K^{\alpha\beta}(u)\,Q_{\alpha\beta}(v) + \bar{\nu}K^\rho_\rho(u)\,Q^\rho_\rho(v))dS \\
&+\frac{E}{3(1-\bar{\nu}^2)}\int_S \left(f^3 - (\frac{h}{2})^3\right)((1-\bar{\nu})Q^{\alpha\beta}(u)\,e_{\alpha\beta}(v) + \bar{\nu}Q^\rho_\rho(u)\,e^\rho_\rho(v))dS \\
&-\frac{E}{4(1-\bar{\nu}^2)}\int_S \left(f^4 - (\frac{h}{2})^4\right)((1-\bar{\nu})Q^{\alpha\beta}(u)\,K_{\alpha\beta}(v) + \bar{\nu}Q^\rho_\rho(u)\,K^\rho_\rho(v))dS \\
&+\frac{E}{5(1-\bar{\nu}^2)}\int_S \left(f^5 - (\frac{h}{2})^5\right)((1-\bar{\nu})Q^{\alpha\beta}(u)\,Q_{\alpha\beta}(v) + \bar{\nu}Q^\rho_\rho(u)\,Q^\rho_\rho(v))dS \\
=&\qquad L(v)
\end{aligned}
$$

The shell is stiffened at intrados if $f = 0$ and at extrados if $g = 0$. The additional terms to $A_1(u, v)$ represent the contribution of stiffeners to virtual strain energy. In the same way, we can also obtain the variational equation for the N model by choosing the kinematic and the transverse distribution function $w(z)$.

6.1.3 Thermoelastic Isotropic Shells

Around the reference temperature T_0, let $\Theta = T - T_0$, then the constitutive law of a linearly thermoelastic shell reads

$$\sigma^{ij}(u, \Theta) = \sigma_0^{ij} + C^{ijkl}\epsilon_{kl}(u) + \Theta M^{ij} \tag{6.4}$$

where σ_0^{ij} is the residual stress, the tangent elastic moduli tensor $C = (C^{ijkl})$ and stress-temperature tensor $M = (M^{ij})$ obtained from the free energy φ satisfy material symmetries and ellipticity conditions. By neglecting the residual stress and considering the free energy of an isotropic, homogeneous, thermoelastic material with Lamé constants $\bar{\lambda}$, $\bar{\mu}$; coefficient of thermal expansion κ, compressibility modulus $E_v = (3\bar{\lambda} + 2\bar{\mu})/3$; heat capacity c and mass density ρ, we have

$$\varphi = \frac{1}{\rho}\left(\frac{\bar{\lambda}}{2}\epsilon_l^l \epsilon_m^m + \bar{\mu}\epsilon^{kl}\epsilon_{kl} - 3\kappa E_v \epsilon_l^l \Theta\right) - c\frac{\Theta^2}{2T_0}, \tag{6.5}$$

$$\sigma = \rho\partial_\epsilon\varphi = \bar{\lambda}\epsilon_l^l G^{\cdot\cdot} + 2\bar{\mu}\epsilon - 3\kappa E_v \Theta G^{\cdot\cdot} \ and \ M = -3\kappa E_v G^{\cdot\cdot} \tag{6.6}$$

Let us consider the thermal strain tensor $\epsilon^{th} = \kappa\Theta G^{\cdot\cdot}$. The stress tensor now reads

$$\sigma = \bar{\lambda}\varepsilon_l^l G^{\cdot\cdot} + 2\bar{\mu}\varepsilon, \quad \varepsilon = \epsilon - \epsilon^{th} \tag{6.7}$$

Considering the hypothesis $\sigma^{33} = 0$, we obtain the 2D thermoelastic constitutive relations

$$\begin{aligned}
\sigma^{\alpha\beta} &= \bar{\Lambda}\varepsilon_\gamma^\gamma G^{\alpha\beta} + 2\bar{\mu}\varepsilon^{\alpha\beta} = \bar{\Lambda}\epsilon_\gamma^\gamma G^{\alpha\beta} + 2\bar{\mu}\epsilon^{\alpha\beta} - 2(\bar{\Lambda} + \bar{\mu})\kappa\Theta G^{\alpha\beta}, \\
&= \bar{\Lambda}\epsilon_\gamma^\gamma G^{\alpha\beta} + 2\bar{\mu}\epsilon^{\alpha\beta} - 3\eta E_v \kappa\Theta G^{\alpha\beta} \\
&= \frac{E\bar{\nu}}{1-\bar{\nu}^2}\epsilon_\gamma^\gamma G^{\alpha\beta} + \frac{E}{1+\bar{\nu}}\epsilon^{\alpha\beta} - \frac{E}{(1-\bar{\nu})}\kappa\Theta G^{\alpha\beta}
\end{aligned} \tag{6.8}$$

$$\bar{\Lambda} = \frac{2\bar{\mu}\bar{\lambda}}{\bar{\lambda} + 2\bar{\mu}} \ and \ \eta = \frac{1 - 2\bar{\nu}}{(1-\bar{\nu})}$$

One can remark the correcting coefficient η, of the compressibility (or volumetric deformation) modulus. The variational equation now comprises contributions of thermal origin. Indeed, in time-independent evolution, the equations are decoupled and the temperature field can be calculated separately. In this case, the N-T model variational equations read

$$\begin{aligned}
A(u, v) &= \int_S (N(u) : e(v) + M(u) : K(v) + M^*(u) : Q(v))dS = L(v) + L_\Theta(v); \\
L_\Theta(v) &= \int_S \int_{\frac{-h}{2}}^{\frac{h}{2}} 3\eta E_v \kappa\Theta G^{\alpha\beta}(e_{\alpha\beta}(v) - zK_{\alpha\beta}(v) + z^2 Q_{\alpha\beta}(v))\psi dz dS
\end{aligned} \tag{6.9}$$

In order to calculate the virtual thermal work $L_\Theta(v)$, we need an hypothesis on through the thickness distribution of $\Theta(x, z)$. Let us consider a Taylor expansion of the temperature field with respect to z, truncated at the first order. We have $\Theta(x, z) = \Theta(x, 0)+z\partial_z\Theta(x, 0) + 0(x, z) \approx \Theta(x, 0) + z\partial_z\Theta(x, 0) = \Theta_m(x) + z\Theta_1(x)$. If temperature is given at the intrados by $\Theta(x, -h/2) = \Theta_-$, we can seek solution of the temperature field in the form $\Theta(x, z) = (1 + 2z/h)\Theta_m(x) - (2z/h)\Theta_-$. Also in the same way, with a given temperature at intrados $\Theta(x, -h/2) = \Theta_-$ and at extrados $\Theta(x, h/2) = \Theta_+$, using a second-order truncated expansion, $\Theta(x, z) \approx \Theta_m(x) + z\Theta_1(x) + z^2\Theta_2(x)/2$, we deduce from the heat equations an admissible transverse distribution of $\Theta(x, z)$.

The additional virtual work due to the residual stress is calculated in the same way as for $L_\Theta(v)$ by replacing $3\eta E_\nu\kappa\Theta G^{\alpha\beta}$ by $-\sigma_0^{\alpha\beta}$. Transverse stresses σ^{i3} are calculated as shown earlier for the N-T model. To obtain thermal virtual work in the N model, we replace $3\eta E_\nu\kappa\Theta G^{\alpha\beta}(e_{\alpha\beta}(v) - zK_{\alpha\beta}(v) + z^2Q_{\alpha\beta}(v))$ by $3E_\nu\kappa\Theta G^{ij}\epsilon_{ij}(v, w\bar{y})$.

6.1.4 Anisotropic Homogeneous Shells

Apart from the idealistic widely used isotropic materials, some other interesting materials present different types of symmetries such as plane symmetry, two perpendicular plane symmetries (orthotropic materials) and transverse isotropy. The number of tangent moduli needed to describe their elastic behaviour in a reference configuration varies from 13 to 2. These materials satisfy some additional particular symmetries other than the natural material symmetries. In general, the tangent moduli tensor $C = (C^{ijkl})$ of an anisotropic material has 21 independent constants. The constitutive law is defined as follows:

$$\sigma^{ij} = C^{ijkl}\epsilon_{kl}, \quad C^{ijkl} = C^{ijlk} = C^{jikl} = C^{klij} \tag{6.10}$$

and $C = (C^{ijkl})$ also satisfies the ellipticity condition, i.e. there exists a constant $a > 0$ such that for any symmetric tensor τ, we have

$$C^{ijkl}\tau_{kl}\tau_{ij} \geqslant a\tau_{kl}\tau_{kl} \tag{6.11}$$

The variational problem with mixed boundary conditions,

$$\begin{cases} \text{Find } u \in U_{ad} \text{ such that} \\ A(u, v) = \int\limits_S \int_{\frac{-h}{2}}^{\frac{h}{2}} C^{ijkl}\epsilon_{kl}(u)\epsilon_{ij}(v)\psi dz dS = L(v)\ \forall v \in U_{ad} \end{cases} \tag{6.12}$$

has a unique solution by applying the lemma of Lax-Milgram. Shell equations for the N-T model are obtained by assuming also the plane stress conditions, i.e.

$$\sigma^{\alpha 3} = C^{\alpha 3\gamma\delta}\epsilon_{\gamma\delta} + 2C^{\alpha 3\gamma 3}\epsilon_{\gamma 3} + C^{\alpha 333}\epsilon_{33} = 0 \tag{6.13}$$

$$\sigma^{33} = C^{33\gamma\delta}\epsilon_{\gamma\delta} + 2C^{33\gamma 3}\epsilon_{\gamma 3} + C^{3333}\epsilon_{33} = 0$$

which is equivalent to

$$2C^{\alpha 3\gamma 3}\epsilon_{\gamma 3} + C^{\alpha 333}\epsilon_{33} = -C^{\alpha 3\gamma\delta}\epsilon_{\gamma\delta};$$

$$2C^{33\gamma 3}\epsilon_{\gamma 3} + C^{3333}\epsilon_{33} = -C^{33\gamma\delta}\epsilon_{\gamma\delta}$$

Let $X_\gamma = \epsilon_{\gamma 3}$, $X_3 = \epsilon_{33}$, $F^\alpha = -C^{\alpha 3\gamma\delta}\epsilon_{\gamma\delta}$, $F^3 = -C^{33\gamma\delta}\epsilon_{\gamma\delta}$, $M^{\alpha\gamma} = 2C^{\alpha 3\gamma 3}$; $M^{\alpha 3} = C^{\alpha 333}$; $M^{33} = C^{3333}$, then we have

$$M^{\alpha\gamma}X_\gamma + M^{\alpha 3}X_3 = F^\alpha; \;\; M^{3\gamma}X_\gamma + M^{33}X_3 = F^3 \tag{6.14}$$

The matrix M^{ij} is symmetric and positive defined. Indeed, by substituting τ such that $\tau_{\alpha\beta} = 0$, $C^{ijkl}\tau_{kl}\tau_{ij} = M^{ij}X_jX_i \geqslant aX_iX_i$. Let (Y_{ij}) be the inverse of (M^{ij}), then

$$\begin{aligned} X_\gamma = \epsilon_{\gamma 3} = Y_{\gamma j}F^j &= -Y_{\gamma\alpha}C^{\alpha 3\varrho\delta}\epsilon_{\varrho\delta} - Y_{\gamma 3}C^{33\varrho\delta}\epsilon_{\varrho\delta} \\ X_3 = \epsilon_{33} = Y_{3j}F^j &= -Y_{3\alpha}C^{\alpha 3\varrho\delta}\epsilon_{\varrho\delta} - Y_{33}C^{33\varrho\delta}\epsilon_{\varrho\delta} \end{aligned} \tag{6.15}$$

and the 2D constitutive law now reads

$$\begin{aligned} \sigma^{\alpha\beta} &= C^{\alpha\beta\varrho\delta}\epsilon_{\varrho\delta} + 2C^{\alpha\beta\gamma 3}\epsilon_{\gamma 3} + C^{\alpha\beta 33}\epsilon_{33} \\ &= C^{\alpha\beta\varrho\delta}\epsilon_{\varrho\delta} - 2C^{\alpha\beta\gamma 3}(Y_{\gamma\nu}C^{\nu 3\varrho\delta} + Y_{\gamma 3}C^{33\varrho\delta})\epsilon_{\varrho\delta} \\ &\quad - C^{\alpha\beta 33}(Y_{3\nu}C^{\nu 3\varrho\delta} + Y_{33}C^{33\varrho\delta})\epsilon_{\varrho\delta} \\ &= A^{\alpha\beta\varrho\delta}\epsilon_{\varrho\delta} \end{aligned} \tag{6.16}$$

We deduce that the tensor $(A^{\alpha\beta\varrho\delta})$ also satisfies all the material symmetries. Let τ be such that $\tau_{i3} = X_i$. From the above calculations, the symmetric positive defined tensors (Y_{ij}) and C^{ijkl}, we deduce that there exists a constant $c > 0$ such that

$$C^{ijkl}\tau_{kl}\tau_{ij} = A^{\alpha\beta\varrho\delta}\tau_{\varrho\delta}\tau_{\alpha\beta} \geqslant a\tau_{kl}\tau^{kl} = a(\tau_{\alpha\beta}\tau^{\alpha\beta} + \tau_{i3}\tau^{i3} + \tau_{3i}\tau^{3i}) \geqslant c\tau_{\alpha\beta}\tau^{\alpha\beta} \tag{6.17}$$

Therefore $(A^{\alpha\beta\varrho\delta})$ is elliptic. Suppose

$$C^{ijkl} = \bar{\lambda}G^{ij}G^{kl} + 2\bar{\mu}G^{ik}G^{lj} \tag{6.18}$$

Then

$$M^{\alpha\beta} = 4\bar{\mu}G^{\alpha\beta}, \quad M^{\alpha 3} = 0, \quad M^{33} = \bar{\lambda} + 2\bar{\mu}; \tag{6.19}$$

$$Y_{\alpha\beta} = \frac{1}{4\bar{\mu}} G_{\alpha\beta}, \quad Y_{\alpha 3} = 0, \quad Y_{33} = \frac{1}{\bar{\lambda} + 2\bar{\mu}} \tag{6.20}$$

which lead to the well-known classical results

$$X_3 = \epsilon_{33} = Y_{3j} F^j = -Y_{33} C^{33\varrho\delta} \epsilon_{\varrho\delta} = -\frac{\bar{\lambda}}{\bar{\lambda} + 2\bar{\mu}} G^{\varrho\delta} \epsilon_{\varrho\delta} = -\frac{\bar{\lambda}}{\bar{\lambda} + 2\bar{\mu}} \epsilon^{\varrho}_{\varrho} \tag{6.21}$$

The variational formulation is identical to that of isotropic homogeneous materials, with the stress and moments resultant tensors N, M and M^* defined using the 2D tangent moduli tensor $(A^{\alpha\beta\varrho\delta})$ defined above. It also has a unique solution for the mixed boundary conditions because all the conditions to apply the Lax-Milgram lemma are satisfied. Governing equations of transverse stresses σ^{i3} remain unchanged. The variational formulation of the N model for an anisotropic material is also obtained in the same way by using the corresponding kinematic.

6.2 Heterogeneous Shells

6.2.1 General Periodic Media

Composite materials are materials whose elastic moduli depend on the material point's position. Locally these materials can be homogeneous, isotropic or anisotropic. They present some interesting properties depending on their internal structures, according to the final use. However, it is not easy to consider their heterogeneity in calculating structures (Fig. 6.2).

In certain theories, homogeneous moduli are obtained by calculating mean values in a Representative Elementary Volume (REV). Certain materials present a "regular" form obtained by the reproduction of a sub-structure. A very frequent case is that of cellular generated media. The moduli of the generic cell are known. Structures thus obtained are periodic media with an ε size generic cell $Y^{\varepsilon} \in \mathbb{R}^p$ $p = 1, 2, 3$. The generic cell can be defined by homothety, from a cell of size 1. We then write $Y^{\varepsilon} = \varepsilon Y$. If $x = \varepsilon y$ then $y = x/\varepsilon$ (Fig. 6.3).

In order to obtain homogeneous moduli, some authors have applied the asymptotic expansion method due to Bensoussan et al., [12], on Euler's equations, by analysing in a cartesian coordinate system, some formal series of the form

$$u^{\varepsilon} = \sum_{n=0}^{\infty} \varepsilon^n u^n(x, y), \quad \sigma^{\varepsilon} = \sum_{n=0}^{\infty} \varepsilon^n \sigma^n(x, y), \quad \epsilon^{\varepsilon} = \sum_{n=0}^{\infty} \varepsilon^n \epsilon^n(x, y) \tag{6.22}$$

On a one-dimensional curvilinear periodic media, Tutek [104] and on a periodic shell structure, Kalamkarov [8]; in order to calculate the homogeneous moduli, these

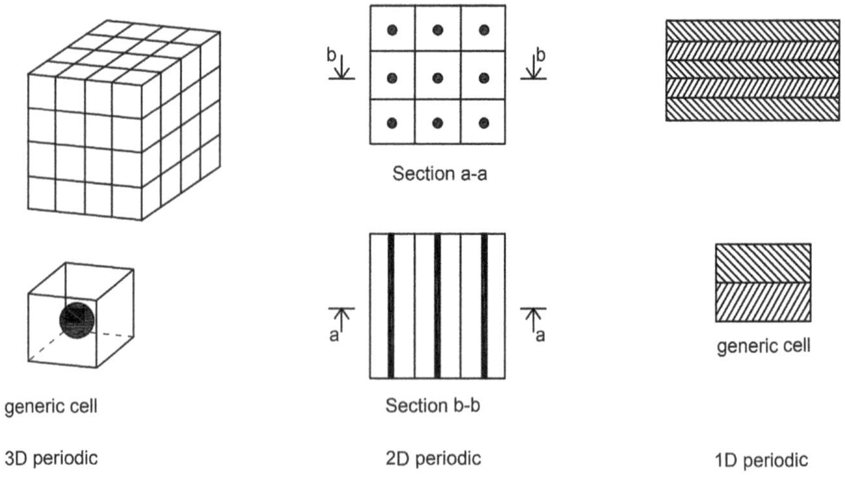

Fig. 6.2 3D, 2D and 1D periodic media

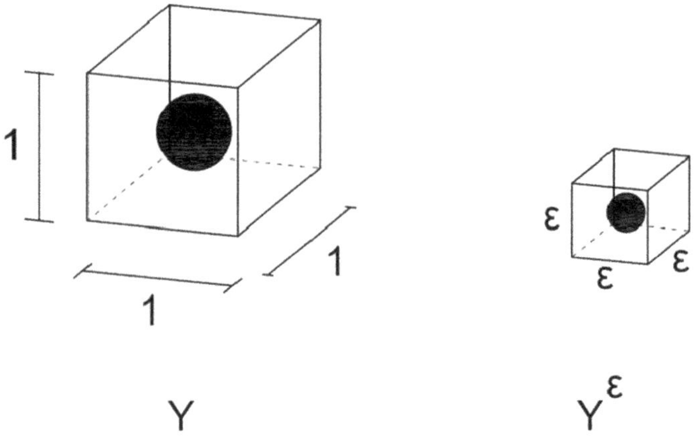

Fig. 6.3 Generic cell

authors considered formal series depending on two parameters (δ being a dimensionless parameter related to the thickness) as follows:

$$\begin{aligned} u^{\varepsilon\delta} &= \sum_{n,p=0}^{\infty} \varepsilon^n \delta^p u^{np}(x, y) \\ \sigma^{\varepsilon\delta} &= \sum_{n,p=0}^{\infty} \varepsilon^n \delta^p \sigma^{np}(x, y) \\ \epsilon^{\varepsilon\delta} &= \sum_{n,p=0}^{\infty} \varepsilon^n \delta^p \epsilon^{np}(x, y) \end{aligned} \tag{6.23}$$

Applied forces were assumed to be periodic and expanded in the same two parameters series. At the cost of a complex and inextricable calculation, by identifying terms of the same power, the homogenized tangent moduli tensor was obtained through the relation $\sigma^0 = C^0\epsilon(u^0)$.
The variational formulation of the model problem for a fixed ε reads

$$\begin{aligned}\int_\Omega \sigma^{ij}\epsilon_{ij}d\Omega &= \int_\Omega C^{ijkl}(x,\frac{x}{\varepsilon})\epsilon_{kl}(U^\varepsilon)\epsilon_{ij}(V)d\Omega \\ &= \int_\Omega \epsilon_{kl}(U^\varepsilon)C^{ijkl}(x,\frac{x}{\varepsilon})\epsilon_{ij}(V)d\Omega \\ &= \int_\Omega U^\varepsilon_{k|l}C^{ijkl}(x,\frac{x}{\varepsilon})V_{i|j}d\Omega = L(V)\end{aligned} \tag{6.24}$$

Let us recall that $x = (x^1, x^2, x^3)$, with $x^3 = z$, $y \in \mathbb{R}^p$, $\frac{x}{\varepsilon} \in \mathbb{R}^p$ and p characterizes the type of periodicity. We can replace $\frac{x}{\varepsilon}$ by $\frac{x^p}{\varepsilon}$ if $p < 3$; for example, for a plane periodicity (on the mid-surface), $\frac{x^p}{\varepsilon} = (\frac{x^1}{\varepsilon}, \frac{x^2}{\varepsilon})$; for a periodic stratified media in the normal direction $\frac{x^p}{\varepsilon} = \frac{x^3}{\varepsilon}$.

Let $\varphi^{kl}(x,\frac{x}{\varepsilon}) = C^{ijkl}(x,\frac{x}{\varepsilon})V_{i|j}$. The variational problem reads

$$\int_\Omega U^\varepsilon_{k|l}\varphi^{kl}(x,\frac{x}{\varepsilon})d\Omega = L(V) \tag{6.25}$$

Let us define the series of functionals by

$$F^\varepsilon : \varphi = (\varphi^{kl}) \longmapsto F^\varepsilon(\varphi) = \int_\Omega U^\varepsilon_{k|l}\varphi^{kl}(x,\frac{x}{\varepsilon})d\Omega \tag{6.26}$$

The homogenization technique consists in characterizing the limit functional F^0. By fixing k and l in F^ε we obtain a continuous functional F^ε_{kl}. We can also write that

$$F^\varepsilon = \sum_{k,l=1}^{3} F^\varepsilon_{kl} \tag{6.27}$$

The method is straightforward unlike the asymptotic expansion methods and is based on the two-scale convergence theorem of NGUETSENG [82] that we shall briefly present below after some notations. We denote the weak convergence and strong convergence in $L^2(\Omega)$ respectively by $\rightharpoonup$ and $\longrightarrow$

$$L^2_{\#}(Y) = \{w \in L^2_{loc}(\mathbb{R}^3)\ Y - periodic\} \tag{6.28}$$

The space $L^2_{loc}(\mathbb{R}^3)$ is the space of real functions which are dx-measurable and square summable in every bounded domain in $\mathbb{R}^3$. The norm in $L^2(\Omega)$ is denoted by $\| \,.\, \|$; $H^1_{\#}(Y)/\mathbb{R}$ is the space of $H^1(Y)\ Y - periodic$ functions with zero mean value.

Two-scale convergence Theorem

Let (u^ε) be a bounded sequence in $L^2(\Omega)$ and let a sub-sequence still denoted (u^ε) be such that $u^\varepsilon \rightharpoonup$, then there exists $u^0\ inL^2(\Omega,\ L^2_{\#}(Y))$ such that $\forall\varphi \in L^2(\Omega,\ L^2_{\#}(Y))$

(i) $\displaystyle\int_\Omega u^\varepsilon \varphi(x, \frac{x}{\varepsilon})dx \longrightarrow \int_{\Omega\times Y} u^0(x, y)\varphi(x, y)dxdy$

(ii) if $u^\varepsilon \longrightarrow u^0\ in\ L^2(\Omega)\ then \int_\Omega u^\varepsilon \varphi(x, \frac{x}{\varepsilon})dx \rightarrow \int_\Omega u^0(x) \int_Y \varphi(x, y)dxdy$

(iii) $if\ \| u^\varepsilon \| \leq c$ and $\| \dfrac{\partial u^\varepsilon}{\partial x_j} \| \leq c$ then $\exists Z \in L^2(\Omega,\ H^1_{\#}(Y)/\mathbb{R})$ such that

$$\int_\Omega \frac{\partial u^\varepsilon}{\partial x_j}\varphi(x, \frac{x}{\varepsilon})dx \longrightarrow \int_{\Omega\times Y} (\frac{\partial u^0(x)}{\partial x_j} + \frac{\partial Z(x, y)}{\partial y_j})\varphi(x, y)dxdy \tag{6.29}$$

This theorem was directly and successfully applied to calculate the homogenized thermal conductivities of a periodic media (see NGUETSENG [82]). We deduce from the demonstration of the theorem that the Lagrange multiplier function $Z(x, y)$ depends on the macroscopic position x and microscopic $\varepsilon y \in \varepsilon Y \subset \mathbb{R}^p$ $\forall p = 1, 2\ or\ 3$. This justifies the notion of two scales and double convergence. Before applying the theorem on the functional F^ε let us recall that in curvilinear media the norms $\| U \|^2 = \displaystyle\int_\Omega (U_{i,j}U^j_{,i} + U_iU^i)dx$ and $\| U \|^2 = \int_\Omega (U_{i|j}U^j_{|i} + U_iU^i)dx$ are equivalent in $(H^1_{\Gamma_0}(\Omega))^3$ (see CIARLET et al. [28]). We deduce that $U^\varepsilon_{k|l}$ is bounded if and only if $U^\varepsilon_{k,l}$ is bounded. The variational problem satisfies all the conditions necessary to apply Lax-Milgram's lemma. Therefore there exists a constant c such that $\| U^\varepsilon \| \leq c$ and $\| U^\varepsilon_{k,l} \| \leq c$. Let U^0 be the limit of the sequence U^ε then $U^\varepsilon_{k,l} \rightharpoonup U^0_{k,l}$ and $U^\varepsilon \longrightarrow U^0$. We have

$$
\begin{aligned}
F^{\varepsilon}(\varphi) = & \int_{\Omega} U^{\varepsilon}_{k|l}\varphi^{kl}(x, \frac{x}{\varepsilon})d\Omega = \int_{\Omega} U^{\varepsilon}_{k|l}C^{ijkl}(x, \frac{x}{\varepsilon})V_{i|j}d\Omega \\
= & \int_{\Omega} (U^{\varepsilon}_{k,l} - \Gamma^{m}_{k,l}U^{\varepsilon}_{m})C^{ijkl}(x, \frac{x}{\varepsilon})V_{i|j}d\Omega \\
= & \int_{\Omega} U^{\varepsilon}_{k,l}C^{ijkl}(x, \frac{x}{\varepsilon})V_{i|j}d\Omega - \int_{\Omega}(\Gamma^{m}_{k,l}U^{\varepsilon}_{m})C^{ijkl}(x, \frac{x}{\varepsilon})V_{i|j}d\Omega
\end{aligned}
\tag{6.30}
$$

We deduce from *point* (ii) of the theorem that

$$
\int_{\Omega} (\Gamma^{m}_{k,l}U^{\varepsilon}_{m})C^{ijkl}(x, \frac{x}{\varepsilon})V_{i|j}d\Omega \longrightarrow \int_{\Omega} (\Gamma^{m}_{k,l}U^{0}_{m})(x)\int_{Y} C^{ijkl}(x, y)V_{i|j}dyd\Omega \tag{6.31}
$$

Because, for each fixed k, the functional F^{ε}_{kl} satisfies the conditions of the theorem, we deduce from (iii) that there exists $Z = (Z_k)$ with $Z_k \in L^2(\Omega,\ H^1_{\#}(Y)/\mathbb{R})$ such that

$$
\int_{\Omega} U^{\varepsilon}_{k,l}C^{ijkl}(x, \frac{x}{\varepsilon})V_{i|j}d\Omega \longrightarrow \int_{\Omega\times Y} (\frac{\partial U^{0}_{k}(x)}{\partial x_l} + \frac{\partial Z_k(x, y)}{\partial y_l})C^{ijkl}(x, y)V_{i|j}dyd\Omega \tag{6.32}
$$

It follows that

$$
\begin{aligned}
F^{\varepsilon} : \varphi = (\varphi^{kl}) \longmapsto \quad & F^{\varepsilon}(\varphi) = \int_{\Omega} U^{\varepsilon}_{k|l}\varphi^{kl}(x, \frac{x}{\varepsilon})d\Omega \\
\longrightarrow \quad & \int_{\Omega\times Y} (\frac{\partial U^{0}_{k}(x)}{\partial x_l} - (\Gamma^{m}_{k,l}U^{0}_{m})(x) \\
& + \frac{\partial Z_k(x, y)}{\partial y_l})C^{ijkl}(x, y)V_{i|j}d\Omega dy \\
= \quad & \int_{\Omega\times Y} (U^{0}_{k|l}(x) + \frac{\partial Z_k(x, y)}{\partial y_l})C^{ijkl}(x, y)V_{i|j}dyd\Omega
\end{aligned}
\tag{6.33}
$$

By replacing in the initial variational equation V by $V = \bar{V}(x) + \varepsilon\phi(x, \frac{x}{\varepsilon})$ with $\phi_i \in L^2(\Omega,\ H^1_{\#}(Y)/\mathbb{R})$ sufficiently smooth, we have

$$
V_{i|j} = \bar{V}_{i|j} + \varepsilon\phi_{i|j} + \frac{\partial \phi_i(x, y)}{\partial y_j} \tag{6.34}
$$

and

$$
\begin{aligned}
F^{\varepsilon}(\varphi) = & \int_{\Omega} U^{\varepsilon}_{k|l} C^{ijkl}(x, \frac{x}{\varepsilon}) V_{i|j} d\Omega \\
= & \int_{\Omega} U^{\varepsilon}_{k|l} C^{ijkl}(x, \frac{x}{\varepsilon})(\bar{V}_{i|j} + \varepsilon \phi_{i|j} + \frac{\partial \phi_i(x, y)}{\partial y_j}) d\Omega \\
= & L(\bar{V}(x) + \varepsilon \phi(x, \frac{x}{\varepsilon}))
\end{aligned}
\tag{6.35}
$$

and at the limit we obtain

$$
\begin{aligned}
\int_{\Omega \times Y} & (U^0_{k|l}(x) + \frac{\partial Z_k(x, y)}{\partial y_l}) C^{ijkl}(x, y)(\bar{V}_{i|j} \\
& + \frac{\partial \phi_i(x, y)}{\partial y_j}) dy d\Omega \qquad = L(\bar{V}(x))
\end{aligned}
\tag{6.36}
$$

We deduce the equation satisfied by $Z = (Z_k)$ by letting $\bar{V}(x) = 0$

$$
\int_Y \frac{\partial Z_k(x, y)}{\partial y_l} C^{ijkl}(x, y) \frac{\partial \phi_i(x, y)}{\partial y_j} dy = -U^0_{k|l}(x) \int_Y C^{ijkl}(x, y) \frac{\partial \phi_i(x, y)}{\partial y_j} dy \tag{6.37}
$$

which, for a fixed x, reads

$$
\int_Y Z_{k,l}(x, y) C^{ijkl}(x, y) \phi_{i,j}(x, y) dy = -U^0_{k|l}(x) \int_Y C^{ijkl}(x, y) \phi_{i,j}(x, y) dy \tag{6.38}
$$

Let $\chi^{mn} = (\chi^{mn}_k)$ with $\chi^{mn}_k \in L^2(\Omega,\ H^1_{\#}(Y)/\mathbb{R})$ be the solution of

$$
\int_Y \chi^{mn}_{k,l}(x, y) C^{ijkl}(x, y) w_{i,j}(y) dy = \int_Y C^{ijmn}(x, y) w_{i,j}(y) dy,\ w_i \in H^1_{\#}(Y)/\mathbb{R} \tag{6.39}
$$

then

$$
Z(x, y) = -U^0_{m|n}(x) \chi^{mn}(x, y) \tag{6.40}
$$

i.e.

$$\begin{aligned}\int_{\Omega\times Y}(U^0_{k|l}(x)+\frac{\partial Z_k(x,y)}{\partial y_l})\times \\ C^{ijkl}(x,y)V_{i|j}d\Omega dy &= \int_{\Omega\times Y}(U^0_{k|l}(x)-U^0_{m|n}(x)\chi^{mn}_{k,l}(x,y))\times \\ &\qquad C^{ijkl}(x,y)V_{i|j}d\Omega dy \\ &= \int_{\Omega}\int_{Y}(C^{ijkl}(x,y)-C^{ijmn}(y)\chi^{kl}_{m,n}(x,y))\times \\ &\qquad U^0_{k|l}(x)V_{i|j}dyd\Omega \\ &= \int_{\Omega}C^{ijkl}_h(x)U^0_{k|l}(x)V_{i|j}d\Omega \\ &= L(V)\end{aligned} \tag{6.41}$$

In the above formula, the homogenized tangent moduli are

$$C^{ijkl}_h(x)=\int_Y(C^{ijkl}(x,y)-C^{ijmn}(x,y)\chi^{kl}_{m,n}(x,y))dy \tag{6.42}$$

Consider the function $\Phi^{mn}=(\Phi^{mn}_k)=(\delta^m_k y^n)$, $\delta^m_k y^n\in H^1_\#(Y)/\mathbb{R}$, $y^n=y^1$ *or* y^2 *or* y^3 and the bilinear form defined in $\left[H^1_\#(Y)/\mathbb{R}\right]^3$ by

$$a(w,\bar{w})=\int_Y C^{ijkl}(x,y)w_{i,j}(y)\bar{w}_{k,l}(y)dy \tag{6.43}$$

then the equation of χ^{mn} becomes

$$\begin{aligned}a(\chi^{mn},w)=\int_Y C^{ijkl}(x,y)\chi^{mn}_{k,l}w_{i,j}(y)dy &= \int_Y C^{ijmn}(x,y)w_{i,j}(y)dy \\ = \quad \int_Y C^{ij\bar{m}\bar{n}}(x,y)\Phi^{mn}_{\bar{m},\bar{n}}(y)w_{i,j}(y)dy &= a(\Phi^{mn},w)\end{aligned} \tag{6.44}$$

which is also equivalent, by the symmetry of $a(.,.)$, to

$$a(\Phi^{mn}-\chi^{mn},w)=a(w,\Phi^{mn}-\chi^{mn})=0 \tag{6.45}$$

We also have $a(-w,\Phi^{mn}-\chi^{mn})=0$ for every w and in particular for $w=\chi^{ij}$, we have

$$a(-\chi^{ij},\Phi^{kl}-\chi^{kl})=0 \tag{6.46}$$

In the same way, we also have

$$
\begin{aligned}
C_h^{ijkl}(x) &= \int_Y (C^{ijkl}(x,y) - C^{ijmn}(x,y)\chi^{kl}_{m,n}(x,y))dy \\
&= \int_Y (C^{\bar{m}\bar{n}mn}(x,y)\Phi^{ij}_{\bar{m},\bar{n}}\Phi^{kl}_{m,n} - C^{\bar{m}\bar{n}mn}(x,y)\Phi^{ij}_{\bar{m},\bar{n}}\chi^{kl}_{m,n}(x,y))dy \\
&= a(\Phi^{ij},\Phi^{kl}) - a(\Phi^{ij},\chi^{kl}) = a(\Phi^{ij},\Phi^{kl}-\chi^{kl})
\end{aligned}
$$

It follows that

$$C_h^{ijkl}(x) = a(\Phi^{ij} - \chi^{ij}, \Phi^{kl} - \chi^{kl})$$

and the homogenized tangent moduli tensor thus obtained satisfies all the material symmetries and ellipticity conditions. The final variational equation now reads

$$\int_\Omega C_h^{ijkl}(x)U^0_{k|l}(x)V_{i|j}d\Omega = \int_\Omega C_h^{ijkl}(x)\epsilon_{kl}(U^0)\epsilon_{ij}(V)d\Omega = L(V)$$

It is therefore sufficient to use the bilinear form $a(.,.)$ to calculate the moduli. This direct method saves tedious calculations of the asymptotic methods in curvilinear media. The homogenized moduli thus obtained are measurable but are different from those obtained by averaging. This shows that even if in every cell the material is homogeneous and isotropic, the homogenized constitutive law is not necessary isotropic nor homogeneous. On the other hand, the microscopic character in the cells, sometimes combined with a sudden variation of the values, is attenuated. The 2D shell model is deduced as in paragraph 6.1.4.

6.2.2 Application on a Two-Component Periodic Stratified Media

We consider a stratified periodic media in the $x^3 = z$ direction. We assume the generic cell is constituted by two homogeneous isotropic components whose moduli are (λ^-, μ^-) and (λ^+, μ^+) or (E^-, ν^-) and (E^+, ν^+) (Fig. 6.4).

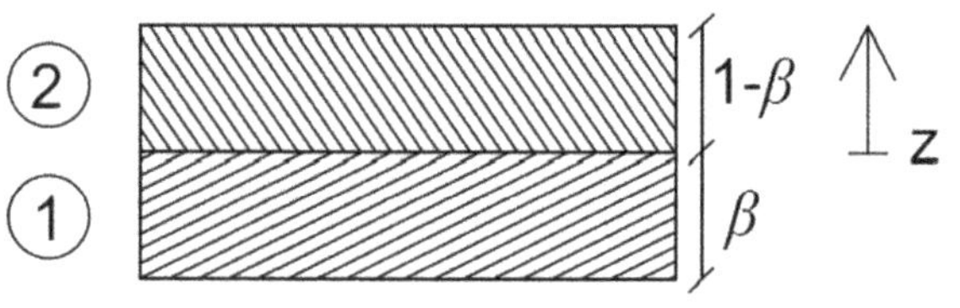

Fig. 6.4 Two-component generic cell

The constitutive law is

$$\sigma^{ij} = (\bar{\lambda} G^{ij} G^{kl} + 2\bar{\mu} G^{ik} G^{jl})\epsilon_{kl} = C^{ijkl}\epsilon_{kl}, \quad (\bar{\lambda}, \bar{\mu}) = (\lambda^-, \mu^-) \tag{6.47}$$

or (λ^+, μ^+),

$$\epsilon_{ij} = (-\frac{\bar{\nu}}{E} G_{ij} G_{kl} + \frac{1+\bar{\nu}}{E} G_{ik} G_{jl})\sigma^{kl} = S_{ijkl}\sigma^{kl}, \quad (E, \bar{\nu}) = (E^-, \nu^-) \tag{6.48}$$

or (E^+, ν^+) and $Y = [0, 1]$

$$\sigma^{ij} = C^{ijkl}(x, \frac{y^3}{\varepsilon})\epsilon_{kl} \tag{6.49}$$

In the fundamental cell or generic cell $(\bar{\lambda}, \bar{\mu}) = (\lambda^-, \mu^-)$ for $0 \le y \le \beta$ and $(\bar{\lambda}, \bar{\mu}) = (\lambda^+, \mu^+)$ for $\beta \prec y \le 1$. The homogenized moduli are

$$C_h^{ijkl}(x) = \int_0^1 (C^{ijkl}(x, y^3) - C^{ijm3}(x, y^3)\chi_{m,3}^{kl}(x, y^3))dy^3 \tag{6.50}$$

and by writing $y = y^3$, we find the functions χ^{kl} by solving the variational equation

$$\int_0^1 C^{i3n3}(x, y)\chi_{n,3}^{kl}(x, y)w_{i,3}(y)dy = \int_0^1 C^{i3kl}(x, y)w_{i,3}(y)dy, \; w_i \in H_\#^1([0, 1])/\mathbb{R} \tag{6.51}$$

which is also equivalent to

$$\int_0^1 (C^{i3n3}(x, y)\chi_{n,3}^{kl}(x, y) - C^{i3kl}(x, y))w_{i,3}(y)dy = 0, \;\; \forall w_i \in H_\#^1([0, 1])/\mathbb{R} \tag{6.52}$$

from which we deduce that

$$\left[C^{i3n3}(x, y)\chi_{n,3}^{kl}(x, y) - C^{i3kl}(x, y)\right]_{,3} = 0 \tag{6.53}$$

It follows that

$$C^{i3n3}(x, y)\chi_{n,3}^{kl}(x, y) = C^{i3kl}(x, y) + A^{ikl}(x) \tag{6.54}$$

or $2\bar{\mu}(y)G^{in}\chi_{n,3}^{kl}(x, y) = 2\bar{\mu}(y)G^{ik}G^{3l} + A^{ikl}(x)$. Multiplying both sides by $\frac{1}{2\bar{\mu}(y)}G_{mi}$ yields $\chi_{m,3}^{kl}(x, y)$ which satisfies the equation

$$\chi^{kl}_{m,3}(x,y) = \delta^k_m G^{3l} + \frac{1}{2\bar{\mu}(y)} G_{mi} A^{ikl}(x) = \delta^k_m G^{3l} + \frac{1}{2\bar{\mu}(y)} A^{kl}_m(x) \tag{6.55}$$

Let $\chi^{kl}_{m,3}(x,y) = f'_m = \dfrac{df_m}{dy}, a_m(x) = \delta^k_m G^{3l}, b_m(x) = A^{kl}_m(x), c(y) = \dfrac{1}{2\bar{\mu}(y)}$ where $c(y) = \dfrac{1}{2\mu^-}$ for $0 \le y \le \beta$ and $c(y) = \dfrac{1}{2\mu^+}$ for $\beta < y \le 1$. The constant $b_m(x)$ should be determined such that the average of f_m be equal to zero, when solving the equation

$$f'_m = a_m(x) + b_m(x)c(y) \tag{6.56}$$

The solution is

$$f_m = a_m(x)y + b_m(x)(\frac{y}{2\mu^-}H(\beta - y) + \frac{y}{2\mu^+}H(y-\beta)) + d(x) \tag{6.57}$$

where the Heaviside function $H(t) = 1$ if $t \ge 0$ and $H(t) = 0$ if $t \prec 0$. By choosing $d(x) = 0$, it follows that the average of f_m is zero if

$$b_m(x) = \frac{-2a_m(x)\mu^-\mu^+}{(1-\beta^2)\mu^- + \beta^2\mu^+} = \frac{-2\mu^-\mu^+\delta^k_m G^{3l}}{(1-\beta^2)\mu^- + \beta^2\mu^+} \tag{6.58}$$

Then

$$\begin{aligned} \chi^{kl}_{m,3}(x,y) &= \delta^k_m G^{3l} - \frac{1}{2\bar{\mu}(y)} \frac{2\mu^-\mu^+}{(1-\beta^2)\mu^- + \beta^2\mu^+}\delta^k_m G^{3l} \\ &= (1 - \frac{1}{\bar{\mu}(y)} \frac{\mu^-\mu^+}{(1-\beta^2)\mu^- + \beta^2\mu^+})\delta^k_m G^{3l} \\ &= \alpha^\star(y)\delta^k_m G^{3l} \end{aligned} \tag{6.59}$$

and

$$\begin{aligned} C^{ijkl}_h(x) &= \int_0^1 (C^{ijkl}(x,y) - \alpha^\star(y)C^{ijm3}(x,y)\delta^k_m G^{3l})dy \\ &= \int_0^1 (C^{ijkl}(x,y) - \alpha^\star(y)C^{ijk3}(x,y)G^{3l})dy \end{aligned} \tag{6.60}$$

We have the following:

$$C^{ijk\alpha}_h(x) = C^{ij\alpha k}_h(x) = \int_0^1 C^{ijk\alpha}(x,y)dy, \tag{6.61}$$

$$C_h^{ijk3}(x) = C_h^{ij3k}(x) = \int_0^1 (1-\alpha^\star(y))C^{ijk3}(x,y))dy, \tag{6.62}$$

$$\begin{aligned} C_h^{3333}(x) = & \int_0^1 (1-\alpha^\star(y))(\bar{\lambda}+2\bar{\mu})dy \\ & = \beta(\lambda^- + 2\mu^-) + (1-\beta)(\lambda^+ + 2\mu^+) \\ & - \int_0^1 \alpha^\star(y)(\bar{\lambda}+2\bar{\mu})(y)dy \end{aligned} \tag{6.63}$$

This result shows that the mean value is greater than the homogenized value and thus averaging is optimistic. From $C^{i\alpha\rho 3} = 0$ we deduce that the moduli $C_h^{i\alpha\rho l}$ are mean values. For a stratified shell with n layers not necessarily periodic we can assume that the shell is extracted from a periodic media whose generic cell is made of the n components and calculate the homogenized moduli as above in a fundamental cell which has n components. 2D shell models are deduced after the tangent moduli tensor $C_h^{ijkl}(x)$ is completely determined. However certain methods consist in integrating the variational equation through the thickness, layer by layer.

6.2.3 A Simplified Calculation Method in a Stratified Shell

Let us consider a stratified shell made of n layers. Each layer of thickness h_m is between z_{m-1} and z_m, $z_0 = -h/2$. In each layer n_m the constitutive law is defined by

$$\sigma_m^{ij} = C_m^{ijkl}(x,z)\epsilon_{kl} \tag{6.64}$$

and the variational equation reads

$$\begin{aligned} \int_\Omega \sigma^{ij}\epsilon_{ij}\psi dz dS = & \int_S \sum_{m=1}^n \int_{z_{m-1}}^{z_m} \sigma_m^{ij}(U)\epsilon_{ij}(V)\psi dz dS \\ = & \int_S \sum_{m=1}^n \int_{z_{m-1}}^{z_m} C_m^{ijkl}(x,z)\epsilon_{kl}(U)\epsilon_{ij}(V)\psi dz dS \\ = & \quad L(V) \end{aligned} \tag{6.65}$$

For the N-T model, we have

$$\int_{\Omega} \sigma^{ij} \epsilon_{ij} \psi dz dS = \int_{\Omega} \sigma^{\alpha\beta} \epsilon_{\alpha\beta} \psi dz dS$$
$$= \int_{s} \sum_{m=1}^{n} \int_{z_{m-1}}^{z_m} C_m^{\alpha\beta\gamma\delta}(x, z)(e_{\gamma\delta}(u) \tag{6.66}$$
$$-zK_{\gamma\delta}(u) + z^2 Q_{\gamma\delta}(u))(e_{\alpha\beta}(v)$$
$$-zK_{\alpha\beta}(v) + z^2 Q_{\alpha\beta}(v))\psi dz dS$$
$$= L(v)$$

For the first-order approximation, $\psi \approx 1$, $C_m^{\alpha\beta\gamma\delta}(x, z) \approx A_m^{\alpha\beta\gamma\delta}(x, 0)$ calculated under the plane strain hypothesis and we have

$$\int_{s} \sum_{m=1}^{n} v \int_{z_{m-1}}^{z_m} C_m^{\alpha\beta\gamma\delta}(x, 0)(e_{\gamma\delta}(u) - zK_{\gamma\delta}(u) + z^2 Q_{\gamma\delta}(u))e_{\alpha\beta}(v)$$
$$+C_m^{\alpha\beta\gamma\delta}(x)(-ze_{\gamma\delta}(u) + z^2 K_{\gamma\delta}(u) - z^3 Q_{\gamma\delta}(u))K_{\alpha\beta}(v) \tag{6.67}$$
$$+C_m^{\alpha\beta\gamma\delta}(x)(z^2 e_{\gamma\delta}(u) - z^3 K_{\gamma\delta}(u) + z^4 Q_{\gamma\delta}(u))Q_{\alpha\beta}(v))dz dS = L(v)$$

Let

$$\int_{s} \sum_{m=1}^{n} N_m^{\alpha\beta}(u)e_{\alpha\beta}(v) + M_m^{\alpha\beta}(u)K_{\alpha\beta}(v) + M_m^{\star\alpha\beta}(u)Q_{\alpha\beta}(v))dS = L(v) \tag{6.68}$$

$$N_m^{\alpha\beta}(u) = C_m^{\alpha\beta\gamma\delta}(x)[(z_m - z_{m-1})e_{\gamma\delta}(u) - \frac{1}{2}(z_m^2 - z_{m-1}^2)K_{\gamma\delta}(u)$$
$$+\frac{1}{3}(z_m^3 - z_{m-1}^3)Q_{\gamma\delta}(u)]$$

$$M_m^{\alpha\beta}(u) = C_m^{\alpha\beta\gamma\delta}(x)[-\frac{1}{2}(z_m^2 - z_{m-1}^2)e_{\gamma\delta}(u) + \frac{1}{3}(z_m^3 - z_{m-1}^3)K_{\gamma\delta}(u)$$
$$-\frac{1}{4}(z_m^4 - z_{m-1}^4)Q_{\gamma\delta}(u)]$$

$$M_m^{\star\alpha\beta}(u) = C_m^{\alpha\beta\gamma\delta}(x)[\frac{1}{3}(z_m^3 - z_{m-1}^3)e_{\gamma\delta}(u) - \frac{1}{4}(z_m^4 - z_{m-1}^4)K_{\gamma\delta}(u)$$
$$+\frac{1}{5}(z_m^5 - z_{m-1}^5)Q_{\gamma\delta}(u)]$$

These resulting forces can be written in the form

$$\begin{bmatrix} N_m(u) \\ M_m(u) \\ M_m^\star(u) \end{bmatrix} = C_m(x) \begin{bmatrix} (z_m - z_{m-1}) & -\frac{1}{2}(z_m^2 - z_{m-1}^2) & \frac{1}{3}(z_m^3 - z_{m-1}^3) \\ -\frac{1}{2}(z_m^2 - z_{m-1}^2) & \frac{1}{3}(z_m^3 - z_{m-1}^3) & -\frac{1}{4}(z_m^4 - z_{m-1}^4) \\ \frac{1}{3}(z_m^3 - z_{m-1}^3) & -\frac{1}{4}(z_m^4 - z_{m-1}^4) & \frac{1}{5}(z_m^5 - z_{m-1}^5) \end{bmatrix} \begin{pmatrix} e(u) \\ K(u) \\ Q(u) \end{pmatrix} \quad (6.69)$$

We rewrite the variational equation as follows:

$$\int_S \sum_{m=1}^{n} (N_m(u) : e(v) + M_m(u) : K(v) + M_m^\star(u) : Q(v))dS = L(v) \quad (6.70)$$

Let us note that if $h_m = z_m - z_{m-1}$ is too small we can apply the mean value theorem (Roll's theorem) to replace $\int_{z_{m-1}}^{z_m} z^p dz \ \ p \geq 0$, by $h_m z_g^p$, where z_g is the centre of the cross section of the layer. More details on N-T composite shell structures can be found in the original works of NGATCHA et al. [78, 79] and on N composite shell structures in NGATCHA et al. [80] (see also NGATCHA et al. [81]).

6.3 Some Semi-Analytic Models

6.3.1 Models with Rigid Normal Direction

In order to solve some particular problems in which transverse shear stresses are significant, some authors have proposed some specific kinematics which account for shear strains without variation of thickness. These kinematics are defined as follows:

$$\begin{cases} U_\alpha(x^1, x^2, z) = \left(1 + \frac{z}{R_\alpha}\right) u_\alpha(x^1, x^2) - \frac{z}{A_\alpha} \partial_\alpha u_3(x^1, x^2) + w(z)\gamma_\alpha(x^1, x^2) \\ U_3(x^1, x^2, z) = \qquad u_3(x^1, x^2) \\ \gamma_\alpha(x^1, x^2) \ = \qquad \frac{1}{A_\alpha} \frac{\partial u_3(x^1, x^2)}{\partial x^\alpha} + f_\alpha(x^1, x^2) \end{cases} \quad (6.71)$$

where R_α is the radius of curvature in the coordinate direction x^α, A_α the norm of the tangent vector to the coordinate line in the covariant basis and the functions $w(z)$ are given by different authors below:

- Reddy's, Levinson's, Murthy's : $w(z) = z(1 - \frac{4z^2}{3h^2})$
- Touratier's model: $w(z) = \frac{h}{\Pi} sin(\frac{\pi z}{h})$
- Ambartsumian's model: $w(z) = \frac{z}{2}(\frac{h^2}{4} - \frac{z^2}{3})$
- Soldatos's model: $w(z) = hsinh(\frac{z}{h}) - zcosh(\frac{1}{2})$
- Reissner: $w(z) = \frac{5z}{4}(1 - \frac{4z^2}{3h^2})$
- Kamara et al: $w(z) = ze^{-2(\frac{z}{h})^2}$
- Mantari et al. $w(z) = sin(\frac{\pi z}{h})e^{\frac{1}{2}cos(\frac{\pi z}{h})} + \frac{\pi}{2h}z$

The models thus derived depend on four independent functions and their strain tensor satisfies the equations $\epsilon_{\alpha 3} \neq 0, \ \ \epsilon_{33} = 0$.

6.3.2 Models with Higher Order Expansion Terms

The kinematics are defined by a series limited at the $N - th$ order as follows:

$$U(x^1, x^2, z) = (U_1(x^1, x^2, z), U_2(x^1, x^2, z), U_3(x^1, x^2, z))$$

defined by

$$U(x^1, x^2, z) = \sum_{n=0}^{n=N} z^n u^n(x^1, x^2) \tag{6.72}$$

It is a classical expansion. We have to look for the N+1 functions $u^n(x^1, x^2)$. Variation through the thickness is assumed to be polynomial. Certain authors have considered trigonometric polynomials defined as follows:

$$U(x^1, x^2, z) = u^0(x^1, x^2) + \sum_{n=1}^{n=N} sin(\frac{n\pi z}{h}) u^n(x^1, x^2) \tag{6.73}$$

or a mixed polynomial with z and the trigonometric function $sine$

$$U(x^1, x^2, z) = u^0(x^1, x^2) + zu^1(x^1, x^2) + \sum_{n=1}^{n=N} sin(\frac{n\pi z}{h}) u^{n+1}(x^1, x^2) \tag{6.74}$$

or z and the trigonometric function $cosine$

$$U(x^1, x^2, z) = u^0(x^1, x^2) + zu^1(x^1, x^2) + \sum_{n=1}^{n=N} cosin(\frac{n\pi z}{h}) u^{n+1}(x^1, x^2) \tag{6.75}$$

or z and the trigonometric functions $sine, cosine$

$$\begin{aligned} U(x^1, x^2, z) = u^0(x^1, x^2) + zu^1(x^1, x^2) + \sum_{n=1}^{n=N} (sin(\frac{n\pi z}{h}) u^{2n}(x^1, x^2) \\ + cosin(\frac{n\pi z}{h}) u^{2n+1}(x^1, x^2)) \end{aligned} \tag{6.76}$$

or z and an exponential expansion

$$U(x^1, x^2, z) = \sum_{n=0}^{n=N} e^{(nz/h)} u^n(x^1, x^2) \tag{6.77}$$

$$U(x^1, x^2, z) = u^0(x^1, x^2) + zu^1(x^1, x^2) + \sum_{n=1}^{n=N} e^{(nz/h)} u^{n+1}(x^1, x^2) \tag{6.78}$$

These models are often used for stratified shells, depend on very many independent functions, are consequently memory greedy and need great computing capacities. Some inextricable problems often arise from boundary conditions inherently imposed on higher order ($n \geq 1$) functions.

6.4 Cosserat Thick Shells

6.4.1 Description

Certain shells cannot be described correctly by thickening following the normal direction of the mid-surface as usual. In a conical shell , for example, the material is distributed in a direction different from the normal of the mid-surface S. The shell occupies the domain $\Omega = \left\{ OM = Om + zD,\ m \in S,\ -\frac{h}{2} \leq z \leq \frac{h}{2},\ D \in \mathbb{R}^3 \right\}$.
The mid-surface S is generated by a straight segment non-parallel to its vertical axis of revolution and D is the unitary directing vector. Let Ω be a shell obtained by thickening the mid-surface following a direction D as shown in Fig. 6.5 and let the deformed configuration be described as follows:

$$\tilde{\Omega} = \left\{ O\tilde{M} = O\tilde{m} + zd,\ \tilde{m} \in \tilde{S},\ -\frac{h}{2} \leq z \leq \frac{h}{2},\ d \in \mathbb{R}^3 \right\}$$

the directing vector d not necessary unitary. We denote by $\{A_1, A_2, A_3\}$, $\{A^1, A^2, A^3\}$, respectively, the covariant and contravariant bases of the mid-surface S and $\{a_1, a_2, a_3\}$, $\{a^1, a^2, a^3\}$ their counterparts on $\tilde{S}$. We shall use the same notations as in the previous chapters.

Let $D(x) = D^i A_i = D_i A^i$, then we have

$$\begin{aligned} D_{,\alpha} &= (\nabla_\alpha D_\rho - D_3 B_{\alpha\rho}) A^\rho + (\partial_\alpha D_3 + D_\rho B^\rho_\alpha) A^3 = D_{i\alpha} A^i \\ &= (\nabla_\alpha D^\rho - D^3 B^\rho_\alpha) A_\rho + (\partial_\alpha D^3 + D^\rho B_{\alpha\rho}) A_3 = D^i_\alpha A_i \end{aligned} \tag{6.79}$$

Consider a vector $u(x) = u^i A_i = u_i A^i$, we also have

$$\begin{aligned} u_{,\alpha} &= (\nabla_\alpha u_\rho - u_3 B_{\alpha\rho}) A^\rho + (\partial_\alpha u_3 + u_\rho B^\rho_\alpha) A^3 = u_{i\alpha} A^i \\ &= (\nabla_\alpha u^\rho - u^3 B^\rho_\alpha) A_\rho + (\partial_\alpha u^3 + u^\rho B_{\alpha\rho}) A_3 = u^i_\alpha A_i \end{aligned} \tag{6.80}$$

$$M = (X, Y, Z) = (R_x \cos\theta, R_x \sin\theta, x)$$

$$R_x = a - x \tan\theta$$

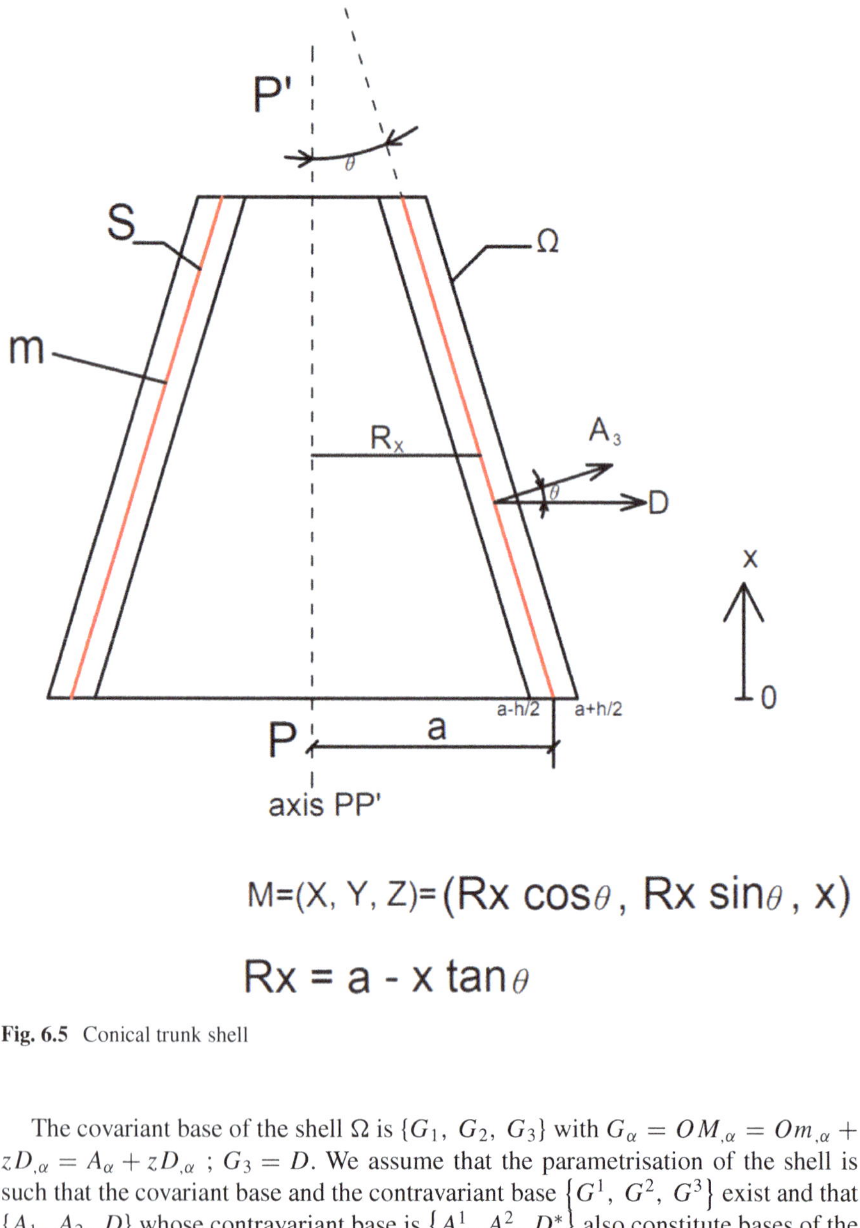

Fig. 6.5 Conical trunk shell

The covariant base of the shell Ω is $\{G_1, G_2, G_3\}$ with $G_\alpha = OM_{,\alpha} = Om_{,\alpha} + zD_{,\alpha} = A_\alpha + zD_{,\alpha}$; $G_3 = D$. We assume that the parametrisation of the shell is such that the covariant base and the contravariant base $\{G^1, G^2, G^3\}$ exist and that $\{A_1, A_2, D\}$ whose contravariant base is $\{A^1, A^2, D^*\}$ also constitute bases of the shell Ω. We also assume that $\{A_1, A_2, A_3\}$ and its dual $\{A^1, A^2, A^3\}$ constitute a base. As in the previous chapters, a vector can be expressed in any base.

We can write the following relations on any vector:

$$U(x,z) = U^i G_i = U_i G^i = \bar{U}^\rho A_\rho + \bar{U}^3 D = \bar{U}_\rho A^\rho + \bar{U}_3 D^* = u^i A_i = u_i A^i$$

We have

$$\bar{U}^\rho A_\rho + \bar{U}^3 D = (\bar{U}^\rho + \bar{U}^3 D^\rho) A_\rho + \bar{U}^3 D^3 A_3 = u^i A_i$$

We therefore deduce that

$$\begin{aligned} u^\rho = \bar{U}^\rho + \bar{U}^3 D^\rho, \quad & u^3 = \bar{U}^3 D^3 = \bar{U}^3 D_3 \\ \bar{U}^3 = \frac{u^3}{D_3}, \quad & \bar{U}^\rho = u^\rho - \frac{D^\rho}{D_3} u^3 \\ \bar{U}_3 = \frac{u_3}{D_3}, \quad & \bar{U}_\rho = u_\rho - \frac{D_\rho}{D_3} u_3 \end{aligned}$$

From the above relations, we choose to express vectors in the bases $\{A_1, A_2, A_3\}$, $\{A^1, A^2, A^3\}$ in the reference configuration or in the bases $\{a_1, a_2, a_3\}$, $\{a^1, a^2, a^3\}$ in the deformed configuration. The directing vector is then $d(x) = d^i a_i = d_i a^i$.

6.4.2 Metric and Strain Tensors

In the reference configuration, we define the covariant components of the metric tensor by

$$\begin{aligned} G_{\alpha\beta} = G_\alpha \cdot G_\beta = & \quad (A_\alpha + z D_{,\alpha}) \cdot (A_\beta + z D_{,\beta}) \\ = & \; A_{\alpha\beta} + z(A_\alpha \cdot D_{,\beta} + A_\beta \cdot D_{,\alpha}) + z^2 D_{,\alpha} \cdot D_{,\beta} \\ = & \quad A_{\alpha\beta} + z(D_{\alpha\beta} + D_{\beta\alpha}) + z^2 D_{i\alpha} D^i_\beta \end{aligned}$$

$$\begin{aligned} G_{\alpha 3} = G_\alpha \cdot D = & \quad (A_\alpha + z D_{,\alpha}) \cdot D \\ = & \; (A_\alpha + z D^i_\alpha A_i) \cdot D_i A^i \\ = & \quad D_\alpha + z D^i_\alpha D_i \end{aligned}$$

$$G_{33} = D \cdot D = D^i D_i$$

On the mid-surface S ($z = 0$) we have

$$\begin{aligned} \left[G_{ij}(x, 0)\right] = & \begin{bmatrix} A_{11} & A_{12} & D_1 \\ A_{21} & A_{22} & D_2 \\ D_1 & D_2 & D \cdot D \end{bmatrix} \\ = & \begin{bmatrix} A_{11} & A_{12} & 0 \\ A_{21} & A_{22} & 0 \\ 0 & 0 & D \cdot D \end{bmatrix} + \begin{bmatrix} 0 & 0 & D_1 \\ 0 & 0 & D_2 \\ D_1 & D_2 & 0 \end{bmatrix} \\ = & \quad M + E = M(I + M^{-1} E) \end{aligned}$$

Therefore

$$\left[G_{ij}(x,0)\right]^{-1} = \left[G^{ij}(x,0)\right] = (I + M^{-1}E)^{-1}M^{-1} \simeq M^{-1} - M^{-1}EM^{-1}$$

$$= \begin{pmatrix} A^{11} & A^{12} & -\dfrac{D^1}{D\cdot D} \\ A^{21} & A^{22} & -\dfrac{D^2}{D\cdot D} \\ -\dfrac{D^1}{D\cdot D} & -\dfrac{D^2}{D\cdot D} & \dfrac{1}{D\cdot D} \end{pmatrix}$$

In the deformed configuration of the shell, the covariant components are

$$\begin{aligned} g_{\alpha\beta} = g_\alpha \cdot g_\beta &= (a_\alpha + zd_{,\alpha}) \cdot (a_\beta + zd_{,\beta}) \\ &= a_{\alpha\beta} + z(a_\alpha \cdot d_{,\beta} + a_\beta \cdot d_{,\alpha}) + z^2 d_{,\alpha} \cdot d_{,\beta} \\ &= a_{\alpha\beta} + z(d_{\alpha\beta} + d_{\beta\alpha}) + z^2 d_{i\alpha} \cdot d^i_\beta \end{aligned}$$

$$g_{\alpha 3} = g_\alpha \cdot d = (a_\alpha + zd_{,\alpha}) \cdot d = (a_\alpha + zd^i_\alpha a_i) \cdot d_i a^i = d_\alpha + zd^i_\alpha d_i$$

$$g_{33} = d \cdot d = d^i d_i$$

Let $\tilde{S}$ be the mid-surface of the deformed shell $\tilde{\Omega}$. We assume it is the image of S in the transformation performed by the displacement vector u, such that $O\tilde{M} = O\tilde{m} + zd,\ \tilde{m} \in \tilde{S},\ \ O\tilde{m} = Om + u$. By letting $d = D + \delta$, we deduce that the displacement vector between the reference (or initial) and the deformed configurations of the shell is $u(x) + z\delta(x) = (u_i(x) + z\delta_i(x))A^i = (u^i(x) + z\delta^i(x))A_i$. It can be remarked that this kinematic reproduces the kinematic of some of the models initially treated: K-L ($\delta_\alpha = -\partial_\alpha u_3,\ \delta_3 = 0$); N-T ($\delta_\alpha = \partial_\alpha u_3 + B^\rho_\alpha u_\rho,\ \delta_3 = 0$); N ($\delta_\alpha = \partial_\alpha u_3 + B^\rho_\alpha u_\rho,\ \delta_3 = \bar{q}(x)\ with\ w(z) = z$); R-M ($\delta_\alpha = \theta_\alpha,\ \delta_3 = 0$).

The small strain tensor $\epsilon = \epsilon_{ij}(x,z)G^i \otimes G^j$ is defined as follows:

$$\begin{aligned} \epsilon_{\alpha\beta}(x,z) &= \frac{1}{2}(g_{\alpha\beta} - G_{\alpha\beta}) \\ &= \frac{1}{2}(a_{\alpha\beta} - A_{\alpha\beta}) + \frac{z}{2}(d_{\alpha\beta} + d_{\beta\alpha} - D_{\alpha\beta} - D_{\beta\alpha}) \\ &\quad + \frac{z^2}{2}(d_{i\alpha}d^i_\beta - D_{i\alpha}D^i_\beta) \end{aligned} \tag{6.81}$$

$$\epsilon_{\alpha 3}(x,z) = \frac{1}{2}(d_\alpha - D_\alpha + z(d^i_\alpha d_i - D^i_\alpha D_i))$$

$$\epsilon_{33}(x,z) = \frac{1}{2}(d^i d_i - D^i D_i)$$

From the previous chapters, we have $\frac{1}{2}(a_{\alpha\beta} - A_{\alpha\beta}) = e_{\alpha\beta} = \frac{1}{2}(\nabla_\alpha u_\beta + \nabla_\beta u_\alpha - 2u_3 B_{\alpha\beta})$. We deduce that $a_{\alpha\beta} = A_{\alpha\beta} + e_{\alpha\beta}$, $a_{..} = A_{..}(I_{..} + \underset{..}{A}^{-1} e_{..})$ and $a^{\alpha\beta} \simeq A^{\alpha\beta} - e^{\alpha\beta}$. In order to express the strain tensor in terms of u and δ we must calculate the following terms:

$$\begin{aligned} d_\alpha = d \cdot a_\alpha = & \quad (D+\delta)\cdot(A_\alpha + u_{,\alpha}) = D_\alpha + \delta_\alpha + D^i u_{i\alpha} + \delta^i u_{i\alpha} \\ & \simeq D_\alpha + \delta_\alpha + D^i u_{i\alpha} = D_\alpha + \delta_\alpha + D_i u^i_\alpha = D_\alpha + \vartheta_\alpha(u,\delta); \end{aligned} \tag{6.82}$$

$$\begin{aligned} d^\alpha = a^{\alpha\rho} d_\rho \simeq & \quad (A^{\alpha\rho} - e^{\alpha\rho}) d_\rho \\ \simeq & \; A^{\alpha\rho} D_\rho + A^{\alpha\rho}\vartheta_\rho - e^{\alpha\rho} D_\rho - e^{\alpha\rho}\vartheta_\rho \\ \simeq & \quad D^\alpha + \vartheta^\alpha \end{aligned} \tag{6.83}$$

$$\begin{aligned} d_3 = d^3 = d \cdot a_3 = & \quad (D+\delta)\cdot(A_3 - u_{3\alpha} A^\alpha) \\ = & \quad D_3 + \delta_3 - D^\alpha u_{3\alpha} + \delta^\alpha u_{3\alpha} \\ \simeq & \; D_3 + \delta_3 - D^\alpha u_{3\alpha} = D_3 + \vartheta_3(u,\delta) \end{aligned} \tag{6.84}$$

$$\begin{aligned} d_{\alpha\beta} = d_{,\beta} \cdot a_\alpha = & \quad (D+\delta)_{,\beta}\cdot(A_\alpha + u_{,\alpha}) \\ = & \; (D_{i\beta} A^i + \delta_{i\beta} A^i)\cdot(A_\alpha + u_{,\alpha}) \\ = & \quad D_{\alpha\beta} + \delta_{\alpha\beta} + D_{i\beta} u^i_\alpha + \delta_{i\beta} u^i_\alpha \\ \simeq & \quad D_{\alpha\beta} + \delta_{\alpha\beta} + D_{i\beta} u^i_\alpha \end{aligned} \tag{6.85}$$

$$\begin{aligned} d_{3\beta} = d_{,\beta} \cdot a_3 = & \quad (D+\delta)_{,\beta}\cdot(A_3 - u_{3\alpha} A^\alpha) \\ = & \; (D_{i\beta} A^i + \delta_{i\beta} A^i)\cdot(A_3 - u_{3\alpha} A^\alpha) \\ = & \quad D_{3\beta} + \delta_{3\beta} - D^\rho_\beta u_{3\rho} - \delta^\rho_\beta u_{3\rho} \\ \simeq & \quad D_{3\beta} + \delta_{3\beta} - D^\rho_\beta u_{3\rho} \end{aligned} \tag{6.86}$$

$$\begin{aligned} d^\rho_\beta = \nabla_\beta d^\rho - d^3 b^\rho_\beta = & \quad \nabla_\beta(a^{\rho\gamma} d_\gamma) - d^3 a^{\rho\gamma} b_{\gamma\beta} \\ = & \quad a^{\rho\gamma}(\nabla_\beta d_\gamma - d_3 b_{\gamma\beta}) \\ \simeq & \; (A^{\rho\gamma} - e^{\rho\gamma})(D_{\gamma\beta} + \delta_{\gamma\beta} + D^i_\beta u_{i\gamma}) \\ \simeq & \quad (D^\rho_\beta + \delta^\rho_\beta + A^{\rho\gamma} D^i_\beta u_{i\gamma}) \end{aligned} \tag{6.87}$$

$$\begin{aligned} d^i d_i = & \quad d^\rho d_\rho + d^3 d_3 = (A^{\rho\gamma} - e^{\rho\gamma}) d_\gamma d_\rho + d_3 d_3 \\ = & \; (A^{\rho\gamma} - e^{\rho\gamma})(D_\gamma + \vartheta_\gamma)(D_\rho + \vartheta_\rho) + (D_3 + \vartheta_3)(D_3 + \vartheta_3) \\ \simeq & \quad D^i D_i + 2 D^i \vartheta_i \end{aligned} \tag{6.88}$$

$$\begin{aligned} d_{\rho\alpha} d^\rho_\beta = & \; (D_{\rho\alpha} + \delta_{\rho\alpha} + D^i_\alpha u_{i\rho})(D^\rho_\beta + \delta^\rho_\beta + A^{\rho\gamma} D^i_\beta u_{i\gamma}) \\ \simeq & \quad D_{\rho\alpha} D^\rho_\beta + D_{\rho\alpha}\delta^\rho_\beta + \delta_{\rho\alpha} D^\rho_\beta + D^\rho_\beta D^i_\alpha u_{i\rho} \\ d_{3\alpha} d^3_\beta = & \quad (D_{3\alpha} + \delta_{3\alpha} - D^\rho_\alpha u_{3\rho})(D_{3\beta} + \delta_{3\beta} - D^\rho_\beta u_{3\rho}) \\ \simeq & \; D_{3\alpha} D^3_\beta + D_{3\alpha}\delta_{3\beta} + D_{3\beta}\delta_{3\alpha} - D_{3\alpha} D^\rho_\beta u^3_\rho - D_{3\beta} D^\rho_\alpha u^3_\rho \end{aligned} \tag{6.89}$$

We calculate $\epsilon_{ij}(u,\delta)$ as follows:

$$\epsilon_{33}(u,\delta) = \frac{1}{2}(d^i d_i - D^i D_i) = D^i \vartheta_i = \varepsilon_3(u,\delta)$$

$$\begin{aligned}\epsilon_{\alpha 3}(u,\delta) = & \frac{1}{2}(d_\alpha - D_\alpha + z(d^i_\alpha d_i - D^i_\alpha D_i)) \\ = & \frac{1}{2}(\vartheta_\alpha + z(D^i_\alpha \vartheta_i + \delta^i_\alpha D_i)) \\ = & \frac{1}{2}(\vartheta_\alpha + z(D^i_\alpha \vartheta_i + \delta_{i\alpha} D^i)) \\ = & \frac{1}{2}(\vartheta_\alpha(u,\delta) + z\varepsilon_\alpha(u,\delta))\end{aligned}$$

Let us evaluate the terms in $\epsilon_{\alpha\beta}(u,\delta)$. It contains $e_{\alpha\beta}(u)$ which is the change in the first fundamental form. We have

$$\begin{aligned}D^i_\beta u_{i\alpha} = & D^\rho_\beta u_{\rho\alpha} + D^3_\beta u_{3\alpha} \\ = & (\nabla_\beta D^\rho - D^3 B^\rho_\beta)(\nabla_\alpha u_\rho - u_3 B_{\alpha\rho}) \\ & +(\partial_\beta D^3 + D^\rho B_{\beta\rho})(\partial_\alpha u_3 + u_\rho B^\rho_\alpha)\end{aligned}$$

This expression contains $-D^3 B^\rho_\beta(\nabla_\alpha u_\rho - u_3 B_{\alpha\rho})$ and

$$\partial_\beta D^3(\partial_\alpha u_3 + u_\rho B^\rho_\alpha) = \nabla_\beta(D^3(\partial_\alpha u_3 + u_\rho B^\rho_\alpha)) - D^3\nabla_\beta(\partial_\alpha u_3 + u_\rho B^\rho_\alpha).$$

Therefore, this term contains $-D^3 B^\rho_\beta(\nabla_\alpha u_\rho - u_3 B_{\alpha\rho}) - D^3\nabla_\beta(\partial_\alpha u_3 + u_\rho B^\rho_\alpha) = -D_3 K_{\beta\alpha}$. Also $D^i_\alpha u_{i\beta}$ contains $-D_3 K_{\alpha\beta}$. We can now write

$$\begin{aligned}\frac{z}{2}(\delta_{\alpha\beta} + D^i_\beta u_{i\alpha} + \delta_{\beta\alpha} + D^i_\alpha u_{i\beta}) = & -zD_3 K_{\alpha\beta}(u) - zD_3 K^c_{\alpha\beta}(u,\delta) \\ = & -zD_3(K_{\alpha\beta}(u) + K^c_{\alpha\beta}(u,\delta)) \\ = & -zD_3 \Bbbk_{\alpha\beta}(u,\delta)\end{aligned}$$

The term $\frac{z^2}{2}(D^\rho_\alpha \delta_{\rho\beta} + D^\rho_\beta \delta_{\rho\alpha} + D^\rho_\beta D^i_\alpha u_{i\rho} + D_{3\alpha}\delta_{3\beta} + D_{3\beta}\delta_{3\alpha} - D_{3\alpha} D^\rho_\beta u^3_\rho - D_{3\beta} D^\rho_\alpha u^3_\rho)$ can be rewritten $z^2 D^2_3 Q^c_{\alpha\beta}(u,\delta)$. It can be remarked that by taking $D = A_3$ and $d = a_3$ then $\delta = -(\partial_\rho u_3 + u_\gamma B^\gamma_\rho)A^\rho = -\theta_\rho A^\rho$ and $\delta_3 = 0$. We have

$$D^\rho_\alpha \delta_{\rho\beta} = B^\rho_\alpha \nabla_\beta \theta_\rho, \quad D_{3\alpha}\delta_{3\beta} = D_{3\beta}\delta_{3\alpha} = 0, \quad D^\rho_\beta \delta_{\rho\alpha} = B^\rho_\beta \nabla_\alpha \theta_\rho \ \textit{and}$$

$$\frac{z^2}{2}(D^\rho_\alpha \delta_{\rho\beta} + D^\rho_\beta \delta_{\rho\alpha}) = z^2 Q_{\alpha\beta}(u)$$

We deduce that the term in z^2 is similar to the change in the third fundamental form. We write

$$\begin{aligned} D_3^2 Q^c_{\alpha\beta} = \ & (D^\rho_\alpha \delta_{\rho\beta} + D^\rho_\beta \delta_{\rho\alpha} + D_{3\alpha}\delta_{3\beta} + D_{3\beta}\delta_{3\alpha})/2 \\ & +(-D_{3\alpha} D^\rho_\beta u^3_\rho - D_{3\beta} D^\rho_\alpha u^3_\rho + D^\rho_\beta D^i_\alpha u_{i\rho})/2 \end{aligned}$$

The last term is neglected in the calculation of the change in the third fundamental form. In order to be consistent with the previous chapters, it will also be left out and finally we obtain

$$\begin{aligned} \epsilon_{\alpha\beta}(u,\delta) &= e_{\alpha\beta}(u) - zD_3 \Bbbk_{\alpha\beta}(u,\delta) + z^2 D_3^2 Q^c_{\alpha\beta}(u,\delta) \\ Q^c_{\alpha\beta}(u,\delta) &= (D^\rho_\alpha \delta_{\rho\beta} + D^\rho_\beta \delta_{\rho\alpha} + D_{3\alpha}\delta_{3\beta} + D_{3\beta}\delta_{3\alpha})/2D_3^2 \\ \Bbbk_{\alpha\beta}(u,\delta) &= -(\delta_{\alpha\beta} + \delta_{\beta\alpha} + D^i_\beta u_{i\alpha} + D^i_\alpha u_{i\beta})/2D_3 \\ e_{\alpha\beta}(u) &= \frac{1}{2}(\nabla_\alpha u_\beta + \nabla_\beta u_\alpha - 2u_3 B_{\alpha\beta}) \end{aligned} \tag{6.90}$$

We remark in the transverse strains $\epsilon_{\alpha 3}(u,\delta)$ and $\epsilon_{33}(u,\delta)$ the terms $\vartheta_\alpha(u,\delta)$ and $\vartheta_3(u,\delta)$ which are found in Cosseratats' theory, without the ε_i. The transverse strains are rewritten as follows:

$$\begin{aligned} \epsilon_{\alpha 3}(u,\delta) &= \frac{1}{2}(\vartheta_\alpha + z\varepsilon_\alpha) \\ \epsilon_{33}(u,\delta) &= \varepsilon_3 \\ \vartheta_\alpha(u,\delta) = \delta_\alpha + D_i u^i_\alpha, \ & \vartheta_3(u,\delta) = \delta_3 - D^\alpha u_{3\alpha} \\ \varepsilon_\alpha(u,\delta) = (D^i_\alpha \vartheta_i + \delta_{i\alpha} D^i), \ & \varepsilon_3(u,\delta) = D^i \vartheta_i \end{aligned} \tag{6.91}$$

6.4.3 Equilibrium Equations

As usual we reconsider Euler's equations of the model problem:

$$\begin{gathered} \sigma^{ij}_{/j} + f^i = 0 \text{ in } \Omega \\ \sigma\vec{n} = p \ on \ \Gamma_- \cup \Gamma_+ \cup \bar{\gamma} \times [-\tfrac{h}{2}, \tfrac{h}{2}]; \ \Gamma_- = S \times \{-\tfrac{h}{2}\}; \\ \Gamma_+ = S \times \{\tfrac{h}{2}\}; \ \gamma \text{ is a part of } \partial S; \\ U = 0 \text{ on } \Gamma_0; \ \Gamma_0 = \bar{\gamma}_0 \times [-\tfrac{h}{2}, \tfrac{h}{2}] \ \bar{\gamma}_0 \\ \text{is the complementary part of } \bar{\gamma} \text{ of } \partial S \\ \sigma^{ij} = (\bar{\lambda} G^{ij} G^{kl} + 2\bar{\mu} G^{ik} G^{jl})\epsilon_{kl}(u,\delta) \\ = A^{ijkl}(x,z)\epsilon_{kl}(u,\delta) \end{gathered} \tag{6.92}$$

No displacement on the border implies that $u = \delta = 0$ on $\bar{\gamma}_0$. We deduce from the strain tensor that components of the displacement vector must satisfy the conditions $u_\alpha, \delta_\alpha \in H^1_{\bar{\gamma}_0}(S)$, $u_3, \delta_3 \in H^2(S) \cap H^1_{\bar{\gamma}_0}(S)$. We deduce from the expression of the strain tensor that $d\epsilon_{kl}(u,\delta) = \epsilon_{kl}(du, d\delta)$. Let $du = v$ and $d\delta = w$. The following approximations will be adopted:

$$\begin{gathered} d\Omega = (G_1, G_2, D)dxdz \simeq (A_1, A_2, D)dxdz \\ = D_3(A_1, A_2, A_3)dxdz = D_3 dSdz \\ G^{ij}(x,z) \simeq G^{ij}(x,0) \\ G^{\alpha\beta}(x,0) = A^{\alpha\beta}, \quad G^{\alpha 3}(x,0) = -D^{\alpha}/D\cdot D, \quad G^{33}(x,0) = 1/D\cdot D \\ A^{ijkl}(x,z) = A^{ijkl}(x,0) = A^{ijkl}(x) \\ \sigma^{ij}(u,\delta) = (\bar{\lambda}G^{ij}(x,0)G^{kl}(x,0) + 2\bar{\mu}G^{ik}(x,0)G^{jl}(x,0))\epsilon_{kl}(u,\delta) \\ = A^{ijkl}(x)\epsilon_{kl}(u,\delta) \end{gathered}$$

Let $U_{ad} = H^1_{\gamma_0}(S) \times H^1_{\gamma_0}(S) \times H^2(S) \cap H^1_{\gamma_0}(S)$. The variational formulation reads

$$\begin{cases} \text{find } (u,\delta) \in U_{ad} \times U_{ad} \text{ such that} \\ \int\limits_{\Omega} \sigma^{ij}(u,\delta)\epsilon_{ij}(v,w)d\Omega = \int\limits_{S}\int\limits_{-h/2}^{h/2} A^{ijkl}(x)\epsilon_{kl}(u,\delta)\epsilon_{ij}(v,w)D_3 dzdS \\ \qquad = \quad L(v,w) \quad \forall (v,w) \in U_{ad} \times U_{ad} \end{cases} \tag{6.93}$$

The external virtual work $L(v,w)$ is expressed as follows:

$$L(v,w) = \int\limits_{S} \left(\bar{p}^i v_i + \tilde{p}^i w_i\right) dS + \int\limits_{\gamma_1} \left(\bar{q}^i v_i + \tilde{q}^3 w_3\right) d\gamma + \int\limits_{\gamma_1} \tilde{m}^{\alpha} w_{\alpha} d\gamma$$

The internal virtual work is

$$\begin{aligned} \int\limits_{\Omega} \sigma^{ij}(u,\delta)\epsilon_{ij}(v,w)d\Omega = \int\limits_{\Omega} \sigma^{\alpha\beta}(u,\delta)\epsilon_{\alpha\beta}(v,w)d\Omega + 2\int\limits_{\Omega} \sigma^{\alpha 3}(u,\delta)\epsilon_{\alpha 3}(v,w)d\Omega \\ + \int\limits_{\Omega} \sigma^{33}(u,\delta)\epsilon_{33}(v,w)d\Omega \end{aligned}$$

By evaluating the different terms as in the previous chapters, we obtain the different tensors $N = (N^{\alpha\beta})$, $M = (M^{\alpha\beta})$, $M^* = (M^{*\alpha\beta})$, $T = (T^{\alpha})$, $\tilde{T} = (\tilde{T}^{\alpha})$, H, such that

$$\begin{aligned} \int\limits_{\Omega} \sigma^{\alpha\beta}(u,\delta)\epsilon_{\alpha\beta}(v,w)d\Omega &= \int\limits_{S}\int\limits_{-h/2}^{h/2} (A^{\alpha\beta\rho\gamma}(x)\epsilon_{\rho\gamma}(u,\delta) \\ &\quad + 2A^{\alpha\beta\rho 3}(x)\epsilon_{\rho 3}(u,\delta) + A^{\alpha\beta 33}(x)\epsilon_{33}(u,\delta))\epsilon_{\alpha\beta}(v,w)D_3 dzdS \\ &= \int\limits_{S} (N^{\alpha\beta}e_{\alpha\beta}(v) + M^{\alpha\beta}\mathbb{k}_{\alpha\beta}(v,w) + M^{*\alpha\beta}Q^c_{\alpha\beta}(v,w))D_3 dS \end{aligned}$$

We also have

$$
\begin{aligned}
2\int_{\Omega}\sigma^{\alpha 3}(u,\delta)\epsilon_{\alpha 3}(v,w)d\Omega = & \int_{S} 2\int_{-h/2}^{h/2}((A^{\alpha 3\rho\gamma}(x)\epsilon_{\rho\gamma}(u,\delta) \\
& +2A^{\alpha 3\rho 3}(x)\epsilon_{\rho 3}(u,\delta)+A^{\alpha 333}(x)\epsilon_{33}(u,\delta))\epsilon_{\alpha 3}(v,w)D_3 dz dS \\
= & \int_{S}(T^{\alpha}\vartheta_{\alpha}(v,w)+\tilde{T}^{\alpha}\varepsilon_{\alpha}(v,w))D_3 dS
\end{aligned}
\tag{6.94}
$$

and

$$
\begin{aligned}
\int_{\Omega}\sigma^{33}(u,\delta)\epsilon_{33}(v,w)d\Omega = & \int_{S}\int_{-h/2}^{h/2}((A^{33\rho\gamma}(x)\epsilon_{\rho\gamma}(u,\delta)+2A^{33\rho 3}(x)\epsilon_{\rho 3}(u,\delta) \\
& +A^{3333}(x)\epsilon_{33}(u,\delta))\epsilon_{33}(v,w)D_3 dz dS \\
= & \int_{S} H\varepsilon_3(v,w)D_3 dS
\end{aligned}
\tag{6.95}
$$

Finally, the variational equation is equivalent to

$$
\begin{cases}
\text{Find } (u,\delta)\in U_{ad}\times U_{ad} \text{ such that} \\
E((u,\delta);(v,w)) = \int_{S}(N^{\alpha\beta}(u,\delta)e_{\alpha\beta}(v)+M^{\alpha\beta}(u,\delta)\Bbbk_{\alpha\beta}(v,w) \\
\quad +M^{*\alpha\beta}(u,\delta)Q^{c}_{\alpha\beta}(v,w))D_3 dS+\int_{S}(T^{\alpha}(u,\delta)\vartheta_{\alpha}(v,w) \\
\quad +\tilde{T}^{\alpha}(u,\delta)\varepsilon_{\alpha}(v,w))D_3 dS+\int_{S} H(u,\delta)\varepsilon_3(v,w)D_3 dS \\
\quad =\int_{S}\left(\bar{p}^{i}v_i+\tilde{p}^{i}w_i\right)dS+\int_{\gamma_1}\left(\bar{q}^{i}v_i+\tilde{q}^{3}w_3\right)d\gamma+\int_{\gamma_1} m^{\alpha}w_{\alpha}d\gamma \\
\quad = L(v,w)\quad \forall (v,w)\in U_{ad}\times U_{ad}
\end{cases}
\tag{6.96}
$$

The final variational problem obtained satisfies all the necessary conditions for the existence of a unique solution because

$$
E((u,\delta);(v,w))=\int_{\Omega}\sigma^{ij}(u,\delta)\epsilon_{ij}(v,w)d\Omega=\int_{\Omega}A^{ijkl}(x)\epsilon_{kl}(u,\delta)\epsilon_{ij}(v,w)d\Omega
$$

and the external virtual work is assumed to be a continuous linear form. Let us remind that, in general, the virtual vector fields v, w belong to the tangent space of U_{ad}. In this example both spaces are equal. The terms $N=(N^{\alpha\beta})$, $M=(M^{\alpha\beta})$, $M^{*}=(M^{*\alpha\beta})$, $T=(T^{\alpha})$, $\tilde{T}=(\tilde{T}^{\alpha})$ and H depend on $e_{\alpha\beta}(u)$, $\Bbbk_{\alpha\beta}(u,\delta)$, $Q^{c}_{\alpha\beta}(u,\delta)$, $\vartheta_i(u,\delta)$, $\varepsilon_i(u,\delta)$.
This model accounts for transverse shear strains and for a homogenous thickness change (ϵ_{33} is z-independent). The N model also accounts for shear strains with a through the thickness distributed pinch strain which depends on z. Heuristic models based on formal higher order expansion terms account for transverse strains but the successive terms hardly satisfy boundary conditions. The R-M model and its various variants account for transverse shear strains but under invariable thickness. The N-T model which is rigorously justified does not account for transverse strains but calculates transverse stresses as reactions due to plane strain constraints. The K-L model which is also rigorously justified and suitable for thin shells does not address transverse strain or stress.

Appendix A
Brief Introduction to Three-Dimensional (3D) Linear Elasticity

The main objective of this appendix is to let the reader revise or acquire minimum notions in 3D linear elasticity, in order to be familiar with some terms, notations and especially variational principles which are frequently used in the text book. To avoid lengthiness the reader will be directed to specialized documents for more details. A short overview of continuum (solid) mechanics will precede the linear elasticity theory in a cartesian coordinates system. We start with definitions, transformation and measurements, i.e. line element, surface element, volume element, strain tensor and some examples. In the next section, from the fundamental equations of dynamics, we shall establish Euler's equations and variational equations of equilibrium. The linear elastic theory, a few examples of problems and a brief introduction to curvilinear media will be devoted to the last section.

A.1 Transformation of a Continuous Media

A.1.1 Definition

A media is said to be a continuum if during its transformation two different configurations (or shapes) can be matched point-wise by a bijection. This means that during transformation no particle is lost (Fig. A.1).

Let M^0 and M denote the position vector of a point in a reference and the deformed configurations, respectively. In a cartesian coordinate system, we denote by $x = (x_1, x_2, x_3)$ and $y = (y_1, y_2, y_3)$ their respective coordinates and suppose they are matched by a function φ such that $y = \varphi(x)$, then the deformation φ is bijective. The displacement vector is defined by $\varphi(x) = x + u(x)$. We suppose the deformation function is differentiable. Then we can calculate $F_{ij} = \partial\varphi_i/\partial x_j$ that we shall also denote $\varphi_{i,j}$. The matrix $F = (F_{ij})$ is called the gradient of the deformation and is also denoted $F = \nabla\varphi = I + \nabla u$ where I is the identity matrix. Because the deformation φ is a bijection $det F \neq 0$. We shall only consider that $det F \succ 0$.

R. Nzengwa, *Plate and Shell Models*,
https://doi.org/10.1007/978-981-97-2780-3

Fig. A.1 Position vector before and after transformation

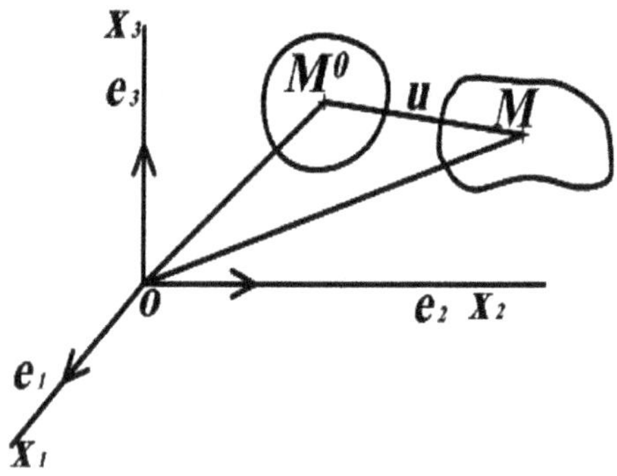

A.1.2 Examples

1. $\varphi(x) = Qx + a \quad Q = \begin{bmatrix} cos\alpha & sin\alpha & 0 \\ -sin\alpha & cos\alpha & 0 \\ 0 & 0 & 1 \end{bmatrix}, \quad a \in \mathbb{R}^3$

2. $\varphi(x) = Fx + a$, F is a constant matrix and $detF \succ 0$

Exercise 1

Determine the displacement vectors u for the above examples. Suppose $cos\alpha \approx 1$, $sin\alpha \approx \alpha$, the cartesian reference is $\{O;\ e_1, e_2, e_3\}$, O is the origin. Calculate $-\alpha e_3 \times OM^0$ and compare with the first displacement, the symbol $\times$ is the vector product. Calculate $\nabla\varphi$ of the above examples.

The gradient of the displacement vector can be decomposed just like any matrix in two parts as follows:

$\nabla u = \frac{1}{2}(\nabla u +^t \nabla u) + \frac{1}{2}(\nabla u -^t \nabla u) = sym(\nabla u) + asym(\nabla u)$

where ${}^t\nabla u$ is the transpose matrix of ∇u, $sym(\nabla u)$ is the symmetric component and $asym(\nabla u)$ the anti-symmetric component.

Transportation

Let Ω^0 be a domain occupied by a continuum media and $\gamma_0 : \lambda \in [0, a] \subset \mathbb{R} \rightarrow \Omega^0$ a curve at least differentiable, of class C^1 for example. Then $\frac{d\gamma_0}{d\lambda}(\lambda) = X^0$ is a tangent vector to the curve. Then $\gamma = \varphi \circ \gamma_0 : \lambda \in [0, a] \rightarrow \Omega$ is the transported curve and $\frac{d\gamma}{d\lambda}(\lambda) = \nabla\varphi(\lambda)X^0 = FX^0 = X$ is the transported tangent vector in the deformed configuration. So $\nabla\varphi = F$ transforms a tangent vector of the reference configuration into a tangent vector in the deformed configuration.

Consider a volume element made of three independent tangent vectors X^0, Y^0 and Z^0, the volume element is the mixed product (X^0, Y^0, Z^0). The transported volume is

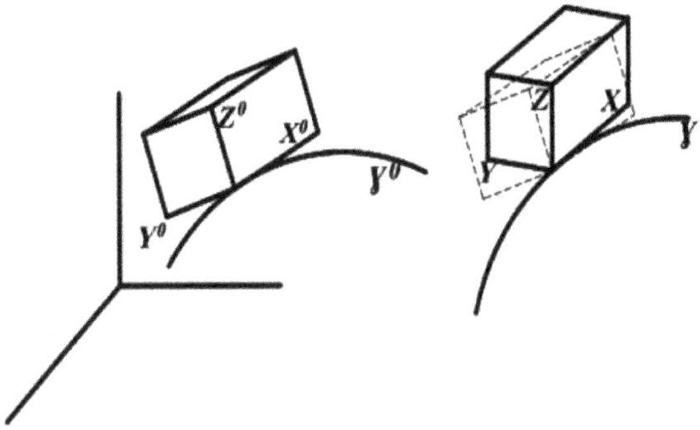

Fig. A.2 Transported objects

$(X, Y, Z) = (FX^0, FY^0, FZ^0) = det F(X^0, Y^0, Z^0) = J(X^0, Y^0, Z^0)$. So J measures the change of volume and the media is incompressible if during transformation $J = 1$.

Let A^0 be an area element constructed with two independent tangent vectors $A^0 = X^0 \times Y^0$. The transported area element is

$$A = X \times Y = FX^0 \times FY^0 \tag{A.1}$$

Let $Z = FZ^0$, then

$$(X, Y, Z) = Z \cdot X \times Y = (FX^0, FY^0, FZ^0) = J(X^0, Y^0, Z^0) = JZ^0 \cdot X^0 \times Y^0 \tag{A.2}$$

$$= JZ^0 \cdot A^0 = JF^{-1}Z \cdot A^0 = Z \cdot J^t F^{-1} A^0$$

Therefore, $Z \cdot X \times Y = Z \cdot J^t F^{-1} A^0$ and it follows that the transported area is $A = X \times Y = J^t F^{-1} A^0$. It should be remarked that $J^t F^{-1} = Cof F$, where $Cof F$ is the cofactor of the matrix F. Let $\| A^0 \| = dS^0$ then we can write $A^0 = \overrightarrow{n_0} dS^0$ where $\overrightarrow{n_0}$ is a unit vector. The transported area vector will also be written as $A = \overrightarrow{n}\, dS = Cof F \overrightarrow{n_0} dS^0$ (Fig. A.2).

Strain tensor

Let X^0 and $X = FX^0$ be, respectively, a tangent vector and its transported counterpart in the deformed configuration, their respective small line elements are defined by $dM^0 = d\lambda X^0$ and $dM = d\lambda X$. The square of their norms is $dM^0 \cdot dM^0$ and $dM \cdot dM = FdM^0 \cdot FdM^0 =^t FFdM^0 \cdot dM^0 = CdM^0 \cdot dM^0$. We also have

$dM \cdot dM - dM^0 \cdot dM^0 = (C - I)dM^0 \cdot dM^0$

The tensors

$C =^t FF$ and $E = \dfrac{1}{2}(C - I) = \frac{1}{2}\left({}^t(I + \nabla u)(I + \nabla u) - I\right) = \dfrac{1}{2}\left(\nabla u + {}^t\nabla u + {}^t\nabla u \nabla u\right)$

called, respectively, the Cauchy-Green tensor and Green-Saint Venant or Green-Lagrange tensor, which are all symmetric tensors, measure the length change of line elements after deformation. There is no change of length if $C = I$ or $E = O$. We also have $det(C) = (det(F))^2 = J^2 \succ 0$ and the Green-Lagrange tensor can be expressed as the sum of a linear part and a nonlinear component by

$$E(u) = \frac{1}{2}\left(\nabla u + {}^t\nabla u\right) + \frac{1}{2}\left({}^t\nabla u \nabla u\right) = \epsilon(u) + q(u) \tag{A.3}$$

The linear part $\epsilon(u) = \frac{1}{2}\left(\nabla u + {}^t\nabla u\right)$ also called small strain or infinitesimal deformation strain is devoted to linear elasticity analysis.

Theorem

$E = O$ is equivalent to φ is a rigid movement, i.e. $\varphi(x) = Qx + b$, $Q^t Q = {}^t QQ = I$, b is a constant vector.

The Green-Lagrange tensor characterizes length change and also angle change between two tangent vectors. In fact, let X^0 and Y^0 be initially perpendicular in the reference configuration, then

$$\frac{\| X \|^2}{\| X^0 \|^2} = \frac{CX^0 \cdot X^0}{\| X^0 \|^2} = (I + 2E)e_1 \cdot e_1 = 1 + 2E_{11}; \tag{A.4}$$

$$\begin{aligned} \frac{\| X \|^2 - \| X^0 \|^2}{\| X^0 \|^2} &= 2E_{11} = \frac{(\| X \| - \| X^0 \|)}{\| X^0 \|} \frac{(\| X \| + \| X^0 \|)}{\| X^0 \|} \\ &\approx \frac{(\| X \| - \| X^0 \|)}{\| X^0 \|}(2 + E_{11}) \approx 2\frac{(\| X \| - \| X^0 \|)}{\| X^0 \|} \end{aligned} \tag{A.5}$$

and

$$E_{11} \approx \frac{(\| X \| - \| X^0 \|)}{\| X^0 \|} \tag{A.6}$$

We also have

$$\begin{aligned} X \cdot Y &= \| X \| \| Y \| \, sin\theta = \left\{\left(CX^0 \cdot X^0\right) \cdot \left(CY^0.Y^0\right)\right\}^{\frac{1}{2}} sin\theta \\ &= \| X^0 \| \| Y^0 \| \, \{(Ce_1 \cdot e_1) \cdot (Ce_2.e_2)\}^{\frac{1}{2}} sin\theta = 2 \| X^0 \| \| Y^0 \| \, E_{12} \end{aligned} \tag{A.7}$$

where $e_1 = X^0 / \| X^0 \|$, $e_2 = Y^0 / \| Y^0 \|$ are unitary vectors and $\pi/2 - \theta$ is the angle between the transported tangent vectors.
Therefore,

$$\{(1 + 2E_{11}) \cdot (1 + 2E_{12})\}^{\frac{1}{2}} sin\theta \approx \theta = 2E_{12} \text{ and } E_{12} = E_{21} = \theta/2 \tag{A.8}$$

Recall that components of $E(x)$ have no unit. So E_{21} is half the total distortion or right angle deviation in the plane generated by the vectors e_1 and e_2. If $E_{ij} = 0$ then right angles remain unchanged in the tangent space generated by Fe_i and Fe_j while

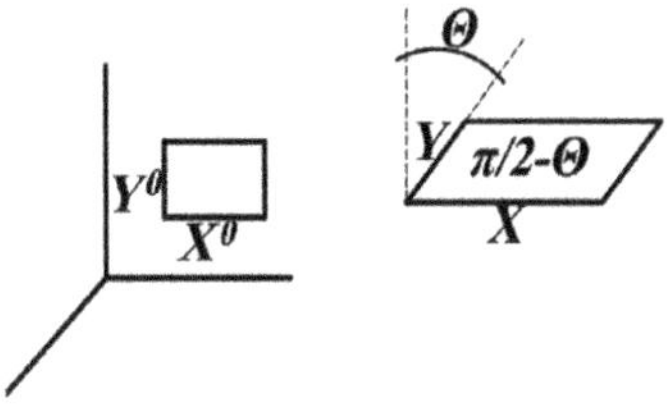

Fig. A.3 Length and angle change

$E_{ii} = 0$ (without index summation) means the movement is rigid in the direction e_i (Fig. A.3).

The tensor$E$ can be decomposed in a spherical strain E^s and a deviatoric E^d as follows:

$$E = E^s + E^d, \ E^s = eI, \ e = trE/3, \ E^d = E - eI \tag{A.9}$$

The spherical strain describes a volumetric deformation, i.e. a uniform elongation if e is positive or a uniform shortening if e is negative. The deviatoric part satisfies the relation $trE^d = 0$ and describes distortion.

Diagonalization of $E(x)$

The Green-Lagrange tensor is symmetric. So it can be diagonalized at any point in an orthonormal base by solving

$$det\,(E - \lambda_i I) = 0 \tag{A.10}$$

and determining the unitary eigenvectors (or principal directions) e_i, solution of $Ee_i = \lambda_i e_i$ (no summation). In these directions, the deformation is locally either a relative elongation ($\lambda_i \succ 0$) or shortening ($\lambda_i \prec 0$) or rigid deformation ($\lambda_i = 0$). In any other direction of director vector $\vec{n}$, the deformation vector $E\vec{n}$ is a combination of distortion and the previous deformations. The deformation vector can be decomposed by $E\vec{n} = e\vec{n} + d\vec{t}$ where $\vec{n}$ and $\vec{t}$ are unitary perpendicular vectors, $e = E\vec{n}.\vec{n}$ is positive (elongation), negative (shortening) or zero (rigid) in the $\vec{n}$ direction and $d = E\vec{n}.\vec{t}$ measures the distortion in the plane perpendicular to $\vec{n}$ (Fig. A.4).

The Eigenvalues are Roots of the Equation

$$-\lambda^3 + E_I\lambda^2 - E_{II}\lambda + E_{III} = 0 \tag{A.11}$$

where $E_I = \lambda_1 + \lambda_2 + \lambda_3 = traceE$, $E_{II} = \lambda_1\lambda_2 + \lambda_1\lambda_3 + \lambda_2\lambda_3 = trace\,(cCof(E)) = \frac{1}{2}\left((trE)^2 - trE^2\right)$ and $E_{III} = det(E) = \left[(trE)^3 - 3trE \cdot trE^2 + 2trE^3\right]/6 = \lambda_1\lambda_2\lambda_3$ are the first, second and third invariants of $E\,(x)$.

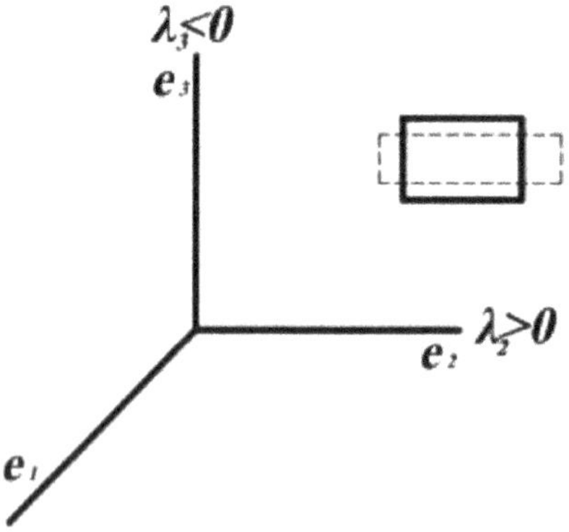

Fig. A.4 Deformation in eigen directions

Plane Strain

Let $x = (x_1, x_2, x_3)$ be the coordinate of a generic point in the reference configuration in an orthonormal base $\{e_1, e_2, e_3\}$. The strain tensor is said to be a plane, perpendicular to the e_3 direction if

$$E = \begin{bmatrix} E_{11} & E_{12} & 0 \\ E_{21} & E_{22} & 0 \\ 0 & 0 & 0 \end{bmatrix} \tag{A.12}$$

It means that there is no dimension change in the e_3 direction and no distortion or right angle deviation in the planes generated by $\{e_1, e_3\}$ and $\{e_2, e_3\}$. On the contrary, there is dimension change and distortion in the plane generated by $\{e_1, e_2\}$. There is automatically plane deformation if the displacement depends only on (x_1, x_2) and the third component is zero $(u = (u_1(x_1, x_2),\ u_2(x_1, x_2), u_3 = 0))$ (Fig. A.5).

Exercise 2

Consider the displacement defined component-wise by $u_i = a_j x_j + d_i$. Calculate $E(u)$

Calculate $E(u)$ of the deformations $\varphi(x) = Qx + a$, and $\varphi(x) = Fx + a$

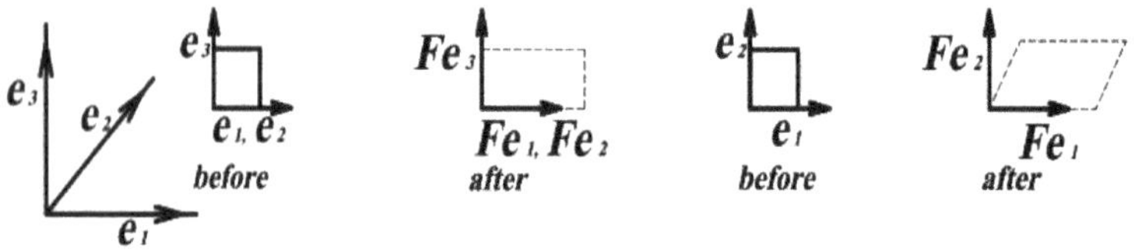

Fig. A.5 Interpretation of plane strain

Calculate $E(u)$ of $u(x) = -\alpha e_3 \times OM^0$

The Divergence Theorem

Let $V : x \in D \subseteq \Omega \to \mathbb{R}^3$ be a sufficiently smooth vector field and $divV = V_{i,i}$ denote the divergence of the vector field, the divergence theorem reads

$$\int_{\partial D} V \cdot \vec{n}\, dS = \int_D divV dx \tag{A.13}$$

$\vec{n} = (n_1, n_2, n_3)$ is the unitary outer normal vector on the border of D.

Let $V = (u, 0, 0)$ with $u : \Omega \to \mathbb{R}$ a sufficiently smooth function, then $divV = u_{,1}$ and from the theorem we deduce that

$$\int_D divV dx = \int_D u_{,1} dx = \int_{\partial D} V \cdot \vec{n}\, dS = \int_{\partial D} u n_1 dS \tag{A.14}$$

Similarly by considering $V = (0, u, 0)$ and $V = (0, 0, u)$ we obtain

$$\int_D u_{,j} dx = \int_{\partial D} u n_j dS \tag{A.15}$$

Applying the same relation to the divergence of a second-order tensor $T = (T_{ij})$ that we denote $divT = T_{ij,j}$ gives

$$\int_D T_{ij,j} dx = \int_{\partial D} T_{ij} n_j dS \tag{A.16}$$

From the above result, we deduce that

$$\int_D u_{,i} v dx = \int_D (uv)_{,i} dx - \int_D u v_{,i} dx = -\int_D u v_{,i} dx + \int_{\partial D} u v n_i dS \text{ and}$$
$$-\int_D T_{ij,j} v_i dx = \int_D T_{ij} v_{i,j} dx - \int_{\partial D} T_{ij} n_j v_i dS \tag{A.17}$$

For further analyses, given two matrices $A = (A_{ij})$ and $B = (B_{ij})$, we define the matrix scalar product by

$$A : B = A_{ij} B_{ij} \tag{A.18}$$

It can be observed that if $A = (A_{ij})$ is symmetric then $A : B = A : symB$ and $A : asymB = 0$. Consequently if $-divT = f$ in Ω then we can write

$$-\int_\Omega T_{ij,j} v_i dx = \int_\Omega T_{ij} v_{i,j} dx - \int_{\partial\Omega} T_{ij} n_j v_i dS = \int_\Omega f.v dx \tag{A.19}$$

and if the tensor is symmetric

$$-\int_{\Omega} divT.vdx = \int_{\Omega} T : \epsilon(v)dx - \int_{\partial\Omega} T\vec{n}.vdS = \int_{\Omega} f.vdx \qquad \text{(A.20)}$$

Example

$$\text{Let } T(x) = \begin{bmatrix} 4x_1^3 + x_2^4 + x_3^2 & x_1^2 + x_2^2 + 2x_3^3 & 4x_1^4 + 2x_2^4 + 3x_3^2 \\ x_1^2 + x_2^4 + x_3 & 4x_1^2 + x_2^4 + x_3^4 & x_1^2 + x_2^5 + 4x_3^2 \\ 2x_1 + 4x_2^5 + x_3^3 & 3x_1^2 + 2x_2^4 + x_3^2 & x_1^2 + 3x_2^2 + 2x_3^2 \end{bmatrix}$$

Then

$$divT(x) = \begin{bmatrix} T_{1j,j} \\ T_{2j,j} \\ T_{3j,j} \end{bmatrix} = \begin{bmatrix} 12x_1^2 + 2x_2 + 6x_3 \\ 2x_1 + 4x_2^3 + 8x_3 \\ 2 + 8x_2^3 + 4x_3 \end{bmatrix}$$

2. Fundamental Equations of Equilibrium

We shall begin by presenting the notion of internal force. Let Ω denote the deformed configuration of a continuum media. Any part or sub-domain D is glued to the rest on the common boundary, by an internal force $\vec{t}$ which resists to any action to separate both parts. The internal forces at the common boundary are opposite on each side of the common boundary and depend on the plane of separation. In other words, at each material point, the internal force vector, also called the Cauchy stress vector, depends on the point and a unitary normal vector $\vec{n}$ to the plane of separation (Fig. A.6).

Axiom

Let Ω denote the deformed configuration of a continuum media, D a sub-domain, S^2 the set of all unitary vectors and $\vec{t}$ the internal force. We admit that there exists a function $t : (x, \vec{n}) \in \bar{\Omega} \times S^2 \to \mathbb{R}^3$ such that

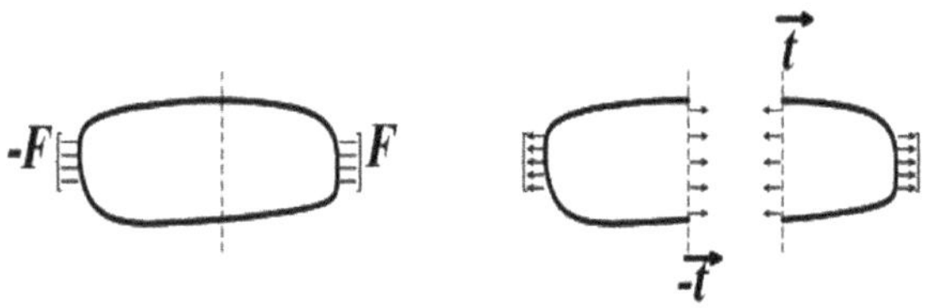

Fig. A.6 Internal force vector

(i) $x \in \partial D$ and $\partial D \cap \partial\Omega$ is empty, then $t(x, \vec{n}) = \vec{t}$, $\vec{n}$ is the unitary outer vector on ∂D.
(ii) $x \in \partial D \cap \partial\Omega$, then $t(x, \vec{n}) = p$, the external pressure due to interaction with the external world and $\vec{n}$ is the outer unitary vector.
(iii) $t(x, \overrightarrow{Qn}) = Qt(x, \vec{n})$ for any rotation matrix Q.

It is thus admitted that any sub-domain of a continuum media is subjected to a body force f and a pressure on its border which is the internal force $\vec{t}$ (see Ciarlet [1-2]). This internal force satisfies the Principle of Material Indifference (PMI). In other words the internal force is objective, i.e. does not depend on the base of observation. The equilibrium of any sub-domain is governed by the equations

$$\int_D f dx + \int_{\partial D} t(x, \vec{n}) dS = O \tag{A.21}$$

$$\int_D ox \times f dx + \int_{\partial D} ox \times t(x, \vec{n}) dS = O \tag{A.22}$$

i.e. the resultant force (the first equation) and the resultant moment (the second equation) are all zero. There is the equilibrium of force and moment. It is proved (see Ciarlet [1]) that there exists a tensor $T(x) = (T_{ij}(x))$ (called the Cauchy stress tensor) such that $t(x, \vec{n}) = T(x)\vec{n}$. The first equation now reads

$$\int_D f dx + \int_{\partial D} t(x, \vec{n}) dS = \int_D f dx + \int_{\partial D} T(x)\vec{n} dS = \int_D (f + divT) dx = O \tag{A.23}$$

for any D and deduce that $f + divT = O$ at each point. It follows from the PMI that $T = QT^t Q$. The second equilibrium equation also reads

$$\int_D ox \times f dx + \int_{\partial D} ox \times t(x, \vec{n}) dS = \int_D ox \times f dx + \int_{\partial D} ox \times T(x)\vec{n} dS = O \tag{A.24}$$

By applying the relations obtained above from the divergence theorem, componentwise, this equation reads

$$\Big[\int_D (x_{i+1} f_{i+2} - x_{i+2} f_{i+1}) dx + \int_D (x_{i+1} T_{i+2j,j} - x_{i+2} T_{i+1j,j}) dx + \int_D (\delta_{i+1j} T_{i+2j} - \delta_{i+2j} T_{i+1j}) dx = 0\Big]_{mod3} \tag{A.25}$$

which is equivalent to

$$\int_D ox \times (f + divT) dx + \int_D (T_{i+2i+1} - T_{i+1i+2})_{mod3} dx = \int_D (T_{i+2i+1} - T_{i+1i+2})_{mod3} dx = 0 \tag{A.26}$$

Recall that $4 = 1\ mod3, 5 = 2\ mod3$. We then deduce that $T_{ij} = T_{ji}$ and conclude that the equilibrium equations are equivalent to

$$divT + f = 0 \text{ in } \Omega \text{ and } T_{ij} = T_{ji} \tag{A.27}$$

This is a system of three equations with six unknowns defined in the deformed configuration which is not known. To overcome this difficulty, another tensor called the first Piola-Kirchhoff tensor is defined using the first equilibrium equation and transport of area vector as follows:

$$\begin{aligned}\int_D f dx + \int_{\partial D} T(x)\vec{n}\, dS &= \int_{D^0} f J dx^0 + \int_{\partial D^0} T(x) Cof F \vec{n_0} d S^0 \\ &= \int_{D^0} f J dx^0 + \int_{\partial D^0} T_0(x)\vec{n_0} d S^0 \\ &= \int_{D^0} (divT_0 + Jf) dx^0 = O\end{aligned} \tag{A.28}$$

It follows that the first Piola-Kirchhoff tensor defined by $T_0 = T(x)Cof F$, which is not a symmetric tensor, satisfies the equation

$$divT_0 + Jf = 0 \tag{A.29}$$

which is a system of three equations with nine unknowns defined in the reference configuration. The tensor is a function of x and F. So we can write $T_0 = T_0(x, F)$. We deduce from the relation $Cof QF = QCof F$ that the PMI implies that $T_0(x, QF) = QT_0(x, F)$. A second Piola-Kirchhoff tensor is defined by $\sigma = F^{-1}T_0$ and satisfies the equation

$$divF\sigma + Jf = 0 \tag{A.30}$$

and

$$^t\sigma =^t T_0{}^t F^{-1} =^t Cof F T^t F^{-1} = J^{-1t} Cof F T Cof F = F^{-1} T_0 = \sigma \tag{A.31}$$

Because σ is symmetric the system is made of three equations and six unknowns. We can also write $\sigma = \sigma(x, F)$ and

$$\sigma(x, QF) = F^{-1t} Q T_0(x, QF) = F^{-1t} Q Q T_0(x, F) = F^{-1} T_0 = \sigma \tag{A.32}$$

which is the Principle of Material Indifference on σ. For any matrix F that satisfies $detF \succ 0$ there exists a rotation matrix R such that $F = RU = VR$, U

and V symmetric and positive defined are the polar decomposition of F. But ${}^tFF = U^2$ which implies that $U = \left({}^tFF\right)^{1/2}$ and $R = F\left({}^tFF\right)^{-1/2}$. Similarly, we obtain $V = \left(F{}^tF\right)^{1/2}$. Let the rotation matrix $Q = {}^t R$, then by the PMI

$$\sigma = \sigma(x, QF) = \sigma(x, U) = \sigma(x, \left({}^tFF\right)^{1/2}) = \sigma(x, (I + 2E)^{1/2}) = \sigma(x, E) \tag{A.33}$$

In infinitesimal deformation (small displacement) it is admitted that

$\sigma \approx T_0 \approx T$, $J \approx 1$; $div\sigma + f \approx O$ in the reference configuration and $E(u) = \epsilon(u)$.

It is easy to guess the form of the stress tensor σ due to some simple loadings. Consider for example a small beam of length L and cross section S subjected to two opposite forces at the tips F and $-F$ in the e_3 direction of an orthonormal base $\{e_1, e_2, e_3\}$. At a point far from the tips if a cut is made with a plane made of a soft material and perpendicular to the e_3 direction, then the two parts will move in the direction of the internal force which equilibrates the external applied force (Fig. A.7). In this example it is clear that the internal force is oriented in the e_3 direction and its component is $\dfrac{F}{S}$. Therefore $t(x, e_3) = \frac{F}{S}e_3$. Similarly we consider planes with respective normals e_1 and e_2. No displacement will occur in these directions. So $t(x, e_1) = t(x, e_2) = O$ and the stress tensor takes the form

$$\sigma = \begin{bmatrix} 0 & 0 & 0 \\ 0 & 0 & 0 \\ 0 & 0 & F/S \end{bmatrix} \tag{A.34}$$

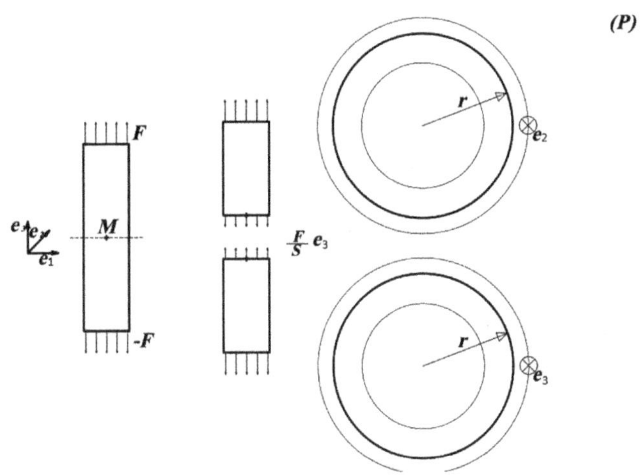

Fig. A.7 Simple loads and stress tensors

We can also consider a ball subjected to a uniform interior pressure $pe_r = pe_1$. In the base $\{e_1, e_2, e_3\}$,e_2 and e_3 being tangent unitary vectors to the coordinate meridian and latitude lines, respectively, if a spherical cut of radius r, a plane cut perpendicular to e_2 and a plane cut perpendicular to e_3, respectively, are made with soft materials, the different movements will be in the e_i direction, i.e. $t(x, e_i) = \alpha_i e_i$. (no summation). The stress tensor will therefore be diagonal

$$\sigma = \begin{bmatrix} \sigma_{11} & 0 & 0 \\ 0 & \sigma_{22} & 0 \\ 0 & 0 & \sigma_{33} \end{bmatrix} \tag{A.35}$$

3. Introduction to Linear Elasticity

In linear elasticity, the second Piola-Kirchhoff stress tensor $\sigma = \sigma(x, \epsilon)$ is sufficient to analyse the equilibrium of a structure under infinitesimal deformation. The indefinite equations (or Euler's equations) written above on regular points (points without jump in the data involved in the analysis) should be completed with boundary conditions which describe how the structure (or media) is linked to the external world. Very many boundary conditions exist but not all can be analysed in a rigorous mathematical setting. The most frequent are the imposed displacement or the imposed traction on parts of the border, rigid or non-rigid contacts with other structures, glued contacts or even heat conditions on the border. We shall restrict ourselves to the genuine mixed problem in which displacement is imposed on a part of the border and traction (pressure) is imposed on the complementary part. The full equations read

$$\begin{gathered} div\sigma + f = 0 \; in \; \Omega^0 \\ u = u_0 \; on \; \Gamma_0 \\ \sigma \vec{n} = p \; on \; \Gamma_1 \end{gathered} \tag{A.36}$$

$$\partial\Omega^0 = \Gamma_0 \cup \Gamma_1, \; mes\Gamma_0 \neq 0, \; mes\Gamma_1 \neq 0$$

where $mes\Gamma_n$ denotes the measure of each portion.

Example on a Dam

Consider a dam of height H, whose border is partitioned as follows:

$$\Gamma_0 = \Gamma_0^b \cup \Gamma_0^{ll} \cup \Gamma_0^{lr} \text{ and } \Gamma_1 = \Gamma_1^w \cup \Gamma_1^v \cup \Gamma_1^e \tag{A.37}$$

The displacement is zero on the bottom base Γ_0^b, the left lateral border Γ_0^{ll} and right lateral border Γ_0^{lr} while Γ_1^w is subjected to water pressure, Γ_1^v to wind pressure and Γ_1^e to pressure due to operating equipment. Their outer unitary normal vectors are, respectively, denoted $\vec{n}_w$, $\vec{n}_v$ and $\vec{n}_e$. Let $p_w(x)$, $p_v(x)$ and $p_e(x)$ denote the external pressures on the borders and $f = -\gamma_s(x)\overrightarrow{k}$ the material weight force per unit volume, then the complete Euler equations read

$$\begin{aligned}
&-div\sigma = -\gamma_s(x)\overrightarrow{k} \ in \ \Omega^0 \\
&u = 0 \ on \ \Gamma_0 = \Gamma_0^b \cup \Gamma_0^{ll} \cup \Gamma_0^{lr} \\
&\sigma\vec{n}_w = p_w \ on \ \Gamma_1^w \\
&\sigma\vec{n}_v = p_v \ on \ \Gamma_1^v \\
&\sigma\vec{n}_e = p_e \ on \ \Gamma_1^e \\
&\partial\Omega^0 = \Gamma_0 \cup \Gamma_1
\end{aligned} \tag{A.38}$$

Euler's equations which are valid for any material cannot be solved because there are three equations for six unknowns. A material constitutive law is necessary in order to complete the indefinite equations. Though in some simple loadings, the stress tensor can bc determined by integrating the above equations, additional equations (Beltrami's equations deduced from compatibility relations and constitutive law) are still needed to calculate the displacement even in plane stress or plane strain problems where stress components can be calculated through Airy stress functions (Fig. A.8).

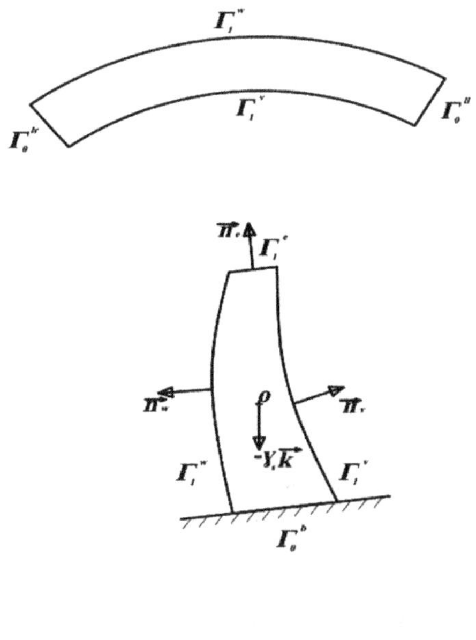

Fig. A.8 Euler's equations on a dam

A.1.3 Constitutive Law

Elastic behaviour is a conservative phenomenon, Therefore $\sigma(x, \epsilon)$, according to the second principle in thermodynamics, is derived from a free energy ψ which does not depend on parameters source of dissipation. The free energy is a convex function of ϵ and

$$\sigma(x, \epsilon) = \rho \partial \psi / \partial \epsilon \tag{A.39}$$

where ρ is the material mass density. We have

$$\sigma(x, \epsilon) = \sigma(x, 0) + \tfrac{\partial \sigma}{\partial \epsilon}(x, 0)\epsilon + 0(\epsilon) = \sigma^{res} + C\epsilon + 0(\epsilon) \tag{A.40}$$

where σ^{res} is the residual stress and C the fourth-order tangent moduli tensor. In a natural reference configuration, the residual stress is zero or is a uniform pressure. At times it is neglected. In any way, it can be considered as an additional force to the volume force. So we shall analyse $\sigma = C\epsilon$.

Because $\sigma_{ij} = \sigma_{ji}$ and $\epsilon_{kl} = \epsilon_{lk}$ it follows that $C_{ijkl} = C_{jikl} = C_{ijlk}$ and because $(C_{ijkl}) = \rho \partial^2 \psi / \partial \epsilon^2$ where the free energy ψ is a strictly convex function, we deduce that $C_{ijkl} = C_{klij}$ and there exists a strictly positive constant c such that $C_{jikl}\epsilon_{kl}\epsilon_{ij} \geq c\epsilon_{lk}\epsilon_{lk}$. This is the ellipticity condition which is necessary for the existence of the engineering constants fourth-order tensor $S = (S_{ijkl}) = (C_{ijkl})^{-1}$ which satisfies $\epsilon = S\sigma$. Because of the material symmetries of (C_{ijkl}), 21 constants are necessary to characterize a material. These constants are reduced to two in a perfect linear elastic homogeneous isotropic material, the ideal material. The tangent modulus tensor and the constitutive law now read

$$C_{ijkl} = \lambda \delta_{ij} \delta_{kl} + 2\mu \delta_{ik} \delta_{jl}, \;\; \sigma_{ij} = \lambda \epsilon_{kk} \delta_{ij} + 2\mu \epsilon_{ij} \tag{A.41}$$

where $\lambda \succ 0, \mu \succ 0$ are the Lamé constants. The constitutive law can also be expressed by using Young's modulus $E \succ 0$ and Poisson's ratio $0 < \nu < \frac{1}{2}$ as follows:

$$\begin{aligned} \sigma &= \frac{E\nu}{(1+\nu)(1-2\nu)}\epsilon_{kk} I + \frac{E}{1+\nu}\epsilon \;\text{ and} \\ \epsilon = S\sigma &= \frac{1+\nu}{E}\sigma - \frac{\nu}{E}\sigma_{kk} I, \; S_{ijkl} = \frac{1+\nu}{E}\delta_{ik}\delta_{jl} - \frac{\nu}{E}\delta_{ij}\delta_{kl} \end{aligned} \tag{A.42}$$

These constants are related as follows:

$$\lambda = \frac{E\nu}{(1+\nu)(1-2\nu)}, \quad \mu = \frac{E}{2(1+\nu)}, \quad E = \mu \frac{3\lambda + 2\mu}{\lambda + \mu} \;\text{ and }\; \nu = \frac{\lambda}{2(\lambda + \mu)}. \tag{A.43}$$

The modulus E characterizes material rigidity to elongation and shortening, μ rigidity to shearing (shear modulus), ν cross-section variation during shortening or elongation and $(3\lambda + 2\mu)/3$ rigidity under volumetric deformation. It is the compressibility modulus. Consider a column of cross section $D_0 \times D_0$ and of height H_0. When subjected to traction/compression on both ends, the new dimensions become $D \times D$ and H. If traction is applied D becomes smaller than D_0 while H becomes greater than H_0. The inverse occurs if compression is applied, i.e. D becomes greater while H becomes smaller. Poisson's ratio is the ratio

$$\nu = \left| \frac{(D - D_0)/D_0}{(H - H_0)/H_0} \right| \tag{A.44}$$

The material is incompressible if $\nu = \frac{1}{2}$. To see this, consider a displacement $u = (D_0 + x_1(D - D_0)/D_0; D_0 + x_2(D - D_0)/D_0; H_0 + x_3(H - H_0)/H_0)$. The initial column is transformed to a column of cross section $D \times D$ and height H. The gradient is (Fig. A.9)

$$\nabla u = \begin{bmatrix} \frac{(D-D_0)}{D_0} & 0 & 0 \\ 0 & \frac{(D-D_0)}{D_0} & 0 \\ 0 & 0 & \frac{(H-H_0)}{H_0} \end{bmatrix} \tag{A.45}$$

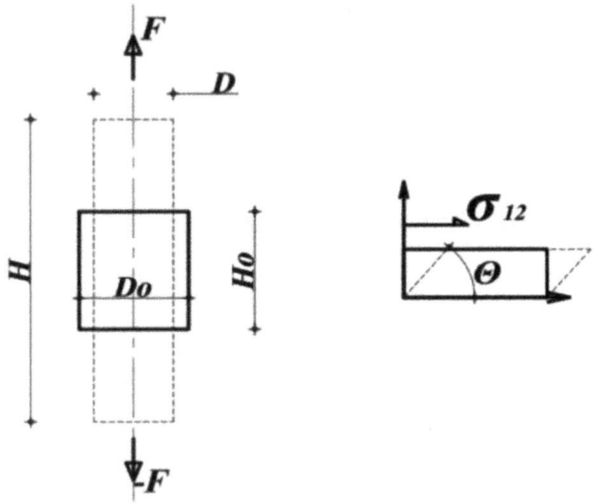

Fig. A.9 Interpretation of Poisson's ratio and shear modulus

From the formula $det(A+B) = detA + CofA : B + A : CofB + detB$ we deduce that $det(I+\nabla u) - 1 \approx divu$ for an infinitesimal transformation. So there is incompressibility if $divu = 0$ which in this case is $2\frac{(D-D_0)}{D_0} + \frac{(H-H_0)}{H_0} = 0$, i.e. $\nu = \left|\frac{(D-D_0)/D_0}{(H-H_0)/H_0}\right| = \frac{1}{2}$.

If $u = \alpha x$ then $\epsilon = \alpha I$ and $\sigma = (3\lambda + 2\mu)\alpha I = \frac{3\lambda+2\mu}{3} 3\alpha\, I = \frac{3\lambda+2\mu}{3} trace\epsilon I$

If σ_{12} is applied on the top end of a parallelopiped then the distortion $\theta = \epsilon_{12} + \epsilon_{21} = 2\epsilon_{12}$. But $\sigma_{12} = 2\mu\epsilon_{12} = \mu(\epsilon_{12} + \epsilon_{21}) = \mu\theta$. These few examples show the role of each modulus.

The complete Euler's equations of the traction-displacement problem now read

$$
\begin{gathered}
-\sigma_{ij,j} = f_i \; in \; \Omega^0 \\
u = u_0 \; on \; \Gamma_0 \\
\sigma \vec{n} = p \; on \; \Gamma_1 \\
\sigma = \frac{E\nu}{(1+\nu)(1-2\nu)}\epsilon_{kk} I + \frac{E}{1+\nu}\epsilon = \lambda\epsilon_{kk} I + 2\mu\epsilon \\
\epsilon(u) = (\nabla u +^t \nabla u)/2 \\
\partial\Omega^0 = \Gamma_0 \cup \Gamma_1, \; mes\Gamma_0 \neq 0, \; mes\Gamma_1 \neq 0
\end{gathered}
\tag{A.46}
$$

In plane strain $\epsilon_{i3} = 0$ or plane stress $\sigma_{i3} = 0$ implies that $\sigma_{33} = \lambda\epsilon_{\alpha\alpha} + (\lambda + 2\mu)\epsilon_{33} = 0$ which leads to $\epsilon_{33} = -\frac{\lambda}{\lambda+2\mu}\epsilon_{\alpha\alpha}$ and the 2D constitutive law reads

$$
\sigma_{\alpha\beta} = \Lambda\epsilon_{\rho\rho}\delta_{\alpha\beta} + 2\mu\epsilon_{\alpha\beta} \tag{A.47}
$$

where $\Lambda = \frac{2\lambda\mu}{\lambda+2\mu}$ or

$$
\sigma_{\alpha\beta} = \frac{E}{1-\nu^2}(\nu\epsilon_{\rho\rho}\delta_{\alpha\beta} + (1-\nu)\epsilon_{\alpha\beta}) \tag{A.48}
$$

and also

$$
\epsilon_{\alpha\beta} = \frac{1+\nu}{E}\sigma_{\alpha\beta} - \frac{\nu}{E}\sigma_{\varrho\varrho}\delta_{\alpha\beta} \tag{A.49}
$$

A.1.4 Existence of Solutions

Let us consider the full partial differential equations of the model problem

$$-(\lambda+2\mu)\nabla divu-\mu\Delta u=f \tag{A.50}$$

$$boundary conditions \tag{A.51}$$

Analytical solutions do not always exist. By multiplying both sides of the equation by an admissible virtual displacement v, i.e.$v=0$ on Γ_0 we obtain by integration

$$-\int_{\Omega^0}\sigma_{ij,j}v_i dx=\int_{\Omega^0}\sigma_{ij}v_{i,j}dx-\int_{\partial\Omega^0}\sigma_{ij}n_j v_i dS=\int_{\Omega^0}f.vdx \tag{A.52}$$

which is equivalent to the variational equation

$$\int_{\Omega^0}\sigma(u):\epsilon(v)dx=\int_{\Omega^0}f.vdx+\int_{\Gamma_1}pvdS=L(v) \tag{A.53}$$

where it is admitted that the various functions are sufficiently smooth. Without loss of generality, let us consider the case where the imposed displacement u_0 is zero, then the space of admissible displacement is $U_{ad}=\{u_i:\Omega^0\longrightarrow\mathbb{R},\ u_i,\ u_{i,j}\in L^2(\Omega^0)\ u_i=0 \text{ on } \Gamma_0\}$ and by letting $\int_{\Omega^0}\sigma(u):\epsilon(v)dx=\int_{\Omega^0}(\lambda\epsilon_{kk}(u)\epsilon_{ll}(v)+2\mu\epsilon_{ij}(u)\epsilon_{ij}(v))$ $dx=a(u,v)$ the variational formulation is rewritten as

$$a(u,v)=L(v) \tag{A.54}$$

$a(.,.)$ is a continuous symmetric bilinear form defined in U_{ad}, i.e. ($\exists c>0,\ |a(u,v)|\leq c\|u\|\|v\|$) and $L(v)$ is a continuous linear form if $f\in L^2(\Omega^0)$ and $p\in L^2(\Gamma_1)$. The solution of the variational equation is the stationary point of the total potential energy $J(u)=\frac{1}{2}a(u,u)-L(u)$ defined in U_{ad} . In order to proof the existence of a solution, we shall make use of Korn's inequality and the Lax-Milgram lemma in a more general form.

Lemma 1 (Korn's Inequality)

Let $|v|=\left\{\int_{\Omega^0}\epsilon_{ll}\epsilon_{ij}(u)\epsilon_{ij}(v))dx\right\}^{1/2}$, and let $\|v\|$ denote the natural induced norm of U_{ad}, i.e. the H^1 norm, then there exists a positive constant c such that $c\ \|v\|\leq|v|$.

In fact, this inequality which is frequently used in linear elasticity (see Duvaut and Lions [4]) shows that the semi-norm $|v|$ is equivalent to the norm $\|v\|$ because of the particular boundary condition imposed on part of the border, i.e. $u=0$ on Γ_0. Also because L is continuous we have $a(u,u)\geq 2\mu\ |v|\succeq c_1\ \|u\|^2$ (i.e. $a(.,.)$ is coercive) and $|L(v)|\leq c_2\ \|v\|$

Lemma 2 (Lax-Milgram)

Let $J:E\to\mathbb{R}$ be a convex functional defined in a Banach space E that satisfies the condition $J(u)\to\infty$ when $\|u\|\to\infty$. Then J has a stationary point, i.e. there exists $u\in E$ such that $\langle J'(u),v\rangle=0$ for any v; $\langle.,.\rangle$ denotes the dual product.

Proof

Let us construct a minimizing sequence (u^n) of the functional $J(.)$ constructed as follows: if u^1 is not a minimizer then we can find u^2 such that $J(u^1) \geq J(u^2)$. If u^2 is not a minimizer then there exists u^3 such that $J(u^2) \geq J(u^3)$ and so on. The sequence is bounded, otherwise there will be a contradiction with $J(u^1) \succeq J(u^n)$ for any n and $J(u^n)$ going to infinity. In a reflexive Banach space as the case herein there exists a sub-sequence that for convenience we still denote (u^n) such that $u^n \rightharpoonup u(\rightharpoonup$ denote weak convergence). Because $J(.)$ is a convex functional $\neq \infty$, $J(u^n) \to J(u)$ and u is a minimizer, i.e. $J(u) \leq J(v)\ \forall v \in E$ But $J(.)$ is convex and therefore Gâteau differentiable. So $\forall v \in E$.

We have

$$\lim_{t \to 0} \frac{J(u+tv) - J(u)}{t} = \langle J'(u), v\rangle \geqq 0 \tag{A.55}$$

and

$$\lim_{t \to 0} \frac{J(u-tv) - J(u)}{t} = -\langle J'(u), v\rangle \geqq 0 \tag{A.56}$$

We deduce that the minimizer satisfies the stationary point condition (the variational equation)

$\langle J'(u), v\rangle = 0$ and the proof of the lemma is complete.

Concerning the total potential energy $J(u) = \frac{1}{2}a(u,u) - L(u)$ defined in U_{ad}, there exist independent positive constants c_i such that

$$J(u) = \frac{1}{2}a(u,u) - L(u) \geqq c_3 \parallel u \parallel^2 - c_2 \parallel u \parallel \tag{A.57}$$

Using Hölder's inequality, $c_2 \parallel u \parallel \leq \frac{(c_2 k)^2}{2} + \frac{\|u\|^2}{2k^2}$. We can choose an integer k such that

$c_3 \parallel u \parallel^2 - c_2 \parallel u \parallel \geq (c_3 - \frac{1}{2k^2}) \parallel u \parallel^2 - \frac{(c_2 k)^2}{2}$

and $(c_3 - \frac{1}{2k^2}) \succ 0$.

It follows that $J(u) = \frac{1}{2}a(u,u) - L(u) \to \infty$ when $\parallel u \parallel \to \infty$. The total potential energy is a convex functional and satisfies the condition of lemma2. Therefore there exists a minimizer u solution to the variational equation

$$a(u,v) = L(v)\ \ \forall v \in U_{ad} \tag{A.58}$$

The solution is unique. Suppose there exists another solution $\overline{u}$. Then

$$a(u, u - \overline{u}) = L(u - \overline{u}) \text{ and } a(\overline{u}, u - \overline{u}) = L(u - \overline{u}) \tag{A.59}$$

which implies that $0 = a(u - \overline{u}, u - \overline{u}) \geq c_1 \parallel u - \overline{u}) \parallel^2$ which leads to $u = \overline{u}$ and the proof is complete. So, to solve a problem, first: the space of admissible displacements must be determined; then follows the total potential energy of the structure and a minimizer or stationary point is characterized.

A.1.5 Application

Consider a vertical bar of length H, fixed on its base and subjected to a torque C at the top end $x_3 = H$. The material's Lamé moduli are λ and μ. The weight of the material is neglected (Fig. A.10).

Euler's equation cannot be easily written because of the boundary condition at the top edge though the Saint Venant principle can be applied. The variational equation is suitable. We can guess an admissible displacement field, for example $u(x) = OM^0 \times \alpha x_3 e_3 = (\alpha x_3 x_2, -\alpha x_3 x_1, 0)$. Then

$$\nabla u = \begin{bmatrix} 0 & \alpha x_3 & \alpha x_2 \\ -\alpha x_3 & 0 & -\alpha x_1 \\ 0 & 0 & 0 \end{bmatrix}, \epsilon(u) = \frac{1}{2}\begin{bmatrix} 0 & 0 & \alpha x_2 \\ 0 & 0 & -\alpha x_1 \\ \alpha x_2 0 & -\alpha x_1 & 0 \end{bmatrix}, \quad (A.60)$$

$$\sigma(u) = \mu \begin{bmatrix} 0 & 0 & \alpha x_2 \\ 0 & 0 & -\alpha x_1 \\ \alpha x_2 0 & -\alpha x_1 & 0 \end{bmatrix},$$

$\sigma(u) : \epsilon(u)/2 = \mu\alpha^2(x_1^2 + x_2^2)$; $L(u) = C\alpha H$,

$J(u) = \frac{1}{2}\int_{\Omega^0}(\lambda\epsilon_{kk}(u)\epsilon_{ll}(u) + 2\mu\epsilon_{ij}(u)\epsilon_{ij}(u))dx - L(u) = \mu\alpha^2 H \int_S (x_1^2 + x_2^2)dS - C\alpha H = \mu\alpha^2 HK - C\alpha H = J(\alpha)$

The solution satisfies

$\langle J'(\alpha), \delta\alpha\rangle = 0 \Leftrightarrow 2\mu\alpha K\delta\alpha = C\delta\alpha$

and

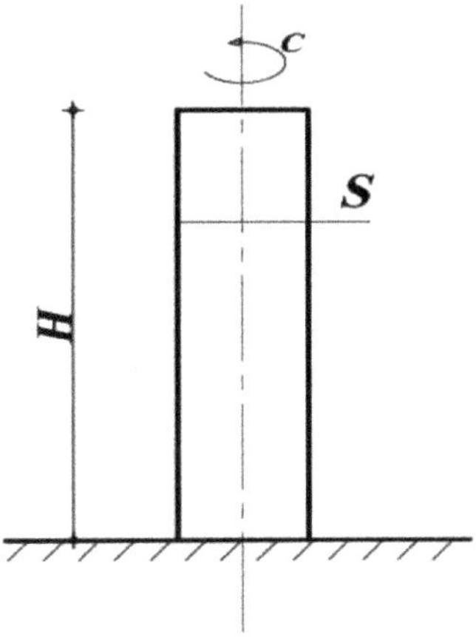

Fig. A.10 Torsion of a bar

$$\alpha = \frac{C}{2\mu K}, u = \frac{C}{2\mu K} \begin{pmatrix} x_3 x_2 \\ -x_3 x_1 \\ 0 \end{pmatrix}, \sigma(u) = \frac{C}{2K} \begin{bmatrix} 0 & 0 & x_2 \\ 0 & 0 & -x_1 \\ x_2 0 & -x_1 & 0 \end{bmatrix}, \quad \text{(A.61)}$$

$$\epsilon(u) = \frac{C}{4\mu K} \begin{bmatrix} 0 & 0 & x_2 \\ 0 & 0 & -x_1 \\ x_2 & -x_1 & 0 \end{bmatrix}$$

The solution is as realistic as the quality of the set of admissible displacements.

A Spherical Tank Under Uniform Pressure We consider a spherical tank under uniform pressure inside and outside. The tank is sufficiently thick in order not to be classified as a shell (Fig. A.11).

The weight is neglected and material moduli are E, ν or λ, μ. We deduce from the symmetry of the problem that the displacement is radial and the stress tensor is diagonal, i.e. in a physical coordinate system $\left(r, S_\theta, S_\varphi\right) = \left(X^1, X^2, X^3\right)$ where S_θ and S_φ are curvilinear meridional and latitude coordinates,

$$u = u\,(r)\,\vec{e_r} \text{ and } \sigma = \begin{bmatrix} \sigma_{11} & 0 & 0 \\ 0 & \sigma_{22} & 0 \\ 0 & 0 & \sigma_{33} \end{bmatrix}$$

Let $\vec{e}$ be a unitary tangent vector of a curve of curvilinear abscissa S, then $\frac{d\vec{e}}{dS} = \frac{1}{R}\vec{n}$ where the radius of curvature R is positive or negative according to the position of the centre of the osculating circle to the curve and the direction of the normal. The radius is positive if the normal points to the centre of the circle. It is otherwise negative. Let M_0 and M be the positions of a point at the reference and deformed configurations, respectively, then $M = M_0 + u$ and the strain tensor is

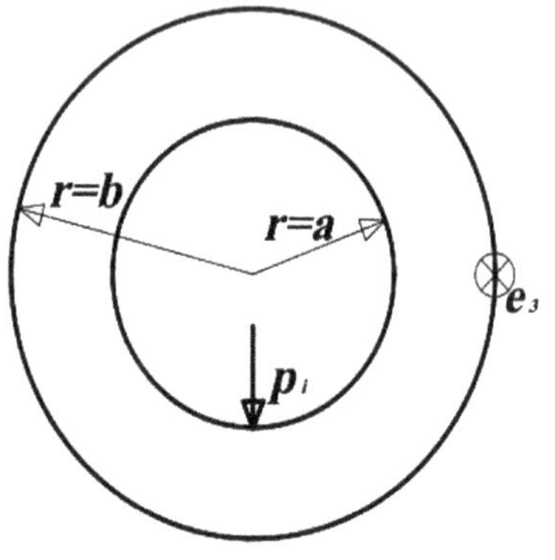

Fig. A.11 Pressurized spherical tank

$$E_{ij}dX^i dX^j = \frac{dM \cdot dM - dM_0 \cdot dM_0}{2} \tag{A.62}$$
$$= du \cdot dM_0 + \frac{du \cdot du}{2} \approx du \cdot dM_0 = \epsilon_{ij} dX^i dX^j = du \cdot dM_0 = du_i dX^i$$

On the sphere we have

$$dM_0 = \begin{pmatrix} dr \\ dS_\theta \\ dS_\varphi \end{pmatrix}, \quad d\left(u(r)\vec{e_r}\right) = du\,\vec{e_r} + ud\vec{e_r} \tag{A.63}$$

and

$$d\left(u\vec{e_r}\right) = u'dr\vec{e_r} + \frac{u\partial\vec{e_r}}{\partial S_\theta}dS_\theta + \frac{u\partial\vec{e_r}}{\partial S_\varphi}dS_\varphi = u'dr\vec{e_r} + \frac{u}{r}dS_\theta\vec{e_\theta} + \frac{u}{r}dS_\varphi\vec{e_\varphi} \tag{A.64}$$

$r \in [a, b]$, $S_\theta \in [0, r\pi]$ or $\theta \in [0, \pi]$, $S_\varphi \in [0, 2\pi r \sin\theta]$ or $\varphi \in [o, 2\pi]$. Therefore, from the constitutive law, $\sigma = \lambda tr\varepsilon I + 2\mu\varepsilon$, we obtain

$$\sigma^{11} = \sigma^{rr} = (\lambda + 2\mu)u' + 2\lambda\frac{u}{r} \quad \text{and} \quad \sigma^{22} = \sigma^{s_\theta s_\theta} = \sigma^{s_\varphi s_\varphi} = \sigma^{33} = 2(\lambda + \mu)\frac{u}{r} + \lambda u' \tag{A.65}$$

Also
$\phi(\epsilon) = \frac{1}{2}\sigma : \epsilon = \frac{1}{2}\left[(\lambda + 2\mu)u'^2 + 4\lambda\frac{uu'}{r} + 4(\lambda + \mu)\left(\frac{u}{r}\right)^2\right] = \phi(u)$

On the internal face ($r = a$) the applied pressure is $p = p\vec{e_r}$. On the external face ($r = b$) atmospheric pressure is neglected. The total potential energy is

$$\begin{aligned} J(u) &= \int_a^b \int_0^\pi \int_0^{2\pi} \phi(u) r^2 sin\theta dr d\theta d\varphi - pu(a)a^2 \int_0^\pi \int_0^{2\pi} sin\theta d\varphi d\theta \\ &= 4\pi\left(\int_a^b \phi(u) r^2 dr - pu(a)a^2\right) \end{aligned} \tag{A.66}$$

Then we have to find a minimizer $u(r)$ of J which is the stationary point of J. We have

$$\delta J(u) = \langle J'(u), \delta u\rangle = 0 \tag{A.67}$$

equivalent to

$$\delta J(u) = \int_a^b \left[r^2(\lambda + 2\mu) u'\delta u' + 2\lambda r u\delta u' + 2\lambda r u'\delta u + 4(\lambda + \mu) u\delta u\right] dr$$
$$= \int_a^b - \left\{\left(r^2(\lambda + 2\mu) u'' + 2r(\lambda + 2\mu) u'\right) - 2\lambda u + 4(\lambda + \mu) u\right\} \delta u = 0 \tag{A.68}$$

Letting $\delta u = 0$ for $r = a$ and $r = b$ and integrating we obtain Euler's equation

$$-r^2 u'' - 2ru' + 2u = 0 \tag{A.69}$$

It follows that

$$u'' + \frac{2u'}{r} - \frac{2u}{r^2} = 0 \tag{A.70}$$

$$u'' + 2\left(\frac{u}{r}\right)' = \left(\frac{1}{r^2}\left(r^2 u\right)'\right)' = 0 \tag{A.71}$$

and

$$\frac{1}{r^2}\left(r^2 u\right)' = \alpha \tag{A.72}$$

Finally

$$u = \gamma_r + \frac{\beta}{r^2} \tag{A.73}$$

The constants γ and β are found by solving boundary equations. On $r = a$, $\sigma\vec{n} = p\vec{e_r}$, i.e. $-\sigma\vec{e_r} = p\vec{e_r}$.

On $r = b$, $\sigma\vec{e_r} = 0$. Finally $\sigma^{rr}(a) = -p$ and $\sigma^{rr}(b) = 0$ are equivalent to

$$(\lambda + 2\mu)\left(\gamma - \frac{2\beta}{a^3}\right) + 2\lambda\left(\gamma + \beta/a^3\right) = -p \tag{A.74}$$
$$(\lambda + 2\mu)\left(\gamma - \frac{2\beta}{b^3}\right) + 2\lambda\left(\gamma + \beta/b^3\right) = 0$$

For the spherical tank we have been able to calculate the strain tensor, the stress tensor and the total potential of a very simple admissible displacement field. This is not obvious in a more general situation without a good mastering of curvilinear media which is the first chapter of the text book.

Bibliography

[1] Philippe G. Ciarlet: Linear and Nonlinear Functional Analysis with Applications. SIAM 2013.

[2] P.G. Ciarlet, Mathematical Elasticity Volume I ;Three-dimensional Elasticity, North Holland, Amsterdam, 1988.

[3] Allan F. Bower; Applied Mechanics of Solids. 2010. CRC press, Taylor and Francis Group.

[4] Duvaut and Lions. Les inequations en mécanique et en physique, Dunod, Paris, 1972.

Appendix B
Summary of Chapters

Chapter 1: Curvilinear Media

B.1 3D Curvilinear Media

B.1.1 Parametrization

$X = (X^1, X^2, X^3) = (X^i)$ or $X = (X_1, X_2, X_3) = (X_i)$

B.1.2 Covariant Base Vectors and Metric

The covariant base is defined as follows:

$$\{G_1, G_2, G_3\}, \quad G_i = \frac{\partial OM(X)}{\partial X^i} = OM_{,i}$$

and the metric is

$$(G_{ij}) = (G_i.G_j); \quad \det(G_{ij}) = G > 0$$

B.1.3 Contravariant Base and Metric Tensor

The contravariant base vector is defined by

R. Nzengwa, *Plate and Shell Models*,
https://doi.org/10.1007/978-981-97-2780-3

$$\{G^1, G^2, G^3\}, \quad G^i.G_j = \delta^i_j, \quad (G^{ij} = (G^i.G^j)$$

and the metric tensor is

$$(G^{ij}) = (G_{ij})^{-1}, \quad G^i = G^{ij}G_j, \quad G_i = G_{ij}G^j$$

B.1.4 Tensor Base

Using the above contravariant and covariant bases, a tensor T is expressed as follows:

$$T = (T_{ij}) = T_{ij}G^i \otimes G^j = T^{ij}G_i \otimes G_j$$

T can also be given by

$$T = T^i_j G_i \otimes G^j$$

From the relations:

$$T = T_{ij}G^i \otimes G^j = T_{ij}G^{ik}G_k \otimes G^{jl}G_l = T_{ij}G^{ik}G^{jl}G_k \otimes G_l$$

and

$$T = T^i_j G_i \otimes G^j = T^i_j G^{ik}G^k \otimes G^j = T^i_j G^i \otimes G_{jl}G_l$$

we deduce the variance change relations:

$$T_{ij} = G_{ik}G_{jl}T^{kl}, \quad T^{ij} = G^{ik}G^{jl}T_{lk} \tag{B.1}$$

and

$$(T^i_j) = G_{jl}T^{il} = G^{ik}T_{jk} \tag{B.2}$$

B.1.5 Covariant Derivation $G_{i,j}$ and Christoffel Symbols Γ^k_{ij}

We have

$G_{i,j} = \Gamma^k_{ij}G_k, \quad \Gamma^k_{ij} = G_{i,j}.G^k = G^{kl}\Gamma_{ijl}, \quad \Gamma^k_{ij} = \Gamma^k_{ji}, \quad \Gamma_{ijl} = \dfrac{1}{2}(G_{il,j} + G_{jl,i} - G_{ij,l})$.

From $G_iG^j = \delta^j_i$ we deduce that $G^j_{,k} = -\Gamma^j_{kl}G^l$.

B.1.6 Covariant Derivation of Tensors

We have

$$V_{,k} = (V_i G^i)_{,k} = V_{i,k} G^i + V_i G^i_{,k} = (V_{i,k} - \Gamma^l_{ki} V_l) G^i = V_{i/k} G^i \tag{B.3}$$

Similarly

$$V^i_{/k} = V^i_{,k} + \Gamma^i_{kl} V^l \tag{B.4}$$

and

$$divV = V^i|_i = V^i_{,i} + \Gamma^i_{il} V^l = \frac{1}{\sqrt{G}} (\sqrt{G} V^i)_{,i}$$

• If $v : \Omega \rightarrow \mathbb{R}$ is the scalar function, then

$$\nabla v = v_{,i} G^i, \quad v_{,ij} = (v_{,i} G^i)_{,j} = (v_{,ij} - \Gamma^k_{ij} v_{,k}) G^i \tag{B.5}$$

which gives

$$\nabla_{ij} v = v_{,ij} - \Gamma^k_{ij} v_{,k} \text{ and } \Delta v = v_{,ii} - \Gamma^k_{ii} v_{,k}$$

• If $T = (T_{ij})$, then

$$(T_{ij})_{,k} = (T_{ij} G^i \otimes G^j)_{,k} = (T_{ij,k} - \Gamma^l_{ki} T_{lj} - \Gamma^l_{kj} T_{il}) G^i \otimes G^j$$

which leads to

- $T_{ij}|_k = T_{ij,k} - \Gamma^m_{ik} T_{mj} - \Gamma^m_{kj} T_{im}, \; T^{ij}|_k = T^{ij}_{,k} + \Gamma^i_{kl} T^{lj} + \Gamma^i_{kl} T^{lj}$

  Similarly
- $T^i_j|_k = T^i_{j,k} + \Gamma^i_{kl} T^l_j - \Gamma^l_{jk} T^i_l$
- $G_{ij}|_k = G^{ij}|_k = 0$

B.1.7 Line Element dl, Volume Element dv and Area Element dS

$$dl^2 = dM.dM = dX^i G_i.dX^j G_j = dX^i dX^j G_{ij} \tag{B.6}$$

$$= dX_i G^i.dX_j G^j = G^{ij} dX_i dX_j \tag{B.7}$$

$$= dX^i G_i.dX_j G^j = dX_i dX^i \tag{B.8}$$

$$
\begin{aligned}
d\mathbf{v} &= (dX^i G_i, dX^j G_j, dX^k G_k) = e^{ijk}(G_i, G_j, G_k)dX \\
&= (G_1, G_2, G_3)dX^1 dX^2 dX^3 = \sqrt{G}dX
\end{aligned} \tag{B.9}
$$

We also have $d\mathbf{v} = \dfrac{1}{\sqrt{G}}(G^1, G^2, G^3)dX_1 dX_2 dX_3$.

The area vector A is

$$
\begin{aligned}
A &= dM \times dM = dX^i G_i \times dX^i G_j = G_i \times G_j dX^i dX^j \\
&= (G_i, G_j, G_k)e_{ijk} dX^i dX^j G^k = \sqrt{G} e_{ijk} dX^i dX^j G^k
\end{aligned} \tag{B.10}
$$

$dS_k = e_{ijk}\sqrt{G}dX^i dX^j$, e_{ijk}, e^{ijk} is the permutation symbols.

B.1.8 Linearized Strain Tensor

The linearized tensor is given by

$$
\epsilon_{ij}(U) = \frac{1}{2}\left(U_i|_j + U_j|_i\right) = \frac{1}{2}\left(U_{i,j} + U_{j,i} - 2\Gamma^l_{ij} U_l\right), \quad U = U_i G^i
$$

B.1.9 Linearized Elastic Stress

$\sigma^{ij} = (\lambda G^{ij} G^{kl} + 2\mu G^{ik} G^{jl})\epsilon_{kl}(U)$, $\sigma^{ij}|_j = (\lambda G^{ij} G^{kl} + 2\mu G^{ik} G^{jl})\epsilon_{kl}(U)|_j$

B.1.10 Divergence Theorem

$\int_\Omega \sigma^{ij}|_j d\Omega = \int_{\partial\Omega} \sigma^{ij} n_j ds$, $\quad \vec{n} = n_j G^j$ is the unit outer normal vector.

B.1.11 Euler's Equation of a Mixed Boundary Value Problem

$$\begin{cases} \sigma^{ij}|_j + f^i = 0 \;\; \text{in} \;\; \Omega \\ \sigma^{ij} n_j = p^i \quad \text{on} \;\; \Gamma_1 \subset \partial\Omega \\ U_i = 0 \quad \text{on} \;\; \Gamma_0 \subset \partial\Omega, mes(\Gamma_0) \neq 0, \quad mes(\Gamma_1) \neq 0, \overline{\Gamma}_0 \cup \overline{\Gamma}_1 = \partial\Omega \\ \sigma^{ij} = \lambda \epsilon_k^k(U) G^{ij} + 2\mu \epsilon^{ij}(U) \end{cases} \tag{B.11}$$

B.1.12 Variational Equation

Find $U \in U_{ad} = \big\{U_i : \Omega \to \mathbb{R}, U_i \in L^2(\Omega), U_i|_j \in L^2(\Omega), U_i = 0, \quad \text{on} \;\; \Gamma_0\big\}$:

$$\int_\Omega \sigma^{ij}(U)\epsilon_{ij}(V)d\Omega = \int_\Omega f^i V_i d\Omega + \int_{\Gamma_1} P^i V_i d\Gamma_1, \forall V \in U_{ad}$$

B.2 2D Curvilinear Media (Surface Geometry)

B.2.1 Parameters, Covariant Base and Metric

Let

$X = (X^\alpha)$ or $X = (X_\alpha)$.

$A_\alpha = Om_{,\alpha}$, $A_3 = A_1 \times A_2/|A_1 \times A_2|$, $\{A_1, A_2, A_3\}$ is the covariant base.

$A_{\alpha\beta} = A_\alpha.A_\beta$, $\;A = \det(A_{\alpha\beta}) > 0$ is the metric tensor.

B.2.2 Contravariant Base and Metric Tensor

$(A^{\alpha\beta}) = (A_{\alpha\beta})^{-1}$, $A^\alpha = A^{\alpha\beta} A_\beta$, $\;A^\alpha.A_\beta = \delta^\alpha_\beta$, $\;A^{\alpha\beta} = A^\alpha.A^\beta$.

B.2.3 Line Element dl, Surface Element dA

We define a line element dl by

$$dl^2 = dm.dm = dX^\alpha A_\alpha.dX^\beta A_\beta = A_{\alpha\beta} dX^\alpha dX^\beta = A^{\alpha\beta} dX_\alpha dX_\beta \tag{B.12}$$

and the surface element by

$$\begin{aligned} dA &= dX^\alpha A_\alpha \times dX^\beta A_\beta = A_\alpha \times A_\beta dX^\alpha dX^\beta \\ &= |A_\alpha \times A_\beta| e_{\alpha\beta} dX^\alpha dX^\beta A^3 = \sqrt{A} e_{\alpha\beta} dX^\alpha dX^\beta A^3 \end{aligned} \tag{B.13}$$

where $e_{\alpha\beta}$ is the permutation symbol.

B.2.4 Fundamental Forms

• $A_{\alpha\beta} = A_\alpha . A_\beta = A_{\beta\alpha}$ is the first fundamental form or metric tensor.
• $B_{\alpha\beta} = A_{\alpha,\beta} . A_3 = -A_{3,\alpha} . A_\beta = B_{\beta\alpha}$ is the second fundamental form or curvature tensor.
• $C_{\alpha\beta} = B_\alpha^\rho B_{\rho\beta} = A^{\rho\gamma} B_{\alpha\gamma} B_{\rho\beta} = B_{\rho\beta} A^{\rho\gamma} B_{\alpha\gamma} = B_\beta^\gamma B_{\alpha\gamma} = C_{\beta\alpha}$ is the third fundamental form or Gauss curvature tensor.

B.2.5 Christoffel Symbols

$$\Gamma_{\alpha\beta}^\rho = A_{\alpha,\beta} . A^\rho = A^{\rho\gamma} \Gamma_{\alpha\beta\gamma},$$

$$\Gamma_{\alpha\beta\rho} = \frac{1}{2} \left(A_{\alpha\rho,\beta} + A_{\beta\rho,\alpha} - A_{\alpha\beta,\rho} \right)$$

B.2.6 Covariant Derivation

$$A_{\alpha,\beta} = \Gamma_{\alpha\beta}^\rho A_\rho + B_{\alpha\beta} A_3, \quad A_{,\beta}^\alpha = -\Gamma_{\beta\rho}^\alpha A^\rho + B_\beta^\alpha A^3 \tag{B.14}$$

noting that $A_3 = A^3$.

$$(V_\alpha A^\alpha)_{,\beta} = V_{\alpha,\beta} A^\alpha + V_\alpha A_{,\beta}^\alpha = V_{\alpha,\beta} A^\alpha - \Gamma_{\alpha\beta}^\rho V_\rho A^\alpha + V_\alpha B_\beta^\alpha A^3 \tag{B.15}$$

We define

$$\nabla_\beta V_\alpha = V_{\alpha,\beta} - \Gamma_{\alpha\beta}^\rho V_\rho, \quad \nabla_\beta V^\alpha = V_{,\beta}^\alpha + \Gamma_{\beta\rho}^\alpha V^\rho \tag{B.16}$$

For $V = V_\alpha A^\alpha + V_3 A^3$ we have

$$\begin{aligned} V_{,\alpha} &= \left(\nabla_\alpha V^\lambda - B_\alpha^\lambda V^3 \right) A_\lambda + \left(V_{,\alpha}^3 + B_{\lambda\alpha} V^\lambda \right) A_3 \\ &= \left(\nabla_\alpha V_\lambda - B_{\alpha\lambda} V_3 \right) A^\lambda + \left(V_{3,\alpha} + B_\alpha^\lambda V_\lambda \right) A^3 \end{aligned} \tag{B.17}$$

B.2.7 Covariant Derivation of Tensors

$$(\overline{T}_{\alpha\beta})_{,\rho} = (\overline{T}_{\alpha\beta}A^{\alpha} \otimes A^{\beta})_{,\rho} = \overline{T}_{\alpha\beta,\rho}A^{\alpha} \otimes A^{\beta} + \overline{T}_{\alpha\beta}A^{\alpha}_{,\rho} \otimes A^{\beta} + \overline{T}_{\alpha\beta}A^{\alpha} \otimes A^{\beta})_{,\rho} \tag{B.18}$$

which leads to

$$\nabla_{\rho}\overline{T}_{\alpha\beta} = \overline{T}_{\alpha\beta,\rho} - \Gamma^{\gamma}_{\alpha\rho}\overline{T}_{\gamma\beta} - \Gamma^{\alpha}_{\gamma\rho}\overline{T}^{\gamma\beta} \tag{B.19}$$

$$\nabla_{\rho}\overline{T}^{\alpha\beta} = \overline{T}^{\alpha\beta}_{,\rho} + \Gamma^{\alpha}_{\rho\gamma}\overline{T}^{\gamma\beta} + \Gamma^{\beta}_{\rho\gamma}\overline{T}^{\alpha\gamma} \tag{B.20}$$

$$\nabla_{\rho}\overline{T}^{\alpha}_{\beta} = \overline{T}^{\alpha}_{\beta,\rho} + \Gamma^{\alpha}_{\rho\gamma}\overline{T}^{\gamma}_{\beta} - \Gamma^{\gamma}_{\rho\beta}\overline{T}^{\alpha}_{\gamma} \tag{B.21}$$

If $v : S \to \mathbb{R}$ is a scalar function then, $\nabla v = v_{,\alpha}A^{\alpha}$,

$$(v_{,\alpha}A^{\alpha})_{,\beta} = (v_{,\alpha\beta} - \Gamma^{\rho}_{\alpha\beta}v_{,\rho})A^{\alpha},$$

$$\nabla_{\alpha\beta}v = v_{,\alpha\beta} - \Gamma^{\rho}_{\alpha\beta}v_{,\rho},$$

$$\Delta v = v_{\alpha\alpha} - \Gamma^{\rho}_{\alpha\alpha}v_{,\rho}.$$

$$div v = \nabla_{\alpha}v^{\alpha} = v^{\alpha}_{,\alpha} + \Gamma^{\alpha}_{\alpha\beta}v^{\beta} = \frac{1}{\sqrt{A}}(\sqrt{A}v^{\alpha})_{,\alpha}$$

B.2.8 Strain Tensor

$$e_{\alpha\beta}(u) = \frac{1}{2}(\nabla_{\alpha}u_{\beta} + \nabla_{\beta}u_{\alpha} - 2u_3B_{\alpha\beta}) = e_{\beta\alpha}$$

B.2.9 Curvature Tensor

$$K_{\alpha\beta}(u) = \nabla_{\alpha}(\nabla_{\beta}u_3 + B^{\rho}_{\beta}u_{\rho}) + B^{\rho}_{\alpha}(\nabla_{\beta}u_{\rho} - u_3B_{\beta\rho}) = K_{\beta\alpha}$$

B.2.10 Gauss Curvature Tensor

$$Q_{\alpha\beta}(u) = \tfrac{1}{2}\left(B_\alpha^\nu \nabla_\beta(\nabla_\nu u_3 + B_\nu^\rho u_\rho) + B_\beta^\nu \nabla_\alpha(\nabla_\nu u_3 + B_\nu^\rho u_\rho)\right)$$

$$= \tfrac{1}{2}\left(B_\alpha^\nu \nabla_\beta \theta_\nu + B_\beta^\nu \nabla_\alpha \theta_\nu\right) = Q_{\beta\alpha}$$

$$\theta_\nu = \nabla_\nu u_3 + B_\nu^\rho u_\rho$$

B.2.11 Divergence Theorem

$v = v_\alpha A^\alpha = v^\alpha A_\alpha$

$$\int_S \nabla_\alpha v^\alpha dS = \int_{\partial S} v^\alpha \nu_\alpha d\gamma$$

where $\vec{\nu} = \nu_\alpha A^\alpha$ is the unit outer normal vector.

Exercise 1.1

For a ball $B(0, R) = \Omega$, with self-weight $f = -\rho g\vec{k}$, write down Euler's equation and the variational equation. The constitutive material is elastic with Lamé constants $\lambda,\ \mu > 0$.

Exercise 1.2

On a sphere of radius R
(i) Calculate the Riemann curvature $R^\rho_{\alpha\beta\gamma}$ deduced from

$$A_{\alpha,\beta\gamma} - A_{\alpha,\gamma\beta} = R^\rho_{\alpha\beta\gamma} A_\rho$$

(ii) Calculate Ricci curvature tensor

$$R_{\alpha\gamma} = R^\rho_{\alpha\rho\gamma}$$

(iii) Calculate the scalar curvature

$$\overline{R} = A^{\alpha\beta} R_{\alpha\beta} = R^\alpha_\alpha$$

(iv) Calculate the tensor

$$E_{\alpha\beta} = R_{\alpha\beta} - \frac{\overline{R}}{2} A_{\alpha\beta}$$

Exercise 1.3

Calculate the total area of a spherical surface of radius R, with the parameters x_α and x^α.

Chapter 2: Shell Theory

B.3 Shell Description

Description of a shell with constant thickness h and mid-surface S:

$$\Omega = \{OM = Om + zA_3, m(x) = m(x^1, x^2) \in S, z \in [-h/2, h/2]\} \qquad (B.22)$$

B.4 Covariant Bases and Metric Tensor

Covariant bases vectors

-If $\chi = \dfrac{h}{2R} < 1$ where $R = \min\{|R_1|, |R_2|\} \neq 0$, then

$$\{G_\alpha, A_3\} = \{OM_{,\alpha}, A_3\} = \{(\delta_\alpha^\rho - zB_\alpha^\rho)A_\rho, A_3\} = \{\mu_\alpha^\rho A_\rho, A_3\} \text{ and } \{A_\alpha, A_3\} \qquad (B.23)$$

constitute two covariant bases of Ω because

$$\det(\mu_\alpha^\rho) = 1 - 2z\overline{H} + z^2 \det(b_\beta^\alpha) \geq (1-\chi)^2 > 0$$

where $\overline{H}$ is the mean curvature and $\det(b_\beta^\alpha)$ is the Gauss curvature

The covariant metric tensor is

$$[G_{ij}] = [G_i.G_j] = \begin{pmatrix} & & 0 \\ & G_{\alpha\beta} & 0 \\ 0 & 0 & 1 \end{pmatrix}$$

$$G_{\alpha\beta} = \mu_\alpha^\rho \mu_\beta^\gamma A_{\rho\gamma} = A_{\alpha\beta} - 2zB_{\alpha\beta} + z^2 C_{\alpha\beta}$$

A vector $U : \Omega \to \mathbb{R}^3$ can be expressed as

$$U = U^\alpha(\mathbf{x}, z)G_\alpha + U^3(\mathbf{x}, z)A_3 = \overline{U}^\alpha(\mathbf{x}, z)A_\alpha + \overline{U}^3(x, z)A_3 \qquad (B.24)$$

B.5 Contravariant Bases

• $\{G^\alpha, A^3\} = \{(\mu^{-1})_\rho^\alpha A^\rho, A^3 = A_3\}$ and $\{A^\alpha, A^3\}$ constitute two contravariant bases of the shell.

• The contravariant metric is

$$[G^{ij}] = [G^i.G^j] = \begin{pmatrix} & & 0 \\ & G^{\alpha\beta} & 0 \\ 0 & 0 & 1 \end{pmatrix} \quad G^{\alpha\beta} = (\mu^{-1})_\rho^\alpha (\mu^{-1})_\gamma^\beta A^{\rho\gamma}$$

• A vector U can be expressed as

$$U = U_\alpha(\mathbf{x}, z)G^\alpha + U_3(\mathbf{x}, z)A^3 = \overline{U}_\alpha(\mathbf{x}, z)A^\alpha + \overline{U}_3(\mathbf{x}, z)A^3 \tag{B.25}$$

B.6 Christoffel Symbols

– In the bases $\{G_i\}$ or $\{G^i\}$, we have

$$\Gamma^k_{ij} = \Gamma^k_{ji} = G_{i,j}G^k = G_{j,i}G^k$$

– In the bases $\{A^i\}$ or $\{A_i\}$, we have

$$\overline{\Gamma}^\rho_{\alpha\beta} = A_{\alpha,\beta}.A^\rho$$

We can define the relation between the 3D and 2D Christoffel symbols by

$$\Gamma^\gamma_{\alpha\beta} = G_{\alpha,\beta}.G^\gamma = \bar{\Gamma}^\gamma_{\alpha\beta} + \left(\mu^{-1}\right)^\gamma_\nu \nabla_\beta \mu^\nu_\alpha$$

$$\Gamma^\alpha_{\beta 3} = -\left(\mu^{-1}\right)^\alpha_\lambda B^\lambda_\beta,\ \Gamma^3_{\alpha\beta} = \mu^\nu_\alpha B_{\nu\beta},\ \Gamma^3_{3\beta} = \Gamma^\alpha_{33} = \Gamma^3_{33} = 0$$

B.7 3D and 2D Tensor Relations

$$T_\alpha = \mu^\nu_\alpha \bar{T}_\nu,\ T^\alpha = (\mu^{-1})^\alpha_\nu \bar{T}^\nu,\ \bar{T}_\alpha = (\mu^{-1})^\nu_\alpha T_\nu,\ \bar{T}^\alpha = (\mu)^\alpha_\nu T^\nu,\ T^3 = T_3 = \bar{T}^3 = \bar{T}_3$$

B.8 Covariant Derivation Relations

From $T_{i/j} = T_{i,j} - \Gamma^k_{ij} T_k$, we deduce the following relations:

$$T_{\alpha/\beta} = T_{\alpha,\beta} - \Gamma^\lambda_{\alpha\beta} T_\lambda - \Gamma^3_{\alpha\beta} T_3 = \mu^\nu_\alpha \left[\nabla_\beta \overline{T}_\nu - B_{\nu\beta} \overline{T}^3\right]$$

$$T^\alpha_{/\beta} = T^\alpha_{,\beta} + \Gamma^\alpha_{\beta\lambda} T^\lambda + \Gamma^\alpha_{\beta 3} T^3,\ T^\alpha_{/\beta} = \left(\mu^{-1}\right)^\alpha_\nu \left[\nabla_\beta \overline{T}^\nu - B^\nu_\beta \overline{T}^3\right]$$

$$T_{\alpha/3} = \mu^\nu_\alpha \overline{T}_{\nu,3};\ T_{3/\alpha} = \overline{T}_{3,\alpha} + B^\lambda_\alpha \overline{T}_\lambda$$

$$T^\alpha_{/3} = \left(\mu^{-1}\right)^\alpha_\nu \overline{T}^\nu_{,3};\ \ldots T^3_{/\alpha} = \overline{T}^3_{,\alpha} + B_{\alpha\lambda} \overline{T}^\lambda$$

$$T^3_{/3} = T_{3/3} = T_{3,3} = \overline{T}^3_{,3} = \overline{T}_{3,3}$$

B.9 Strain Tensor

$\epsilon = \left(\epsilon_{ij}\right) = \frac{1}{2}\left(U_{i/j} + U_{j/i}\right)$, i.e.

$$\epsilon_{\alpha\beta} = \frac{1}{2}\left(U_{\alpha/\beta} + U_{\beta/\alpha}\right) = \frac{1}{2}\left[\mu^\nu_\alpha\left(\nabla_\beta\overline{U}_\nu - B_{\alpha\beta}\overline{U}_3\right) + \mu^\nu_\beta\left(\nabla_\alpha\overline{U}_\nu - B_{\nu\alpha}\overline{U}_3\right)\right]$$

$$\epsilon_{\alpha 3} = \frac{1}{2}\left(U_{\alpha/3} + U_{3/\alpha}\right) = \frac{1}{2}(\mu^\nu_\alpha\overline{U}_{\nu,3} + \overline{U}_{3,\alpha} + B^\nu_\alpha\overline{U}_\nu)$$

$$\epsilon_{33} = U_{3/3} = U_{3,3} = \overline{U}_{3,3}$$

B.10 Plane Strain and Kinematic

$\epsilon_{i3}(U) = 0$ leads to the kinematic given by

$$U_\alpha = \mu^\rho_\alpha\bar{U}_\rho = u_\alpha(x) - z\left(u_{3,\alpha} + 2B^\tau_\alpha u_\tau(x)\right) + z^2\left(B^\tau_\alpha B^\nu_\tau u_\nu(x) + B^\tau_\alpha u_{3,\tau}(x)\right) \tag{B.26}$$

$$= u_\alpha(x) - z\vartheta_\alpha(x) + z^2\varphi_\alpha(x) \tag{B.27}$$

$$U_3 = u_3(x) \tag{B.28}$$

$u_i : \overline{S} \to \mathbb{R}$

B.11 Plane Stress-Strain Constitutive Relations

In this model, the plane stress-strain constitutive relations still hold, i.e.

$$\sigma^{\alpha\beta} = \frac{2\bar{\mu}\bar{\lambda}}{\bar{\lambda} + 2\bar{\mu}}\epsilon^\rho_\rho G^{\alpha\beta} + 2\bar{\mu}\epsilon^{\alpha\beta} = \bar{\Lambda}\epsilon^\rho_\rho G^{\alpha\beta} + 2\bar{\mu}\epsilon^{\alpha\beta}$$

$$G^{\alpha\beta} = \left(\mu^{-1}\right)^\alpha_\nu \cdot \left(\mu^{-1}\right)^\beta_\rho A^{\nu\rho}$$

where the Lamé constant $\bar{\mu}$, $\bar{\lambda} > 0$ and where $\bar{\Lambda} = \dfrac{2\bar{\mu}\bar{\lambda}}{\bar{\lambda} + 2\bar{\mu}}$.

B.12 Equilibrium of a Shell

The equilibrium of a shell subjected to a body force f, a surface force on part of its border p and clamped on another part of its border (model problem) is characterized by the equations:

$$\begin{cases} \sigma^{ij}_{/j} + f^i = 0 \text{ in } \Omega \\ \sigma\vec{n} = p \;\; on \;\; \Gamma_- \cup \Gamma_+ \cup \bar{\gamma} \times [-\frac{h}{2}, \frac{h}{2}];\; \Gamma_- = S \times \left\{-\frac{h}{2}\right\} \\ \Gamma_+ = S \times \left\{\frac{h}{2}\right\};\; \gamma \text{ is a part of } \partial S \\ U = 0 \text{ on } \Gamma_0;\; \Gamma_0 = \bar{\gamma}_0 \times [-\frac{h}{2}, \frac{h}{2}], \\ \bar{\gamma}_0 \text{ is the complementary part of} \bar{\gamma} \text{ of } \partial S \end{cases} \tag{B.29}$$

It is the 3D elasticity of Euler's equation in curvilinear media. The variational equation reads

$$\int_\Omega \sigma^{ij}(U)\epsilon_{ij}(V)d\Omega = P_{ext}(V) \tag{B.30}$$

$$\begin{aligned} d\Omega &= (G_1,\, G_2,\, A_3)\, dx^1dx^2dz = \sqrt{G}dxdz \\ &= \left(det(\mu^\alpha_\beta)\right)\sqrt{A}dxdz = \psi(x, z)dSdz; \end{aligned} \tag{B.31}$$

$$\epsilon_{\alpha\beta}(V) = e_{\alpha\beta}\,(v) - zK_{\alpha\beta}\,(v) + z^2 Q_{\alpha\beta}\,(v) \tag{B.32}$$

$$e_{\alpha\beta}\,(v) = \frac{1}{2}\left(\nabla_\alpha v_\beta + \nabla_\beta v_\alpha - 2v_3 B_{\alpha\beta}\right) = \bar{e}_{\alpha\beta}$$

$$K_{\alpha\beta}\,(v) = \nabla_\alpha B^\rho_\beta v_\rho + B^\rho_\alpha \nabla_\beta v_\rho + B^\rho_\beta \nabla_\alpha v_\rho + \nabla_\alpha \nabla_\beta v_3 - B^\rho_\alpha B_{\rho\beta} v_3 = \bar{K}_{\alpha\beta}$$

$$Q_{\alpha\beta}(v) = \frac{1}{2}\left(B_\alpha^\delta \nabla_\beta(\nabla_\delta v_3 + B_\delta^\rho v_\rho) + B_\beta^\delta \nabla_\alpha(\nabla_\delta v_3 + B_\delta^\rho v_\rho)\right) = \bar{Q}_{\alpha\beta}$$

$v_i: \quad x = (x^1, x^2) \to v_i(x) \in \mathbb{R}$. From

$$\sigma^{\alpha\beta} = \bar{\Lambda}\epsilon_\rho^\rho G^{\alpha\beta} + 2\bar{\mu}\epsilon^{\alpha\beta} = (\bar{\Lambda}G^{\alpha\beta}G^{\gamma\rho} + \bar{\mu}(G^{\alpha\rho}G^{\beta\gamma} + G^{\beta\rho}G^{\alpha\gamma}))\epsilon_{\rho\gamma}(U)$$

$$G^{\alpha\beta} = \left(\mu^{-1}\right)_\nu^\alpha \cdot \left(\mu^{-1}\right)_\rho^\beta A^{\nu\rho} \tag{B.33}$$

It follows that

$$\begin{aligned}\int_\Omega \sigma^{ij}(U)\epsilon_{ij}(V)d\Omega = & \int_\Omega \sigma^{\alpha\beta}(U)\epsilon_{\alpha\beta}(V)d\Omega \\ = & \int_S dS \int_{\frac{-h}{2}}^{\frac{h}{2}} \psi(x,z)\,[\bar{\Lambda}G^{\alpha\beta}G^{\rho\gamma} \\ & + \bar{\mu}\left(G^{\alpha\rho}G^{\beta\gamma} + G^{\beta\rho}G^{\alpha\gamma}\right)]\epsilon_{\rho\gamma}(U)\,\epsilon_{\alpha\beta}(V)\,dz \\ = & A(u,v) = L(v)\end{aligned} \tag{B.34}$$

where

$$\begin{aligned}L(v) = & \int\limits_S \left(\bar{P}^\alpha(x)v_\alpha(x) + \bar{P}^3(x)v_3(x)\right)dS + \int\limits_{\gamma_1} (q^\alpha(x)v_\alpha(x) \\ & + q^3(x)v_3(x))d\bar{\gamma} + \int\limits_{\gamma_1} m^\alpha(x)\bar{\theta}_\alpha(x)d\bar{\gamma}\end{aligned} \tag{B.35}$$

with

$$q^\alpha = \int_{\frac{-h}{2}}^{\frac{h}{2}} p^\alpha dz - \int_{\frac{-h}{2}}^{\frac{h}{2}} zB_\tau^\alpha p^\tau dz; \quad q^3 = \int_{\frac{-h}{2}}^{\frac{h}{2}} p^3 dz;$$

$$m^\alpha = \int_{\frac{-h}{2}}^{\frac{h}{2}} zp^\alpha dz - \int_{\frac{-h}{2}}^{\frac{h}{2}} z^2 B_\tau^\alpha p^\tau dz; \quad m = m^\alpha A_\alpha = m_\alpha A^\alpha$$

$$\bar{\theta}_\alpha = -\left(\nabla_\alpha v_3 + B_\alpha^\gamma v_\gamma\right)$$

B.13 Variational Formulation

Let $A_n^{\alpha\beta\rho\gamma} = \int_{\frac{-h}{2}}^{\frac{h}{2}} \psi z^n \left(\bar{\Lambda} G^{\alpha\beta} G^{\rho\gamma} + \bar{\mu}\left(G^{\alpha\rho} G^{\beta\gamma} + G^{\beta\rho} G^{\alpha\gamma}\right)\right) dz;$

$n = 0, 1, 2, 3, 4.$
Then

$$\begin{aligned}
\int_\Omega \sigma^{\alpha\beta}(U)\epsilon_{\alpha\beta}(V)d\Omega = & \int_S \left[A_0^{\alpha\beta\rho\gamma} e_{\rho\gamma} - A_1^{\alpha\beta\rho\gamma} K_{\rho\gamma} + A_2^{\alpha\beta\rho\gamma} Q_{\rho\gamma}\right] \overline{e}_{\alpha\beta} \\
& + \left[-A_1^{\alpha\beta\rho\gamma} e_{\rho\gamma} + A_2^{\alpha\beta\rho\gamma} K_{\rho\gamma} - A_3^{\alpha\beta\rho\gamma} Q_{\rho\gamma}\right] \overline{K}_{\alpha\beta} \\
& + \left[A_2^{\alpha\beta\rho\gamma} e_{\rho\gamma} - A_3^{\alpha\beta\rho\gamma} K_{\rho\gamma} + A_4^{\alpha\beta\rho\gamma} Q_{\rho\gamma}\right] \overline{Q}_{\alpha\beta} dS \\
& = \int_S (N_0^{\alpha\beta} - N_1^{\alpha\beta} + N_2^{\alpha\beta})\overline{e}_{\alpha\beta} + (-N_1^{\alpha\beta} + N_2^{\alpha\beta} \\
& - \quad N_3^{\alpha\beta})\overline{K}_{\alpha\beta} + (N_2^{\alpha\beta} - N_3^{\alpha\beta} + N_4^{\alpha\beta})\overline{Q}_{\alpha\beta})dS \\
& = \quad \int_\Omega N^{\alpha\beta}\overline{e}_{\alpha\beta} + M^{\alpha\beta}\overline{K}_{\alpha\beta} + M^{*\alpha\beta}\overline{Q}_{\alpha\beta})dS
\end{aligned} \tag{B.36}$$

Therefore the variational formulation also reads

$$\int_\Omega \sigma : \epsilon d\Omega = A(u, v) = \int_S \left(N : \overline{e} + M : \overline{K} + M^* : \overline{Q}\right) dS = L(v) \tag{B.37}$$

We thus deduce indefinite Euler's equations in S:

$$\begin{cases}
(M^{\alpha\rho} + M^{*\lambda\rho} B_\lambda^\alpha)\nabla_\rho(B_\alpha^\beta) - \nabla_\alpha\left(N^{\alpha\beta} + 2M^{\alpha\lambda} B_\lambda^\beta + M^{*\rho\alpha} B_\rho^\lambda B_\lambda^\beta\right) = \bar{P}^\beta \\
\left(N^{\alpha\beta} + M^{\alpha\lambda} B_\lambda^\beta\right) B_{\alpha\beta} - \nabla_\alpha \nabla_\beta (M^{\alpha\beta} + M^{*\lambda\beta} B_\lambda^\alpha) = -\bar{P}^3
\end{cases} \tag{B.38}$$

and the boundary conditions on the free border $\bar{\gamma}_1$:

$$\begin{cases}\left(N^{\alpha\beta}+M^{\alpha\lambda}B_{\lambda}^{\beta}\right)\nu_{\beta}-(M^{\gamma\beta}+M^{*\lambda\beta}B_{\lambda}^{\gamma})\nu_{\beta}t_{\gamma}t^{\rho}B_{\rho}^{\alpha}=q^{\alpha};\\ -\nabla_{\beta}(M^{\alpha\beta}+M^{*\lambda\beta}B_{\lambda}^{\alpha})\nu_{\alpha}-\partial_{\vec{t}}((M^{\gamma\beta}+M^{*\lambda\beta}B_{\lambda}^{\gamma})\nu_{\beta}t_{\gamma})+\partial_{\vec{t}}m^{\nu}=q^{3}\\ (M^{\alpha\beta}+M^{*\lambda\beta}B_{\lambda}^{\alpha})\nu_{\alpha}\nu_{\beta}+m^{t}=0\end{cases} \tag{B.39}$$

on the clamped border $\bar{\gamma_0}$

$$U_i=0 \text{ and } \bar{\theta}_{\nu}=0$$

also equivalent to

$$u_{\alpha}=0,\, u_3=0 \text{ and } \partial_{\vec{\nu}}u_3=0$$

Calculations of Transverse Stresses

We deduce from the equilibrium equations $(div\sigma+f=0)$ and the boundary conditions that the transverse stresses satisfy the equation

$$\begin{cases}\frac{d}{dz}\sigma^{\alpha3}+2\Gamma_{\beta3}^{\alpha}\sigma^{\beta3}+\Gamma_{\beta3}^{\beta}\sigma^{\alpha3} & =-\left(\sigma,_{\beta}^{\alpha\beta}+\Gamma_{\beta\tau}^{\alpha}\sigma^{\tau\beta}+\Gamma_{\beta\tau}^{\beta}\sigma^{\alpha\tau}\right)-f^{\alpha}\\ \sigma^{\alpha3}\left(-h/2\right)=-p_{-}^{\alpha},\quad \sigma^{\alpha3}\left(h/2\right) & =p_{+}^{\alpha}\end{cases} \tag{B.40}$$

$$\begin{cases}\frac{d}{dz}\sigma^{33}+\Gamma_{\alpha3}^{\alpha}\sigma^{33} & =-\left(\sigma,_{\alpha}^{3\alpha}+\Gamma_{\alpha\tau}^{3}\sigma^{\tau\alpha}+\Gamma_{\beta\tau}^{\beta}\sigma^{3\tau}\right)-f^{3}\\ \sigma^{33}\left(-h/2\right)=-p_{-}^{3},\ \sigma^{33}\left(h/2\right) & =p_{+}^{3}\end{cases} \tag{B.41}$$

where
$\Gamma_{\beta3}^{\alpha}=-\left(\mu^{-1}\right)_{\lambda}^{\alpha}B_{\beta}^{\lambda};\ \Gamma_{\beta3}^{\beta}=-\left(\mu^{-1}\right)_{\lambda}^{\beta}B_{\beta}^{\lambda};\ \Gamma_{\beta\tau}^{\alpha}=\bar{\Gamma}_{\beta\tau}^{\alpha}+\left(\mu^{-1}\right)_{\nu}^{\alpha}\nabla_{\beta}\mu_{\tau}^{\nu};\ \Gamma_{\alpha\tau}^{3}=\mu_{\alpha}^{\nu}B_{\nu\tau}$.

It is proven in Nzengwa and Tagne [87] that these special differential equations have unique solutions. Their solutions with the initial conditions satisfy the final conditions.

B.14 Best First-Order Model for Thick Shells

The variational formulation

$$\int_S (N(u):e(v)+M(u):K(v)+M^{*}(u):Q(v))=L\left(v\right) \tag{B.42}$$

can be rewritten in the form

$$\mathrm{A}(u, v) = L(v) \text{ with } \mathrm{A}(u, v) = A_1(u, v) + \sum_{n \geq 2} B^n A^n(u, v) = \mathrm{L}(v) \tag{B.43}$$

By neglecting terms greater than one as calculated above, we obtain

$$N^{\alpha\beta} = \frac{Eh}{1 - \bar{\nu}^2}\left[(1 - \bar{\nu}) e^{\alpha\beta}(u) + \bar{\nu} e^{\rho}_{\rho}(u) A^{\alpha\beta}\right]$$

$$M^{\alpha\beta} = \frac{Eh^3}{12\left(1 - \bar{\nu}^2\right)}\left[(1 - \bar{\nu}) K^{\alpha\beta}(u) + \bar{\nu} K^{\rho}_{\rho}(u) A^{\alpha\beta}\right]$$

$$\bar{N}^{\alpha\beta} = \frac{Eh^3}{12\left(1 - \bar{\nu}^2\right)}\left[(1 - \bar{\nu}) Q^{\alpha\beta}(u) + \bar{\nu} Q^{\rho}_{\rho}(u) A^{\alpha\beta}\right]$$

$$\bar{M}^{\alpha\beta} = \frac{Eh^3}{12\left(1 - \bar{\nu}^2\right)}\left[(1 - \bar{\nu}) e^{\alpha\beta}(u) + \bar{\nu} e^{\rho}_{\rho}(u) A^{\alpha\beta}\right]$$

$$\overline{\overline{M}}^{\alpha\beta} = \frac{Eh^5}{80\left(1 - \bar{\nu}^2\right)}\left[(1 - \bar{\nu}) Q^{\alpha\beta}(u) + \bar{\nu} Q^{\rho}_{\rho}(u) A^{\alpha\beta}\right]$$

Terms with even power on h disappear because the integration of z to the odd power in the interval $-h/2$ to $h/2$ is zero. The corresponding variational equation which also has a unique solution (Nzengwa et al.1999) reads

$$\begin{cases} Find\ u \in U_{ad} = H^1_{\gamma_o}(S) \times H^1_{\gamma_o}(S) \times H^2_{\gamma_o}(S) \\ A_1(u, v) = \int_S \left[\left(N + \overline{N}\right) : e(v) + M : K(v) + \left(\overline{M} + \overline{\overline{M}}\right) : Q(v)\right] dS \\ \qquad = \int_S [\mathcal{N} : e(v) + M : K(v) + \mathcal{M} : Q(v)]\, dS = L(v) \end{cases} \tag{B.44}$$

with $\mathcal{N} = N + \overline{N},\ \mathcal{M} = \overline{M} + \overline{\overline{M}}$

$$\sigma^{\alpha\beta} = \frac{1}{h} N^{\alpha\beta} - \frac{12z}{h^3} M^{\alpha\beta} + \frac{80}{h^5} z^2 \overline{\overline{M}}^{\alpha\beta} \tag{B.45}$$

σ^{i3} are calculated by integrating the differential equations established above for transverse stresses.

Chapter 3: The Dynamic Equations

B.15 The N-T Model Variational Equation

$\epsilon_{i3} = 0$

$$\begin{aligned} U_\alpha(x,z,t) &= u_\alpha(x,t) - z(u_{3,\alpha}(x,t) + 2B^\rho_\alpha u_\rho(x,t)) \\ &\quad + z^2\left(B^\tau_\alpha B^\nu_\tau u_\nu(x,t) + B^\tau_\alpha u_{3,\tau}(x,t)\right) \end{aligned} \tag{B.46}$$

$$U_3(x,z,t) = u_3(x,t)$$

and the 3D variational equation reads

$$\int_\Omega \rho \frac{\partial^2 U}{\partial t^2} V d\Omega + \int_\Omega \sigma(U) : \epsilon(V)\, d\Omega = L(V) \tag{B.47}$$

By using the expressions of U, V in u and v, it becomes

$$B\left(\frac{d^2}{dt^2}u, v\right) + A\left(u(x,t), v(x)\right) = L(v) \tag{B.48}$$

with initial conditions $u(x,o)$; $\dot{u}(x,o)$ and

$$\begin{cases} B\left(\frac{d^2}{dt^2}u, v\right) = \int_\Omega \rho \frac{\partial^2 U}{\partial t^2} V d\Omega = \frac{d^2}{dt^2} B(u,v); \\ A(u(x,t), v(x)) = \int_S [N(u(x,t)) : e(v) + M(u(x,t)) : K(v) \\ \qquad + M^*(u(x,t)) : Q(v)] dS \\ \frac{d^2}{dt^2} B(u,v) = \int_S \left(J^\beta(\ddot{u})\, v_\beta + J^3(\ddot{u})\, v_3\right) dS \end{cases} \tag{B.49}$$

$$\begin{aligned} L(v) &= \int_S \left(\bar{P}^\alpha(x,t) v_\alpha(x) + \bar{P}^3(x,t) v_3(x)\right) dS + \int_{\gamma_1} (q^\alpha(x,t) v_\alpha(x) \\ &\quad + q^3(x,t) v_3(x)) d\bar{\gamma} + \int_{\gamma_1} m^\alpha(x,t) \bar{\theta}_\alpha(x) d\bar{\gamma} \end{aligned} \tag{B.50}$$

$$\begin{cases} \dfrac{d}{dz}\sigma^{\alpha 3}(x,t) + 2\Gamma^{\alpha}_{\beta 3}\sigma^{\beta 3} + \Gamma^{\beta}_{\beta 3}\sigma^{\alpha 3} & = -\left(\sigma,^{\alpha\beta}_{\beta} + \Gamma^{\alpha}_{\beta\tau}\sigma^{\tau\beta} + \Gamma^{\beta}_{\beta\tau}\sigma^{\alpha\tau}\right) - f^{\alpha} + \rho\dfrac{\partial^2}{\partial t^2}U_{\alpha} \\ \sigma^{\alpha 3}(x,-h/2,t) = -p^{\alpha}_{-}, & \\ \sigma^{\alpha 3}(x,h/2,t) = p^{\alpha}_{+} & \end{cases} \tag{B.51}$$

$$\begin{cases} \dfrac{d}{dz}\sigma^{33}(x,t) + \Gamma^{\alpha}_{\alpha 3}\sigma^{33} & = -\left(\sigma,^{3\alpha}_{\alpha} + \Gamma^{3}_{\alpha\tau}\sigma^{\tau\alpha} + \Gamma^{\beta}_{\beta\tau}\sigma^{3\tau}\right) - f^{3} + \rho\dfrac{\partial^2}{\partial t^2}U_{3} \\ \sigma^{33}(x,-h/2,t) = -p^{3}_{-}, & \\ \sigma^{33}(x,h/2,t) = p^{3}_{+} & \end{cases} \tag{B.52}$$

$$\frac{d^2}{dt^2}B_1(u(x,t),v) = \int_S \left(I^{\beta}(\ddot{u})\,v_{\beta} + I^{3}(\ddot{u})\,v_3\right)dS; \tag{B.53}$$

$$\begin{aligned} I^{\beta}(\ddot{u}) = \quad & \rho h A^{\alpha\beta}\ddot{u}_{\alpha} + \rho\{\tfrac{h^3}{12}\{A^{\alpha\tau}B^{\beta}_{\nu}B^{\nu}_{\tau}\ddot{u}_{\alpha} \\ & +2A^{\alpha\tau}B^{\beta}_{\tau}\left(\partial_{\alpha}\ddot{u}_3 + 2B^{\nu}_{\alpha}\ddot{u}_{\nu}\right) + A^{\alpha\beta}(B^{\nu}_{\tau}B^{\tau}_{\alpha}\ddot{u}_{\nu} \\ & +B^{\tau}_{\alpha}\partial_{\tau}\ddot{u}_3)\} + \tfrac{h^5}{80}A^{\delta\alpha}B^{\beta}_{\gamma}B^{\gamma}_{\delta}(B^{\nu}_{\tau}B^{\tau}_{\alpha}\ddot{u}_{\nu} + B^{\tau}_{\alpha}\partial_{\tau}\ddot{u}_3)\} \end{aligned}$$

$$\begin{aligned} I^{3}(\ddot{u}) = \rho h\ddot{u}_3 + \rho\frac{h^3}{12}\partial_{\beta}\left(A^{\alpha\tau}B^{\beta}_{\tau}\ddot{u}_{\alpha} + A^{\alpha\beta}\left(\partial_{\alpha}\ddot{u}_3 + 2B^{\tau}_{\alpha}\ddot{u}_{\tau}\right)\right) \\ -\rho\frac{h^5}{80}\partial_{\beta}\left(A^{\alpha\tau}B^{\beta}_{\tau}\left(B^{\nu}_{\delta}B^{\delta}_{\alpha}\ddot{u}_{\nu} + B^{\nu}_{\alpha}\partial_{\nu}\ddot{u}_3\right)\right) \end{aligned}$$

The inertia I^3 contains the classical term $\rho h\ddot{u}_3$ and the term proposed by Morozov (1967)

$$\rho\frac{h^3}{12}\partial_{\beta}\left(A^{\alpha\beta}\partial_{\alpha}\ddot{u}_3\right) = \rho\frac{h^3}{12}\,\triangle\,\ddot{u}_3 \tag{B.54}$$

in a physical base ($A^{\alpha\beta} = \delta^{\alpha\beta}$). Euler's equations are obtained by adding to the static equations inertial terms and initial conditions. The equations now read

$$I^{\alpha}(\ddot{u}) + (M^{\alpha\rho} + M^{*\lambda\rho}B^{\alpha}_{\lambda})\nabla_{\rho}(B^{\beta}_{\alpha}) - \nabla_{\alpha}\left(N^{\alpha\beta} + 2M^{\alpha\lambda}B^{\beta}_{\lambda} + M^{*\rho\alpha}B^{\lambda}_{\rho}B^{\beta}_{\lambda}\right) = \bar{P}^{\beta}; \tag{B.55}$$

$$I^{3}(\ddot{u}) + (N^{\alpha\beta} + M^{\alpha\lambda}B^{\beta}_{\lambda})B_{\alpha\beta} - \nabla_{\alpha}\nabla_{\beta}(M^{\alpha\beta} + M^{*\lambda\beta}B^{\alpha}_{\lambda}) = -\bar{P}^{3} \tag{B.56}$$

to be completed with boundary and initial conditions.

B.16 Best First-Order Equations

By developing N, M and M^* as in the previous chapter

$N(u) = \mathcal{N}(u) + ...,\ M(u) = \mathcal{M}^{o}(u) + ...,\ M^{*}(u) = \mathcal{M}(u) + ...$

We have

$J^{\beta}(u) = I^{\beta}(u) + ...,\ J^{3}(u) = I^{3}(u) + ...$

$$B_1(u, v) = \int_S [I^{\alpha}(u)v_{\alpha} + I^{3}(u)v_3]ds$$

$$B_2(u, v) = \int_S \rho h(u^{\alpha}v_{\alpha} + u^{3}v_3)ds$$

$$B_3(u, v) = \int_S \rho h u^{3}v_3 ds$$

We can define the following approximate variational equation

$A_1(u, v) = B_1(u, v),\ A_1(u, v) = B_2(u, v).$

B.17 Euler's Equation

The dynamic Euler's equation is obtained by adding the inertia terms as follows:

$J^{\beta}(\ddot{u}) + ... = p^{\beta},$

$J^{3}(\ddot{u}_3) + ... = -p^{3}$

The inertia terms can be replaced by $I^{\beta}(\ddot{u})$, $I^{3}(\ddot{u})$ or $\bar{I}^{\alpha}(\ddot{u})$, $\bar{I}^{3}(\ddot{u})$ and σ^{i3} are calculated as in Chap. 2.

B.18 Free Vibration

The 3D dynamic problem writes

$$\begin{cases} \rho \frac{\partial^2}{\partial t^2} U(x,z,t) - div\sigma = 0 \text{ in } \Omega \\ \sigma \vec{n} = 0 \text{ on } \Gamma = \partial\Omega \setminus \Gamma_0 \\ U(x,z,t) = 0 \text{ on } \Gamma_0 \end{cases} \quad \text{(B.57)}$$

By setting $U(x,z,t) = cos(\sqrt{\Lambda}t)U(x,z)$ or $U(x,z,t) = sin(\sqrt{\Lambda}t)U(x,z)$. The problem (B.57) is equivalent to

$$-div\sigma(U(x,z)) = \Lambda\rho U(x,z) \text{ in } \Omega \quad \text{(B.58)}$$

or in its variational form:

$$\int_{\Omega} \sigma(U(x,z)) : \epsilon(V(x,z))\, d\Omega = \Lambda \int_{\Omega} \rho U(x,z).V(x,z) d\Omega \quad \text{(B.59)}$$

equivalent to

$$A(u(x), v(x)) = \Lambda \int_{S} \left(J^{\beta}(u) v_{\beta} + J^{3}(u) v_3\right) dS = \Lambda B(u,v) \quad \text{(B.60)}$$

or

$$A(u,v) = \Lambda B_2(u,v) = \Lambda(u,v)$$

$$\text{or} \quad A_1(u,v) = \Lambda B_2(u,v) = \Lambda(u,v)$$

B.19 N Shell Model

The "N" model accounts for transverse strains and stress. The kinematic from the main hypothesis $rot\phi = 0$, $\phi_{\alpha} = 2\epsilon_{\alpha 3}$, $\phi_3 = \epsilon_{33}$ implies that

$$U_{\alpha} = u_{\alpha} - z\left(\partial_{\alpha} u_3 + 2B_{\alpha}^{\rho} u_{\rho}\right) + z^2\left(B_{\alpha}^{\rho} B_{\rho}^{\tau} u_{\tau} + B_{\alpha}^{\rho} \partial_{\rho} u_3\right), \quad U_3 = u_3 + q \quad \text{(B.61)}$$

where q is the stretching function. The strain components are

$$\epsilon_{\alpha\beta} = e_{\alpha\beta} - zK_{\alpha\beta} + z^2 Q_{\alpha\beta} + q\Upsilon_{\alpha\beta}, \quad \text{(B.62)}$$

$$\epsilon_{\alpha 3} = \frac{1}{2}\partial_{\alpha} q, \quad \text{(B.63)}$$

$$\epsilon_{33} = \partial_z q(x,z), \quad \text{(B.64)}$$

$$\Upsilon_{\alpha\beta} = -\frac{1}{2}\left(\mu_{\alpha}^{\rho} B_{\rho\beta} + \mu_{\beta}^{\rho} B_{\rho\alpha}\right) \quad \text{(B.65)}$$

where $\Upsilon_{\alpha\beta}$ is the warping tensor and the stress-strain relation reads

$$\sigma^{\cdot\cdot} = \bar{\lambda}\epsilon_l^l G^{\cdot\cdot} + 2\bar{\mu}\epsilon^{\cdot\cdot} \tag{B.66}$$

If $q(x, z, t) = w(z)\bar{q}(x, t)$ where $w(z)$ is a transverse distribution function, the general dynamic variational equation obtained by considering the kinematic:

$$\begin{aligned} U_\alpha(x, z, t) &= u_\alpha(x, t) - z\left(\partial_\alpha u_3(x, t) + 2B_\alpha^\delta u_\delta(x, t)\right), \\ &+z^2\left(B_\alpha^\delta B_\delta^\tau u_\tau(x, t) + B_\alpha^\delta \partial_\delta u_3(x, t)\right) \\ U_3(x, z, t) &= \qquad u_3(x, t) + q(x, z, t) \end{aligned} \tag{B.67}$$

reads

$$\begin{cases} \text{find } (u(x, t), \bar{q}(x, t)) \in U_{ad}, \forall t \\ \\ B^w\left((\ddot{u}, \ddot{\bar{q}}); (v, \bar{y})\right) + A^w\left((u, \bar{q}); (v, \bar{y})\right) = L^w(v, \bar{y}) \end{cases} \tag{B.68}$$

where

$$B^w\left((\ddot{u}, \ddot{\bar{q}}); (v, \bar{y})\right) = \int_S \left(J^\alpha(\ddot{u})v_\alpha + J^3(\ddot{u}, \ddot{\bar{q}})v_3 + J^4(\ddot{u}, \ddot{\bar{q}})\bar{y}\right) dS \tag{B.69}$$

with

$$\begin{aligned} J^3(\ddot{u}, \ddot{\bar{q}}) &= \qquad J^3(\ddot{u}) + \ddot{\bar{q}}\textstyle\int_{-\frac{h}{2}}^{\frac{h}{2}} w\psi dz \\ J^4(\ddot{u}, \ddot{\bar{q}}) &= \ddot{\bar{q}}\int_{\frac{-}{h}2}^{\frac{h}{2}} w^2\psi dz + \ddot{u}_3\int_{\frac{-}{h}2}^{\frac{h}{2}} w\psi dz \end{aligned}$$

and

$$L^w(v, \bar{y}) = \int_S \left(\bar{p}^i v_i + \bar{p}^4\bar{y}\right) dS + \int_{\gamma_1} \left(q^i v_i + q^4\bar{y}\right) d\gamma + \int_{\gamma_1} m^\alpha\bar{\theta}_\alpha(v) d\gamma$$

where

$$\bar{p}^4 = \int_{-\frac{h}{2}}^{\frac{h}{2}} f^3 w\psi dz + w(h/2)p_+^3 - w(-h/2)p_-^3, \quad q^4 = \int_{-\frac{h}{2}}^{\frac{h}{2}} p^3 w dz$$

Exercise 3.1

Write the dynamic equations of a cylindrical internal pressurized thick shell, of radius R, clamped on the border. The pressure p is uniform and approximate inertial $I_\alpha,\ I_3$ are considered.

Chapter 4: Thin Shells

B.20 Assumptions

The theory of thin shell is based on the Kirchhoff-Love model obtained under the following hypotheses:

- $\chi \ll 1, \quad (\chi < 0.15)$;
- The covariant base vectors are $G_\alpha \simeq A_\alpha,\ G_3 = A_3,\ G^\alpha \simeq A^\alpha,\ G^3 = A^3$;
- $d\Omega = dSdz$;
- $\epsilon_{i3}(U) = 0$ (Plane strain).
  The kinematic is given by $U(x,z) = U_\alpha(x,z)A^\alpha + U_3(x,z)A^3$; which implies that the displacement

$$U_\alpha(x,z) = u_\alpha(x) - zu_{3,\alpha}(x), \quad U_3(x,z) = u_3(x), \quad u(x) = u_\alpha(x)A^\alpha + u_3(x)A^3 \tag{B.70}$$

- $\sigma^{i3} = 0$ (plane stress) which implies that

$$\begin{aligned}\sigma^{\alpha\beta} &= (\Lambda A^{\alpha\beta}A^{\rho\gamma} + 2\mu A^{\alpha\rho}A^{\beta\gamma})\epsilon_{\rho\gamma}, \quad \epsilon_{\rho\gamma}(u) = e_{\rho\gamma}(u) - zK_{\rho\gamma}(u) \\ &= \frac{E}{1-\nu^2}(\nu e^\rho_\rho A^{\alpha\beta} + (1-\nu)e^{\alpha\beta}) - z\frac{E}{1-\nu^2}(\nu K^\rho_\rho A^{\alpha\beta} + (1-\nu)K^{\alpha\beta})\end{aligned} \tag{B.71}$$

B.21 Variational Equations of the Model Problem

From the 3D mixed problem the variational equation.

$$\begin{cases}\text{Find} \quad u \in H^1_{\gamma_0} \times H^1_{\gamma_0} \times H^2_{\gamma_0} = U_{ad} \\ \displaystyle\int_S (N(u) : e(v) + M(u) : K(v))dS = L(v), \forall v \in U_{ad}\end{cases} \tag{B.72}$$

where

$$N^{\alpha\beta} = \frac{Eh}{1-\nu^2}(\nu e^\rho_\rho A^{\alpha\beta} + (1-\nu)e^{\alpha\beta}), \quad M^{\alpha\beta} = \frac{Eh^3}{12(1-\nu^2)}(\nu K^\rho_\rho + (1-\nu)K^{\alpha\beta})$$

The functions $L(v), e,\ K$ are defined in the previous chapters.

The equilibrium is also characterized by the minimizer in U_{ad} of the total potential energy:

$$J(u) = \frac{1}{2}\int_S N(u) : e(u)dS + \frac{1}{2}\int_S M(u) : K(u)dS - L(u) = W_m + W_f - W_e \tag{B.73}$$

where $W_m(u)$ is the membrane strain energy, W_f is the flectional deformation (bending) energy and W_e is the external energy.

B.22 Euler's Equation

Euler's equations are similar to the equations in Chap. 2 without terms related to $Q(u)$ and M^*.

B.23 Membrane Theory

Membrane theory is adopted if W_f is relatively insignificant (i.e. $\frac{W_f}{W_m} = \mathcal{O}(h)$),

$$J(u) = \frac{1}{2}\int_S N(u) : e(u)dS - L(u) \tag{B.74}$$

$$\begin{gathered} \nabla_\alpha N^{\alpha\beta} + P^\beta = 0, \quad N^{\alpha\beta} B_{\alpha\beta} + P^3 = 0, \quad \text{on } S \\ N^{\alpha\beta}\nu_\beta = q^\alpha \quad \text{on the free border} \\ u_i = 0, \quad \partial_\alpha w = 0 \quad \text{on the clamped border} \end{gathered} \tag{B.75}$$

B.24 Flectional Theory

The Flectional theory is obtained when $(\frac{W_m}{W_f} = \mathcal{O}(h))$

$$J(u) = \frac{1}{2}\int_S M(u) : K(u)dS - L(u) \tag{B.76}$$

$$\nabla_\alpha \nabla_\beta M^{\alpha\beta} = p^3 \tag{B.77}$$

These equations are completed with boundary conditions deduced from the general boundary conditions established in Chap. 2 as follows:

- (i) on the free border ($\overrightarrow{\nu}$ unit outer normal vector)

$$\begin{cases} \left(M^{\alpha\lambda}B^{\beta}_{\lambda}\right)\nu_\beta - (M^{\gamma\beta} = q^\alpha; \\ -\nabla_\beta(M^{\alpha\beta})\nu_\alpha - \partial_{\overrightarrow{t}}((M^{\gamma\beta})\nu_\beta t_\gamma) + \partial_{\overrightarrow{t}} m^\nu = q^3 \\ (M^{\alpha\beta})\nu_\alpha\nu_\beta + m^t = 0 \end{cases} \tag{B.78}$$

- (ii) on the clamped border

$$\partial_{\overrightarrow{\nu}} w = 0, w = 0 \tag{B.79}$$

The procedure to compute $\nabla_\alpha\nabla_\beta M^{\alpha\beta}$ is given as follows:

$$\nabla_\alpha\nabla_\beta M^{\gamma\delta} = \nabla_\alpha U^{\gamma\delta}_{\beta} = U^{\gamma\delta}_{\beta,\alpha} - \Gamma^{\rho}_{\beta\alpha}U^{\gamma\delta}_{\rho} + \Gamma^{\gamma}_{\alpha\rho}U^{\rho\delta}_{\beta} + \Gamma^{\delta}_{\alpha\rho}U^{\gamma\rho}_{\beta} \tag{B.80}$$

where

$$\begin{aligned} U^{\gamma\delta}_{\beta,\alpha} = {} & M^{\gamma\delta}_{,\beta\alpha} + \Gamma^{\gamma}_{\beta\rho}M^{\rho\delta}_{,\alpha} + \Gamma^{\delta}_{\beta\rho}M^{\gamma\rho}_{,\alpha} + \Gamma^{\gamma}_{\beta\rho,\alpha}M^{\rho\delta} + \Gamma^{\delta}_{\beta\rho,\alpha}M^{\gamma\rho} \\ & = D_\alpha D_\beta M^{\gamma\delta} + \Gamma^{\gamma}_{\beta\rho}M^{\rho\delta}_{,\alpha} + \Gamma^{\delta}_{\beta\rho}M^{\gamma\rho}_{,\alpha} + \Gamma^{\gamma}_{\beta\rho,\alpha}M^{\rho\delta} + \Gamma^{\delta}_{\beta\rho,\alpha}M^{\gamma\rho} \end{aligned} \tag{B.81}$$

and

$$\begin{aligned} \nabla_\alpha\nabla_\beta M^{\gamma\delta} = {} & D_\alpha D_\beta M^{\gamma\delta} + \Gamma^{\gamma}_{\beta\rho}M^{\rho\delta}_{,\alpha} + \Gamma^{\delta}_{\beta\rho}M^{\gamma\rho}_{,\alpha} + \Gamma^{\gamma}_{\beta\rho,\alpha}M^{\rho\delta} + \Gamma^{\delta}_{\beta\rho,\alpha}M^{\gamma\rho} \\ & -\Gamma^{\rho}_{\beta\alpha}\nabla_\beta M^{\gamma\delta} + \Gamma^{\gamma}_{\alpha\rho}\nabla_\beta M^{\rho\delta} + \Gamma^{\delta}_{\alpha\rho}\nabla_\beta M^{\gamma\delta} \end{aligned} \tag{B.82}$$

In a rectangular plate, $\nabla_\alpha = D_\alpha$, $\overrightarrow{\nu} = \overrightarrow{i}$ or $\overrightarrow{j}$ and the same for $\overrightarrow{t}$.

B.25 Mixed Theory

Mixed theory is considered if both energies are non-negligible and a simplified kinematic is admissible. Otherwise the full K-L model is applied.

B.26 Reissner-Mindlin Model for Moderate Thick Plates

The Reissner-Mindlin model for moderate thick plates $U(x, z) = U_\alpha A^\alpha + U_3 A^3$ where $U_\alpha(x, z) = u_\alpha(x) + z\theta_\alpha(x)$ and $U_3(x) = u_3(x)$. The function θ_α does not depend on u which means there are five unknown functions. This model does not account for thickness change because $\epsilon_{33} = 0$.

B.27 Isotropic Plate

In this case a pure bending theory is relevant. In a cartesian coordinates system, for a homogeneous isotropic material, we have

$$\begin{cases} D_\alpha D_\beta M^{\alpha\beta} - P \quad = 0 \text{ on } S \\ \\ \text{boundary conditions} \end{cases} \tag{B.83}$$

$$M_{\alpha\beta} = D\left[(1-\bar{\nu})\, K_{\alpha\beta} + \bar{\nu} tr K \delta_{\alpha\beta}\right], \; K_{..} = \begin{bmatrix} w_{,xx} & w_{,xy} \\ \\ w_{,xy} & w_{,yy} \end{bmatrix}, \; D = \frac{Eh^3}{12\left(1-\bar{\nu}^2\right)} \tag{B.84}$$

which is equivalent to

$$\begin{cases} D\left(\frac{\partial^4 w}{\partial x^4} + 2\frac{\partial^4 w}{\partial x^2 \partial y^2} + \frac{\partial^4 w}{\partial y^4}\right) = P \text{ on } S \\ \\ \text{boundary conditions} \end{cases} \tag{B.85}$$

or

$$\begin{cases} D\Delta\Delta w = P \\ \\ \text{boundary conditions} \end{cases} \tag{B.86}$$

Under simple supported border, we have $w = 0$ and $m^s = 0$;
under clamped border, $w = 0$ and $\partial_{\vec{\nu}} w = 0$.

The variational equation now reads

$$\int_S M^{\alpha\beta} D_\alpha D_\beta v dS + \int_C D_\beta M^{\alpha\beta} \nu_\alpha v dC - \int_C M^{\alpha\beta} \nu_\beta D_\alpha v dC = \int_S P v dS = \langle P, v\rangle \tag{B.87}$$

$w,\ v \in H^2(S) \cap H_0^1(S)$ for the simple supported border and $w,\ v \in H_0^2(S)$ for the clamped border. But $M^{\alpha\beta}\nu_\beta = M^{\alpha\nu} = M^{..}\vec{\nu} = M^{\nu\nu}\vec{\nu} + M^{\nu t}\vec{t}$. Therefore $M^{\alpha\beta}\nu_\beta D_\alpha v = (M^{\nu\nu}\vec{\nu} + M^{\nu t}\vec{t}) \cdot \nabla v = M^{\nu\nu}\partial_{\vec{\nu}} v + M^{\nu t}\partial_{\vec{t}} v$. For the simple supported border, $\partial_{\vec{t}} v = 0$ and $M^{\nu\nu} = 0$ since $m^s = 0$. For the clamped border $\partial_{\vec{\nu}} v =$

0 and $\partial_{\vec{t}} v = 0$. Therefore, in both situations $M^{\alpha\beta}\nu_\beta D_\alpha v = 0$ and $D_\beta M^{\alpha\beta}\nu_\alpha v = 0$ on the border C. The variational formulation now reads

$$\int_S M^{\alpha\beta} D_\alpha D_\beta v dS = \int_S P v dS = \langle P, v \rangle \tag{B.88}$$

$$\begin{aligned}\int_S D\left((1-\bar{\nu})K^{\alpha\beta}v_{,\alpha\beta} + \bar{\nu}K^\rho_\rho v_{,\gamma\gamma}\right) dS = \int_S D((1-\bar{\nu})w_{,\alpha\beta}v_{,\alpha\beta} \\ +\bar{\nu}w_{,\rho\rho}v_{,\gamma\gamma})dS \\ = \langle P, v\rangle\end{aligned} \tag{B.89}$$

or

$$\int_S D\Delta w \Delta v dS = \langle P, v \rangle \tag{B.90}$$

B.28 Orthotropic Plate (Huber's Equation)

Huber's equation is

$$\begin{cases} D_x \dfrac{\partial^4 w}{\partial x^4} + 2H\dfrac{\partial^4 w}{\partial x^2 y^2} + D_y \dfrac{\partial^4 w}{\partial y^4} = P \text{ on } S \\ \text{Boundary condition} \end{cases} \tag{B.91}$$

B.29 Von Karman Equations

$$D_\alpha N^{\alpha\beta} + p^\beta = 0 \text{ on } S \tag{B.92}$$

$$D_\alpha D_\beta M^{\alpha\beta} - b_{\alpha\beta}N^{\alpha\beta} - p^3 = 0 \text{ on } S \tag{B.93}$$

Under simple supported boundary conditions without applied bending moment, or under clamped edge, the variational equation reads

$$\begin{cases} \text{Find } (w_1, w_2, w_3) \text{ such that} \\ \displaystyle\int_S M^{\alpha\beta}\partial_\alpha\partial_\beta \bar{w}_3 dS - \int_S N^{\alpha\beta} b_{\alpha\beta}\bar{w}_3 dS + \int_S N^{\alpha\beta}\partial_\alpha \bar{w}_\beta dS = \int_S p^i \bar{w}_i dS \end{cases} \tag{B.94}$$

$\bar{w}_3, w_3, \in H^2(S) \cap H_0^1(S)$ for simple support and $\bar{w}_3 \in H_0^2(S)$ for clamped edge, $w_\alpha, \bar{w}_\alpha \in H_0^1(S)$.

In practice, it is frequent to deal with plates without membrane (or in-plane) loads p^β. The stresses $N^{\alpha\beta}$ may be due to curvature created by the deformation or pre-stresses applied on the edges.

By introducing Airy's stress functions on $N^{\alpha\beta} = N_{\alpha\beta}$, we obtain new equations.

Let Φ denote the Airy stress function. Then $N_{xx} = \Phi_{,yy}$; $N_{yy} = \Phi_{,xx}$; $N_{xy} = -\Phi_{,xy}$. By replacing $N_{\alpha\beta}$ by these relations in the above equation we obtain the equations

$$\Delta\Delta\Phi = Eh(w_{,xy}^2 - w_{,xx}w_{,yy}) \tag{B.95}$$

and

$$D\Delta\Delta w - \Phi_{,yy}w_{,xx} - \Phi_{,xx}w_{,yy} + 2\Phi_{,xy}w_{,xy} = p^3 \tag{B.96}$$

which are the Von Karman equations. For some particular boundary conditions, under some simplifying hypotheses on Φ and w, an analytic approximate solution of the Von Karman equations can be developed in series.

Exercise 4.1

(i) Calculate D_x, D_y and H and write the equation under a load $p(x, y)$ on the grid beam desk (see Fig. (4.19)).
(ii) Let $w(0, y) = w(a, y) = 0$. Calculate the reaction at $(0, y)$ and at (a, y).
(iii) Calculate the rotation of the cross section at $(0, y)$ and (a, y).

Exercise 4.2

Consider the simply supported desk subject to pre-stress $p(x)$ on its border $x = 0$ and $x = a$.

The displacement is $w(x, y) = a_{mn}w_{mn}(x, y)$, $w_{mn}(x, y) = \sin\dfrac{m\pi x}{a}\sin\dfrac{n\pi y}{b}$.
(i) Write down Von Karman's equation.
(ii) Calculate a_{mn}.

Exercise 4.3

Consider the vertical cylindrical tank of medium radius R, thickness h filled with a liquid of density ρ.
(i) Calculate $\sigma^{\alpha\beta}$ on the lateral part.
(ii) Calculate the max of M^{rr} on the disc.
(iii) Determine the thickness of a material whose maximum bearing stress is $210Mpa$.

Chapter 5: Numerical Methods

The main objective of this chapter is to present succinctly finite element methods (FEM) to discretize the continuous variational equations of N-T and N shell models. These FEM also address the K-L and R-M models which are included in the N-T model. The N model variational equations contain the N-T variational equations and also account for transverse strains and thickness change. We begin by rewriting all the various variational equations considered herein for self-containess of numerical development of the model mixed boundary value problem that has been treated throughout the text.

B.30 N-T Model ($\epsilon_{i3} = 0$)

Recall that in static analysis, the kinematics of the N-T model read

$$U_\alpha(x,z) = u_\alpha(x) - z(u_{3,\alpha}(x) + 2B^\rho_\alpha u_\rho(x)) + z^2\left(B^\tau_\alpha B^\nu_\tau u_\nu(x) + B^\tau_\alpha u_{3,\tau}(x)\right);$$
$$U_3(x,z) = u_3(x) \tag{B.97}$$

The variational solutions are found in the space $U_{ad} \subset \left(H^1(S)\right)^2 \times H^2(S)$.

The strain and stress tensors read

$$\epsilon_{\alpha\beta}(U) = \epsilon_{\alpha\beta}(\bar{U}) = e_{\alpha\beta}(u) - zK_{\alpha\beta}(u) + z^2 Q_{\alpha\beta}(u);\ \epsilon_{\alpha3} = 0;\ \epsilon_{33} = 0$$

and

$$\sigma^{\alpha\beta} = \bar{\Lambda}\epsilon^\rho_\rho G^{\alpha\beta} + 2\bar{\mu}\epsilon^{\alpha\beta} = (\bar{\Lambda}G^{\alpha\beta}G^{\gamma\rho} + \bar{\mu}(G^{\alpha\rho}G^{\beta\gamma} + G^{\beta\rho}G^{\alpha\gamma}))\epsilon_{\rho\gamma}(U)$$

where

$$G^{\alpha\beta} = \left(\mu^{-1}\right)^\alpha_\nu \cdot \left(\mu^{-1}\right)^\beta_\rho A^{\nu\rho} \tag{B.98}$$

Transverse stresses are obtained by solving the equations

$$\begin{cases} \dfrac{d}{dz}\sigma^{\alpha3} + 2\Gamma^\alpha_{\beta3}\sigma^{\beta3} + \Gamma^\beta_{\beta3}\sigma^{\alpha3} = -\left(\sigma,^{\alpha\beta}_\beta + \Gamma^\alpha_{\beta\tau}\sigma^{\tau\beta} + \Gamma^\beta_{\beta\tau}\sigma^{\alpha\tau}\right) - f^\alpha + \rho\dfrac{\partial^2}{\partial t^2}U_\alpha \\ \sigma^{\alpha3}(-h/2) = -p^\alpha_-, \\ \sigma^{\alpha3}(h/2) = p^\alpha_+ \end{cases} \tag{B.99}$$

$$\begin{cases} \dfrac{d}{dz}\sigma^{33} + \Gamma^{\alpha}_{\alpha 3}\sigma^{33} & = -\left(\sigma,^{3\alpha}_{\alpha} + \Gamma^{3}_{\alpha\tau}\sigma^{\tau\alpha} + \Gamma^{\beta}_{\beta\tau}\sigma^{3\tau}\right) - f^3 + \rho\dfrac{\partial^2}{\partial t^2}U_3 \\ \sigma^{33}(-h/2) = -p^3_-, & \\ \sigma^{33}(h/2) = p^3_+ & \end{cases} \tag{B.100}$$

We also have

$$d\Omega = (G_1,\ G_2,\ A_3)\,dx^1dx^2dz = \sqrt{G}dxdz$$
$$= \left(det(\mu^{\alpha}_{\beta})\right)\sqrt{A}dxdz = \psi(x,z)dSdz; \tag{B.101}$$

$$\int_{\Omega}\sigma^{ij}(U)\epsilon_{ij}(V)d\Omega = \int_{\Omega}\sigma^{\alpha\beta}(U)\epsilon_{\alpha\beta}(V)d\Omega$$
$$= \int_S dS\int_{\frac{-h}{2}}^{\frac{h}{2}}\psi(x,z)\,[\bar{\Lambda}G^{\alpha\beta}G^{\rho\gamma}\bar{\mu}\left(G^{\alpha\rho}G^{\beta\gamma} + G^{\beta\rho}G^{\alpha\gamma}\right)]\epsilon_{\rho\gamma}(U)\,\epsilon_{\alpha\beta}(V)\,dz$$
$$= A(u,v) \tag{B.102}$$

The variational equation reads

$$\begin{cases}\text{Find } u \in U_{ad}\\ A(u,v) = L(v), \forall v \in U_{ad}\end{cases} \tag{B.103}$$

where

$$L(v) = \int_S f^i v_i dS + \int_{\gamma_1} q^i v_i d\gamma + \int_{\gamma} m^{\alpha}\overline{\theta}_{\alpha}d\gamma \tag{B.104}$$

with

$$q^{\alpha} = \int_{\frac{-h}{2}}^{\frac{h}{2}} p^{\alpha}dz - \int_{\frac{-h}{2}}^{\frac{h}{2}} zB^{\alpha}_{\tau}p^{\tau}dz;\ \ q^3 = \int_{\frac{-h}{2}}^{\frac{h}{2}} p^3dz;$$

$$m^{\alpha} = \int_{\frac{-h}{2}}^{\frac{h}{2}} zp^{\alpha}dz - \int_{\frac{-h}{2}}^{\frac{h}{2}} z^2B^{\alpha}_{\tau}p^{\tau}dz;\ \ m = m^{\alpha}A_{\alpha} = m_{\alpha}A^{\alpha}$$

We also have

$$\bar{\theta}_{\alpha} = -\left(\nabla_{\alpha}v_3 + B^{\gamma}_{\alpha}v_{\gamma}\right),\ \ \overline{\theta}_{\nu} = \overline{\theta}.\overrightarrow{\nu},\ \ \overline{\theta}_t = \theta.\overrightarrow{t}$$

The external work density due to applied moment on the border is

$$m^{\alpha}\bar{\theta}_{\alpha} = m^{t}\bar{\theta}_{\nu} - m^{\nu}\bar{\theta}_{t} \tag{B.105}$$

The best first-order variational equation is established under the hypotheses: $G^{\alpha\beta} \approx A^{\alpha\beta}$, $\psi(x, z) \simeq 1$, $d\Omega \simeq dSdz$.

The variational formulation reads

$$\begin{cases} \text{Find } u \in U_{ad} \\ A_1(u, v) = L(v), \quad \forall v \in U_{ad} \end{cases} \tag{B.106}$$

B.31 N Model

In static analysis, the kinematic reads

$$U_{\alpha} = u_{\alpha} - z\left(\partial_{\alpha}u_3 + 2B^{\rho}_{\alpha}u_{\rho}\right) + z^2\left(B^{\rho}_{\alpha}B^{\tau}_{\rho}u_{\tau} + B^{\rho}_{\alpha}\partial_{\rho}u_3\right), \ U_3 = u_3 + q \tag{B.107}$$

where q is the stretching function.

Herein

$$q(x, z) = w(z)\bar{q}(x), \tag{B.108}$$

where $w(z)$ is the transverse distribution function.

The strain tensor reads

$$\begin{aligned}
\epsilon_{\alpha\beta} &= e_{\alpha\beta} - zK_{\alpha\beta} + z^2 Q_{\alpha\beta} + q\Upsilon_{\alpha\beta} \\
\epsilon_{\alpha 3} &= \frac{1}{2}\partial_{\alpha}q = \frac{1}{2}w(z)\partial_{\alpha}\bar{q} \\
\epsilon_{33} &= \partial_z w(z)\bar{q} \\
\Upsilon_{\alpha\beta} &= -\frac{1}{2}\left(\mu^{\rho}_{\alpha}B_{\rho\beta} + \mu^{\rho}_{\beta}B_{\rho\alpha}\right)
\end{aligned}$$

The constitutive law now reads

$$\sigma = \bar{\lambda}\epsilon^{l}_{l}G^{\cdot\cdot} + 2\bar{\mu}\epsilon = \sigma\left(\epsilon\left(u\right)\right) + \sigma\left(\epsilon\left(q\right)\right) \tag{B.109}$$

$$\sigma^{ij}(u, q) = \bar{\lambda}\left(\epsilon^{\alpha}_{\alpha}\left(u\right) + \epsilon^{l}_{l}\left(q\right)\right)G^{ij} + 2\bar{\mu}\left(\epsilon^{ij}\left(u\right) + \epsilon^{ij}\left(q\right)\right) = C^{ijkl}\epsilon_{kl}(u, q) \tag{B.110}$$

and the moduli tensor $C = \left(C^{ijkl}\right)$ satisfies all material symmetries and the ellipticity condition.

The best first-order variational equation is rearranged as

$$\begin{cases} \text{find } (u(x), \bar{q}(x)) \in U_{ad} \times H^1_{\gamma_0} \\ A^w\left((u,\bar{q});(v,\bar{y})\right) = L^w(v,\bar{y}) \forall (v,\bar{y}) \in U_{ad} \times H^1_{\gamma_0} \end{cases} \tag{B.111}$$

where

$$\begin{aligned}
A^w\left((u,\bar{q});(v,\bar{y})\right) = & \int_S \int_{-\frac{h}{2}}^{\frac{h}{2}} \sigma^{ij}(u,q)\epsilon_{ij}(v,w\bar{y})\psi dz dS \\
= & \int_S \int_{-\frac{h}{2}}^{\frac{h}{2}} [A^{\alpha\beta\delta\tau}\epsilon_{\delta\tau}(u)\epsilon_{\alpha\beta}(v) + wA^{\alpha\beta\delta\tau}\Upsilon_{\delta\tau}\epsilon_{\alpha\beta}(u)\bar{y} \\
& +\bar{\lambda}w'G^{\alpha\beta}\epsilon_{\alpha\beta}(u)\bar{y} + (\bar{\lambda}G^{\alpha\beta}w' + wA^{\alpha\beta\delta\tau}\Upsilon_{\delta\tau}) \\
& \bar{q}\epsilon_{\alpha\beta}(v)]\psi dz dS + \int_S \int_{-\frac{h}{2}}^{\frac{h}{2}} [(w^2A^{\alpha\beta\delta\tau}\Upsilon_{\delta\tau}\Upsilon_{\alpha\beta} \\
& +\bar{\lambda}ww'G^{\alpha\beta}\Upsilon_{\alpha\beta})\bar{q}\bar{y} + (\bar{\lambda}+2\bar{\mu})(w')^2\bar{q}\bar{y} \\
& +\bar{\mu}G^{\alpha\beta}w^2\partial_\beta\bar{q}\partial_\alpha\bar{y}]\psi dz dS \\
= & \int_S (N^{\alpha\beta}(u)e_{\alpha\beta}(v) + M^{\alpha\beta}(u)K_{\alpha\beta}(v) \\
& +M^{*\alpha\beta}(u)Q_{\alpha\beta}(v))dS + \int_S (D^{\alpha\beta}_{0w}e_{\alpha\beta}(u) - D^{\alpha\beta}_{1w}K_{\alpha\beta}(u) \\
& +D^{\alpha\beta}_{2w}Q_{\alpha\beta}(u))\bar{y}dS + \int_S (E^{\alpha\beta}_{0w}e_{\alpha\beta}(u) - E^{\alpha\beta}_{1w}K_{\alpha\beta}(u) \\
& +E^{\alpha\beta}_{2w}Q_{\alpha\beta}(u))\bar{y}dS + \int_S \bar{q}(F^{\alpha\beta}_{0w}(\Upsilon)e_{\alpha\beta}(v) - F^{\alpha\beta}_{1w}(\Upsilon)K_{\alpha\beta}(v) \\
& +F^{\alpha\beta}_{2w}(\Upsilon)Q_{\alpha\beta}(v))dS + \int_S (\bar{q}\bar{y}(F^{33}_{w0}(\Upsilon) + F^{33}_{w1}(\Upsilon) + F^{33}_{w2})dS \\
& +\int_S (I^{\alpha\beta}_{ww}\partial_\beta\bar{q}\partial_\alpha\bar{y}dS
\end{aligned} \tag{B.112}$$

The terms $N^{\alpha\beta}(u),\ M^{\alpha\beta}(u),\ M^{*\alpha\beta}(u)$ are defined as earlier with $\bar{\lambda}$ in place of $\bar{\Lambda} = 2\bar{\mu}\bar{\lambda}/(\bar{\lambda}+2\bar{\mu})$. The additional coefficients are defined as follows:

$$D_{0w}^{\alpha\beta} = \int_{-\frac{h}{2}}^{\frac{h}{2}} \left[wA^{\alpha\beta\delta\tau}\Upsilon_{\delta\tau}\right]\psi dz\ \ D_{1w}^{\alpha\beta} \quad = \quad \int_{-\frac{h}{2}}^{\frac{h}{2}} z\left[wA^{\alpha\beta\delta\tau}\Upsilon_{\delta\tau}\right]\psi dz;$$

$$D_{2w}^{\alpha\beta} = \int_{-\frac{h}{2}}^{\frac{h}{2}} z^2\left[wA^{\alpha\beta\delta\tau}\Upsilon_{\delta\tau}\right]\psi dz;$$

$$E_{0w}^{\alpha\beta} = \int_{-\frac{h}{2}}^{\frac{h}{2}} \left[\bar{\lambda}G^{\alpha\beta}w'\right]dz,\ \ E_{1w}^{\alpha\beta} = \int_{-\frac{h}{2}}^{\frac{h}{2}} z\left[\bar{\lambda}G^{\alpha\beta}w'\right]dz$$

,

$$E_{2w}^{\alpha\beta} = \int_{-\frac{h}{2}}^{\frac{h}{2}} z^2\left[\bar{\lambda}G^{\alpha\beta}w'\right]dz$$

$$F_{0w}^{\alpha\beta}(\Upsilon) = \int_{-\frac{h}{2}}^{\frac{h}{2}} \left[wA^{\alpha\beta\delta\tau}\Upsilon_{\delta\tau} + \bar{\lambda}w'G^{\alpha\beta}\right]dz$$

$$F_{1w}^{\alpha\beta}(\Upsilon) = \int_{-\frac{h}{2}}^{\frac{h}{2}} z\left[wA^{\alpha\beta\delta\tau}\Upsilon_{\delta\tau} + \bar{\lambda}w'G^{\alpha\beta}\right]dz$$

$$F_{2w}^{\alpha\beta}(\Upsilon) = \int_{-\frac{h}{2}}^{\frac{h}{2}} z^2\left[wA^{\alpha\beta\delta\tau}\Upsilon_{\delta\tau} + \bar{\lambda}w'G^{\alpha\beta}\right]dz$$

$$F_{w0}^{33}(\Upsilon) = \int =_{-\frac{h}{2}}^{\frac{h}{2}} \left[(w^2A^{\alpha\beta\delta\tau}\Upsilon_{\delta\tau}\Upsilon_{\alpha\beta} + \bar{\lambda}ww'G^{\alpha\beta}\Upsilon_{\alpha\beta})\right]dz,$$

$$F_{w1}^{33}(\Upsilon) = \qquad \int_{-\frac{h}{2}}^{\frac{h}{2}} \left[\bar{\lambda}wG^{\alpha\beta}\Upsilon_{\alpha\beta}w'\right]dz,$$

$$F_{w2}^{33}(\Upsilon) = \qquad \int_{-\frac{h}{2}}^{\frac{h}{2}} \left[(\bar{\lambda}+2\bar{\mu})\left(w'\right)^2\right]dz$$

$$I_{ww}^{\alpha\beta} = \int_{-\frac{h}{2}}^{\frac{h}{2}} \left[\bar{\mu}G^{\alpha\beta}w^2\right]\psi dz$$

This variational equation can be rearranged in order to highlight the impact of the transverse deformation on the resultant forces and moments. We rewrite the equation as follows:

$$\int_S ((N^{\alpha\beta}(u) + N_w^{\alpha\beta}(u,\bar{q}))e_{\alpha\beta}(v) + (M^{\alpha\beta}(u) + M_w^{\alpha\beta}(u,\bar{q}))K_{\alpha\beta}(v)$$

$$+(M^{*\alpha\beta}(u) + M_w^{*\alpha\beta}(u,\bar{q}))Q_{\alpha\beta}(v))dS + \int_S (T_w^{\alpha}(u,\bar{q})\partial_\alpha \bar{y}$$

$$+T_w^3(u,\bar{q})\bar{y})dS = L^w(v,\bar{y})$$

$$N_w^{\alpha\beta}(u,\bar{q}) = \bar{q}(F_{0w}^{\alpha\beta}(\Upsilon), \;\; M_w^{\alpha\beta}(u,\bar{q}) = -\bar{q}F_{1w}^{\alpha\beta}(\Upsilon),$$

$$M_w^{*\alpha\beta}(u,\bar{q}) = \bar{q}F_{2w}^{\alpha\beta}(\Upsilon)$$

B.32 Triangularization Mesh

The mid-surface S of the shell is meshed with triangles having straight line or curved line edges. The triangles are characterized by the maximum edge length of all the triangles denoted by h and the mesh is denoted by $\mathcal{T}_h$. Flat elements (element with very very small area $\triangle$) are not accepted.

B.33 Shape Functions and Nodal Vector of Degree of Freedom (dof)

Let T_e, $e \in \mathcal{T}_h$ be a triangle whose summit-nodes have the coordinates $M_1 = (x_1, y_1)$, $M_2 = (x_2, y_2)$ and $M_3 = (x_3, y_3)$. Let P_1 denote spaces of polynomials of degree 1, generated by the bases $\{1, x, y\}$ or the barycentric functions $\{\eta^1, \eta^2, \eta^3\}$ defined as follows:

$$\begin{cases} \eta^1(x,y) = & \dfrac{1}{2\,\triangle}[(y_3 - y_2)(x_2 - x) - (x_3 - x_2)(y_2 - y)] \\ \eta^2(x,y) = & \dfrac{1}{2\,\triangle}[(y_1 - y_3)(x_3 - x) - (x_1 - x_3)(y_3 - y)] \\ \eta^3(x,y) = & \dfrac{1}{2\,\triangle}[(y_2 - y_1)(x_1 - x) - (x_2 - x_1)(y_1 - y)] \end{cases} \tag{B.113}$$

and $\eta^1 + \eta^2 + \eta^3 = 1$. The space of polynomials of degree 2 is denoted by P_2 and $\{1, x, y, x^2, xy, y^2\}$ or $\{\eta^1, \eta^2, \eta^3, 4\eta^1\eta^2, 4\eta^1\eta^3, 4\eta^2\eta^3\}$ constitute two equivalent bases.

The values of a function u at nodes of the triangle are denoted by u^1, u^2, u^3, u^4, u^5, u^6.

If $u \in P_1$ then $u = a_1x + a_2y + a_3 = u^1\eta^1 + u^2\eta^2 + u^3\eta^3$ If $u \in P_2$ then $u = a_1x + a_2y + a_3x^2 + a_4xy + a_5y^2 + a_6 = u^1\eta^1 + u^2\eta^2 + u^3\eta^3 + u^4(4\eta^1\eta^2) + u^5(4\eta^1\eta^3) + u^6(4\eta^2\eta^3)$.

We define specific finite element to be used to interpolate vector fields restricted in a triangle element.

1. The element T_{GR}^{12} used for the gradient recovery method is defined as follows: $u_1, u_2 \in P_1$ and $u_3 \in P_2$

$u_1 = a_1x + a_2y + a_3, u_2 = a_4x + a_5y + a_6, u_3 = a_7x^2 + a_8xy + a_9y^2 + a_{10}x + a_{11}y + a_{12}$.

$$\begin{pmatrix} u_1 \\ u_2 \\ u_3 \end{pmatrix} = \begin{bmatrix} x & y & 1 & 0 & 0 & 0 & 0 & 0 & 0 & 0 & 0 & 0 \\ 0 & 0 & 0 & x & y & 1 & 0 & 0 & 0 & 0 & 0 & 0 \\ 0 & 0 & 0 & 0 & 0 & 0 & x^2 & xy & y^2 & x & y & 1 \end{bmatrix} \begin{pmatrix} a_1 \\ a_2 \\ a_3 \\ a_4 \\ a_5 \\ a_6 \\ a_7 \\ a_8 \\ a_9 \\ a_{10} \\ a_{11} \\ a_{12} \end{pmatrix} \tag{B.114}$$

In the barycentric base, we write

$$\begin{pmatrix} u_1 \\ u_2 \\ u_3 \end{pmatrix} = \begin{bmatrix} \eta^1 & \eta^2 & \eta^3 & 0 & 0 & 0 & 0 & 0 & 0 & 0 & 0 & 0 \\ 0 & 0 & 0 & \eta^1 & \eta^2 & \eta^3 & 0 & 0 & 0 & 0 & 0 & 0 \\ 0 & 0 & 0 & 0 & 0 & 0 & \eta^1 & \eta^2 & \eta^3 & 4\eta^1\eta^2 & 4\eta^1\eta^3 & 4\eta^2\eta^3 \end{bmatrix} \begin{pmatrix} u_1^1 \\ u_1^2 \\ u_1^3 \\ u_2^1 \\ u_2^2 \\ u_2^3 \\ u_3^1 \\ u_3^2 \\ u_3^3 \\ u_3^4 \\ u_3^5 \\ u_3^6 \end{pmatrix} \tag{B.115}$$

Let ${}^tU_e = \left(u_1^1, u_1^2, u_1^3, u_2^1, u_2^2, u_2^3, u_3^1, u_3^2, u_3^3, u_3^4, u_3^5, u_3^6\right)$.

U_e is 12 dof nodal element vector.

2. The element T_{SD}^{18} is used for strain deformation approach. The local shape functions should reproduce three rotations and three translations per node. They should account for uncoupled states due to shearing, membrane deformation and bending. So locally

u can be decoupled as $u = u^0 + u^p$ where u^0 captures rigid body motion (i.e. $\epsilon(u^0) = 0,\ \ u^0 \neq 0$). For a cylindrical shell with radius R and open angle $2\theta_0$, the following shape functions fulfil the conditions.

$$^tU_e = \left(u_1^1, u_2^1, u_3^1, \beta_1^1, \beta_2^1, \gamma^1, u_1^2,\ u_2^2,\ u_3^2,\ \beta_1^2,\ \beta_2^2, \gamma^2,\ u_1^3,\ u_2^3,\ u_3^3,\ \beta_1^3,\ \beta_2^3,\ \gamma^3\right) \quad \text{(B.116)}$$

where $\beta_\alpha = u_{3,\alpha} + B_\alpha^\rho u_\rho$, $\gamma = u_{1,2} - u_{2,1} - u_{3,12}$.

Letting $y = \varphi$.

The solution of these equations on a cylindrical shell with an open angle $2\theta_0$ is $u^0 = (u_{10},\ u_{20},\ u_{30})$, to be completed by a particular solution $u^p = (u_{1p},\ u_{2p},\ u_{3p})$ as follows:

$$\begin{pmatrix} u_{10} \\ u_{20} \\ u_{30} \end{pmatrix} = \begin{bmatrix} 1 & R(cos\theta - cos\theta_0) & -Rsin\theta & 0 & 0 & 0 \\ 0 & xsin\theta & xcos\theta & -Rsin^2\theta & -sin\theta & cos\theta \\ 0 & -xcos\theta & xsin\theta & Rsin\theta cos\theta & cos\theta & sin\theta \end{bmatrix} \begin{pmatrix} a_1 \\ a_2 \\ a_3 \\ a_4 \\ a_5 \\ a_6 \end{pmatrix} \quad \text{(B.117)}$$

$$\begin{pmatrix} u_{1p} \\ u_{2p} \\ u_{3p} \end{pmatrix} = \begin{bmatrix} Rx & R\theta & Rx\theta & 0 & 0 & 0 & 0 & 0 & 0 & 0 & 0 & 0 \\ 0 & 0 & 0 & \theta & x\theta & 0 & 0 & 0 & 0 & 0 & 0 & 0 \\ 0 & 0 & 0 & 0 & 0 & x^2 & x\theta & \theta^2 & x^3 & x^2\theta & x\theta^2 & \theta^3 \end{bmatrix} \begin{pmatrix} a_1 \\ a_2 \\ a_3 \\ a_4 \\ a_5 \\ a_6 \\ a_7 \\ a_8 \\ a_9 \\ a_{10} \\ a_{11} \\ a_{12} \end{pmatrix} \quad \text{(B.118)}$$

such that $u = u^0 + u^p$

B.34 Shift Lagrange Elements

- T_{SL}^9 with 9 dof for the N-T model.
  Shape functions are defined using shift parameters X and Y defined as follows:

$$X(x) = x + \beta, \quad Y(y) = y + \beta, \quad \alpha, \ \beta \in \mathbb{R}$$

$$\begin{aligned} u_1(X, Y) = & \quad a_1 + a_2 X + a_3 Y, \\ u_2(X, Y) = & \quad a_4 + a_5 X + a_6 Y, \\ u_3(X, Y) = & \quad a_7 X^2 + a_8 XY + a_9 Y^2 \end{aligned} \tag{B.119}$$

- T_{SL}^{12} with 12 dof for the N model
  $\bar{q} = a_{10} + a_{11} X + a_{12} Y$ yielding 12dof for the model N
  ${}^tU_e = (u_1^1, u_1^2, u_1^3, \bar{q}^1, u_2^1, u_2^2, u_2^3, \bar{q}^2, u_3^1, u_3^2, u_3^3, \bar{q}^3)$

B.35 The Classical T_{mN}^{15} Element for the N Model

We have

$u_\alpha \in P_1$, $u_3 \in P_2$ and $\bar{q} \in P_1$,
${}^tU_e = (u_1^1, u_2^1, u_3^1, \bar{q}^1, u_1^2, u_2^2, u_3^2, \bar{q}^2, u_1^3, u_2^3, u_3^3, \bar{q}^3, u_3^4, u_3^5, u_3^6)$
or
${}^tU^e = (u_1^1, u_1^2, u_1^3, u_2^1, u_2^2, u_2^3, u_3^1, u_3^2, u_3^3, u_3^4, u_3^5, u_3^6, \bar{q}^1, \bar{q}^2, \bar{q}^3)$

B.36 Some Useful Relations

Let $x = (x^1, x^2)$ be the coordinate system on the mid-surface S. Let $m \in S$ be a generic point and $T \subset S$ a subdomain. Let f be a scalar function defined on T, then

$$\int_T f(m)dS = \int_{\bar{T}_x} \bar{f}\sqrt{A}dx^1dx^2 = \int_{\bar{T}_x} \bar{f}(x)dx \tag{B.120}$$

where $\bar{T}_x$ is the domain of $x \in \mathbb{R}^2$ and $A = \det(A_{\alpha\beta})$.

On a sphere for example $(\theta, \varphi) = (x^1, x^2) = x$. Consider a portion of the sphere of radius R, $\theta \in [\theta_1, \theta_2]$, $\varphi \in [\varphi_1, \varphi_2]$ and $f(x) = 1 = \|\vec{n}\|$. Then $dS = R^2 \sin\theta d\theta d\varphi$, and

$$\int_T f dS = \int_{[\theta_1,\theta_2]\times[\varphi_1,\varphi_2]} R\sin\theta d\theta d\theta = R^2(\varphi_2 - \varphi_1)(\cos\theta_1 - \cos\theta_2) \tag{B.121}$$

If $\varphi_1 = 0$, $\varphi_2 = 2\pi$, $\theta_1 = 0$, $\theta_2 = \pi$, then the answer is $4\pi R^2$, which is the surface of the sphere. Therefore, through the parametrization the integral is performed on a subdomain in R^2. The subdomain is a triangle T_x in R^2 if T is curved triangle on S.

B.37 The Barycentric Coordinates System

Consider a triangle T with summit-nodes denoted M_1, M_2, M_3 with respective coordinate (x_1, y_1), (x_2, y_2), (x_3, y_3). Let $M(x, y) \in T$ then, there exist $\eta^1, \eta^2, \eta^3 \in \mathbb{R}$ such that $x = x_1\eta^1 + x_2\eta^2 + x_3\eta^3 = x_i\eta^i$, $y = y_i\eta^i$ and $\eta^1 + \eta^2 + \eta^3 = 1$.

The numbers η^i constitute the barycentric coordinates of a point $M \in T$ and verify

$$\eta^1(x, y) = \frac{1}{J}(y_{32}x_2x - x_{32}y_2y), \tag{B.122}$$

$$\eta^2(x, y) = \frac{1}{J}(y_{13}x_3x - x_{13}y_3y), \tag{B.123}$$

$$\eta^3(x, y) = \frac{1}{J}(y_{21}x_1x - x_{21}y_1y) \tag{B.124}$$

where $J = x_{12}y_{23} - x_{23}y_{12}$ and $x_{ij} = x_i - x_j$, $y_{ij} = y_i - y_j$, $x_ix = x_i - x$, $y_iy = y_i - y$. So we can rewrite

$$x = x_{13}\eta^1 + x_{23}\eta^2 + x_3 = x(\eta^1, \eta^2), \tag{B.125}$$

$$y = y_{13}\eta^1 + y_{23}\eta^2 + y_3 = y(\eta^1, \eta^2) \tag{B.126}$$

We can observe that the relations are linear. So

$$x_{,\alpha\beta} = y_{,\alpha\beta} = 0,\ \alpha, \beta = \eta^1 \text{ or } \eta^2 \tag{B.127}$$

$$\eta^1_{,\alpha\beta} = \eta^2_{,\alpha\beta} = 0,\ \alpha, \beta = x \text{ or } y \tag{B.128}$$

$$\begin{aligned} \frac{\partial\eta^1}{\partial x^1} &= \frac{1}{J}y_{12}x_2 = a_{11}, \\ \frac{\partial\eta^1}{\partial x^2} &= -\frac{1}{J}x_{32}y_2 = a_{12}, \\ \frac{\partial\eta^2}{\partial x^1} &= \frac{1}{J}y_{13}x_3 = a_{21}, \\ \frac{\partial\eta^2}{\partial x^2} &= -\frac{1}{J}x_{13}y_3 = a_{22} \end{aligned} \tag{B.129}$$

B.38 Relations Between Derivations

Let $x = (x^1, x^2) = (x, y)$. A function $u(x) = u(x(\eta^1, \eta^2)) = \bar{u}(\bar{\eta})$, $\bar{\eta} = (\eta^1, \eta^2)$.

$$\begin{aligned}\frac{\partial u}{\partial x^1} &= \frac{\partial \bar{u}}{\partial \eta^1}\frac{\partial \eta^1}{\partial x^1} + \frac{\partial \bar{u}}{\partial \eta^2}\frac{\partial \eta^2}{\partial x^1} = a_{11}\frac{\partial \bar{u}}{\partial \eta^1} + a_{21}\frac{\partial \bar{u}}{\partial \eta^2} \\ \frac{\partial \bar{u}}{\partial x^2} &= \frac{\partial \bar{u}}{\partial \eta^1}\frac{\partial \eta^1}{\partial x^2} + \frac{\partial \bar{u}}{\partial \eta^2}\frac{\partial \eta^2}{\partial x^2} = a_{12}\frac{\partial \bar{u}}{\partial \eta^1} + a_{22}\frac{\partial \bar{u}}{\partial \eta^2}\end{aligned} \tag{B.130}$$

$$\begin{aligned}\frac{\partial^2 u}{\partial x^1 \partial x^1} &= a_{11}^2\frac{\partial^2 \bar{u}}{\partial \eta^1 \partial \eta^1} + (a_{11}a_{21})\frac{\partial^2 \bar{u}}{\partial \eta^1 \partial \eta^2} + a_{21}^2\frac{\partial^2 u}{\partial \eta^2 \partial \eta^2} \\ \frac{\partial^2 u}{\partial x^1 \partial x^2} &= a_{21}a_{11}\frac{\partial^2 \bar{u}}{\partial \eta^1 \partial \eta^1} + (a_{11}a_{12} + a_{22}a_{21})\frac{\partial^2 \bar{u}}{\partial \eta^1 \partial \eta^2} + a_{12}a_{22}\frac{\partial^2 \bar{u}}{\partial \eta^2 \partial \eta^2} \\ \frac{\partial^2 u}{\partial x^2 \partial x^2} &= a_{12}^2\frac{\partial^2 \bar{u}}{\partial \eta^1 \partial \eta^1} + (a_{12}a_{22})\frac{\partial^2 \bar{u}}{\partial \eta^1 \partial \eta^2} + a_{22}^2\frac{\partial^2 \bar{u}}{\partial \eta^2 \partial \eta^2}\end{aligned} \tag{B.131}$$

$$\begin{pmatrix} u \\ \frac{\partial u}{\partial x^1} \\ \frac{\partial u}{\partial x^2} \\ \frac{\partial^2 u}{\partial x^1 \partial x^1} \\ \frac{\partial^2 u}{\partial x^1 \partial x^2} \\ \frac{\partial^2 u}{\partial x^2 \partial x^2} \end{pmatrix} = \underbrace{\begin{bmatrix} 1 & 0 & 0 & 0 & 0 & 0 \\ 0 & a_{11} & a_{21} & 0 & 0 & 0 \\ 0 & a_{12} & a_{22} & 0 & 0 & 0 \\ 0 & 0 & 0 & a_{11}^2 & a_{11}a_{21} & a_{21}^2 \\ 0 & 0 & 0 & a_{21}a_{11} & a_{11}a_{12} + a_{22}a_{21} & a_{12}a_{22} \\ 0 & 0 & 0 & a_{12}^2 & a_{12}a_{22} & a_{22}^2 \end{bmatrix}}_{B_f} \begin{pmatrix} \bar{u} \\ \frac{\partial \bar{u}}{\partial \eta^1} \\ \frac{\partial \bar{u}}{\partial \eta^2} \\ \frac{\partial^2 \bar{u}}{\partial \eta^1 \partial \eta^1} \\ \frac{\partial^2 \bar{u}}{\partial \eta^1 \partial \eta^2} \\ \frac{\partial^2 \bar{u}}{\partial \eta^2 \partial \eta^2} \end{pmatrix} \tag{B.132}$$

$$\underbrace{\begin{pmatrix} u_1 \\ u_{1,1} \\ u_{1,2} \\ u_2 \\ u_{2,1} \\ u_{2,2} \\ u_3 \\ u_{3,1} \\ u_{3,2} \\ u_{3,11} \\ u_{3,12} \\ u_{3,22} \end{pmatrix}}_{D_e} = \begin{bmatrix} \begin{matrix} 1 & 0 & 0 & 0 & 0 & 0 \\ 0 & a_{11} & a_{21} & 0 & 0 & 0 \\ 0 & a_{12} & a_{22} & 0 & 0 & 0 \\ 0 & 0 & 0 & 1 & 0 & 0 \\ 0 & 0 & 0 & 0 & a_{11} & a_{21} \\ 0 & 0 & 0 & 0 & a_{12} & a_{22} \end{matrix} & O \\ O & B_f \end{bmatrix} \underbrace{\begin{pmatrix} \bar{u}_1 \\ \bar{u}_{1,1} \\ \bar{u}_{1,2} \\ \bar{u}_2 \\ \bar{u}_{2,1} \\ \bar{u}_{2,2} \\ \bar{u}_3 \\ \bar{u}_{3,1} \\ \bar{u}_{3,2} \\ \bar{u}_{3,11} \\ \bar{u}_{3,12} \\ \bar{u}_{3,22} \end{pmatrix}}_{D_{ee}} = \tag{B.133}$$

$$D_e = \begin{bmatrix} B_m & O \\ O & B_f \end{bmatrix} D_{ee}, \quad = B^e_{mf} D_{ee} \tag{B.134}$$

where B^e_{mf} is a 12×12 matrix.

So

$$E^e_e = \begin{pmatrix} e_{11} \\ e_{22} \\ 2e_{12} \end{pmatrix} = C_e D_e = C_e B^e_{mf} D_{ee} \tag{B.135}$$

$$E^e_K = \begin{pmatrix} K_{11} \\ K_{22} \\ 2K_{12} \end{pmatrix} = C_K D_e = C_K B^e_{mf} D_{ee}$$

$$E^e_Q = \begin{pmatrix} Q_{11} \\ Q_{22} \\ 2Q_{12} \end{pmatrix} = C_Q D_e = C_Q B^e_{mf} D_{ee}$$

In the above equations, the vector D_{ee} is given by

$$D_{ee} = F_e(\bar{\eta}) U_e \tag{B.136}$$

where the 12×12 matrix F_e reads

$$F_e = \begin{bmatrix} \begin{matrix} \eta^1 & \eta^2 & \eta^3 & 0 & 0 & 0 \\ 1 & 0 & -1 & 0 & 0 & 0 \\ 0 & 1 & -1 & 0 & 0 & 0 \\ 0 & 0 & 0 & \eta^1 & \eta^2 & \eta^3 \\ 0 & 0 & 0 & 1 & 0 & -1 \\ 0 & 0 & 0 & 0 & 1 & -1 \end{matrix} & O \\ O & \begin{matrix} \eta^1 & \eta^2 & \eta^3 & 4\eta^1\eta^2 & 4\eta^1\eta^3 & 4\eta^2\eta^3 \\ 1 & 0 & -1 & 4\eta^2 & 4\eta^3 - 4\eta^1 & -4\eta^2 \\ 0 & 1 & -1 & 4\eta^1 & -4\eta^1 & 4\eta^3 - 4\eta^2 \\ 0 & 0 & 0 & 0 & -8 & 0 \\ 0 & 0 & 0 & 4 & -4 & -8 \\ 0 & 0 & 0 & 0 & 0 & -8 \end{matrix} \end{bmatrix} \tag{B.137}$$

The nodal strain is given by

$$\epsilon_e = \begin{pmatrix} \epsilon_{11} \\ \epsilon_{22} \\ 2\epsilon_{12} \end{pmatrix} = \left(C_e B^e_{mf} F^e(\eta) - z C_K B^e_{mf} F^e + z^2 C_Q B^e_{mf} F^e \right) U_e \tag{B.138}$$

where C_e, C_K, C_Q are 3×12 matrices and read

$$C_e = \begin{bmatrix} -\Gamma_{11}^1 \; 1 \; 0 \; -\Gamma_{11}^2 \; 0 \; 0 \; -B_{11} \; 0 \; 0 \; 0 \; 0 \; 0 \\ -\Gamma_{22}^1 \; 0 \; 0 \; -\Gamma_{22}^2 \; 0 \; 1 \; -B_{22} \; 0 \; 0 \; 0 \; 0 \; 0 \\ -2\Gamma_{12}^1 \; 0 \; 1 \; -2\Gamma_{12}^2 \; 1 \; 0 \; -2B_{12} \; 0 \; 0 \; 0 \; 0 \; 0 \end{bmatrix} \tag{B.139}$$

$$C_K = \begin{bmatrix} K_1 & B_1^1 & 0 & K_2 & B_1^2 & 0 & B_1^1 B_{11} + B_1^1 B_{22} & -\Gamma_{11}^1 & -\Gamma_{11}^2 & 1 & 0 & 0 \\ K_3 & 0 & B_2^1 & K_4 & B_2^1 & B_2^2 + B_1^2 & -B_2^1 B_{12} - B_2^2 B_{22} & -\Gamma_{22}^1 & -\Gamma_{22}^2 & 0 & 0 & 1 \\ K_5 & 2B_2^1 & 2B_1^1 & K_6 & 2B_2^2 & 2B_1^2 & -2B_1^1 B_{12} - 2B_1^2 B_{22} & -2\Gamma_{12}^1 & -2\Gamma_{12}^2 & 0 & -2 & 0 \end{bmatrix} \tag{B.140}$$

where

$K_1 = \nabla_1 B_1^1 - B_1^1 \Gamma_{11}^1 - B_1^2 \Gamma_{11}^1,$

$K_2 = \nabla_1 B_1^1 - B_1^1 \Gamma_{11}^1 - B_1^2 \Gamma_{11}^1,$

$K_3 = \nabla_2 B_2^1 - B_2^1 \Gamma_{12}^1 - B_2^1 \Gamma_{21}^1 - B_2^2 \Gamma_{12}^2,\ K_4 = B_2^1 \Gamma_{21}^2 + \nabla_2 B_2^2 - B_2^1 \Gamma_{21}^2 - B_2^2 \Gamma_{22}^2,$

$K_5 = 2\nabla_1 B_2^1 - 2B_2^1 \Gamma_{12}^1 - 2B_2^2 \Gamma_{12}^1 - 2B_1^1 \Gamma_{12}^1 - 2B_1^2 \Gamma_{22}^1$

$K_6 = -2\nabla_1 B_2^2 - 2B_2^1 \Gamma_{11}^2 - 2B_2^2 \Gamma_{12}^2 - 2B_1^1 \Gamma_{12}^2 - 2B_1^2 \Gamma_{22}^2$

$$C_Q = \begin{bmatrix} Q_1 & B_1^1 B_1^1 + B_1^2 B_2^1 & 0 & Q_2 & B_1^1 B_1^2 + B_2^1 B_2^2 & 0 & 0 \ldots \\ \ldots - B_1^1 \Gamma_{11}^1 - B_1^2 \Gamma_{12}^1 & -B_1^1 \Gamma_{11}^2 - B_1^2 \Gamma_{12}^2 & B_1^1 & B_1^2 & 0 & & \\ Q_3 & 0 & B_1^1 B_2^1 + B_2^1 B_1^2 & Q_4 & 0 & B_2^1 B_1^2 + B_2^2 B_2^2 & 0 \ldots \\ \ldots - B_2^1 \Gamma_{12}^1 - B_2^2 \Gamma_{22}^1 & -B_2^1 \Gamma_{22}^2 - B_2^2 \Gamma_{22}^1 & 0 & B_2^1 & B_2^2 & & \\ Q_5 & B_2^1 B_1^2 + B_2^2 B_1^1 & B_1^1 B_1^1 + B_2^1 B_1^2 & Q_6 & B_1^2 B_1^2 + B_2^1 B_2^2 & B_1^1 B_1^2 + B_2^1 B_2^2 & 0 \ldots. \\ \ldots. Q_7 & Q_8 & Q_9 & B_2^2 + B_1^2 & B_1^1 & B_1^2 & \end{bmatrix} \tag{B.141}$$

$Q_1 = B_1^1 \nabla_1 B_1^1 - B_1^1 B_1^1 \Gamma_{11}^1 - B_1^1 B_1^2 \Gamma_{21}^1 - B_1^1 B_2^1 \Gamma_{11}^1 - B_2^1 B_2^2 \Gamma_{12}^1,$

$Q_2 = B_1^1 \nabla_1 B_1^2 - B_1^1 B_1^1 \Gamma_{11}^2 - B_1^1 B_1^2 \Gamma_{12}^2 - B_1^2 B_2^1 \Gamma_{11}^2 - B_2^1 B_2^2 \Gamma_{12}^2 + B_1^2 \nabla B_2^2,$

$Q_3 = B_2^1 \nabla_2 B_1^1 + B_2^1 B_1^1 \Gamma_{11}^1 - B_2^1 B_1^2 \Gamma_{22}^1 - B_2^2 B_2^1 \Gamma_{12}^2 - B_2^2 B_2^2 \Gamma_{12}^1 + B_2^2 \nabla B_2^1,$

$Q_4 = -B_2^1 B_1^1 \Gamma_{11}^2 + B_2^1 \nabla_2 B_1^2 - B_2^1 B_1^1 \Gamma_{22}^2 - B_2^2 B_2^1 \Gamma_{12}^2 + B_2^2 \nabla_2 B_2^2 - B_2^2 B_2^2 \Gamma_{22}^2,$

$Q_5 = -B_1^1 B_1^1 \Gamma_{12}^1 - B_1^1 B_1^2 \Gamma_{22}^1 + B_1^1 \nabla_1 B_1^1 + B_1^2 \nabla_2 B_2^1 - B_1^2 B_2^1 \Gamma_{22}^1 - B_1^2 B_2^2 \Gamma_{22}^1 +$
$B_1^2 \nabla_1 B_1^1 - B_1^2 B_1^1 \Gamma_{11}^1 - B_1^2 B_1^1 \Gamma_{12}^1 - \nabla_1 B_1^1 B_2^2 - B_2^2 B_1^1 \Gamma_{11}^1 - B_2^2 B_1^2 \Gamma_{12}^1$

$Q_6 = -B_1^1 B_1^1 \Gamma_{12}^2 - B_1^1 B_1^2 \Gamma_{22}^2 + B_1^1 \nabla_2 B_1^2 + B_1^2 \nabla_2 B_2^2 - B_1^2 B_2^1 \Gamma_{22}^2 - B_1^2 B_2^2 \Gamma_{22}^2 +$
$B_1^2 \nabla_1 B_1^2 - B_1^2 B_1^1 \Gamma_{11}^2 - B_2^2 B_1^1 \Gamma_{12}^2 - B_2^2 B_1^1 \Gamma_{11}^2 - B_1^2 B_2^2 \Gamma_{12}^2 + B_2^2 \nabla_1 B_1^2$

$Q_7 = -B_1^2 \Gamma_{22}^1 - B_2^2 \Gamma_{11}^1 - B_1^1 \Gamma_{12}^1 - B_1^2 \Gamma_{11}^1$

$Q_8 = -B_1^2 \Gamma_{22}^1 - B_1^1 \Gamma_{12}^1 - B_2^2 \Gamma_{11}^1 - B_1^2 \Gamma_{11}^1,$

$Q_9 = -B_1^1 \Gamma_{12}^2 - B_1^2 \Gamma_{22}^2 - B_1^2 \Gamma_{11}^2 - B_2^2 \Gamma_{11}^2.$

The constitutive laws of stress tensor in the N-T model read

$$\sigma^{\alpha\beta} = (\Lambda A^{\alpha\beta} A^{\gamma\delta} + 2\mu A^{\alpha\gamma} A^{\beta\delta}) \epsilon_{\gamma\delta}(U) \tag{B.142}$$

where the exact metric tensor $G^{\alpha\beta}$ has been replaced by $A^{\alpha\beta}$ in order to implement the best first-order variational equation. In the vector form, we have

$$\begin{pmatrix} \sigma^{11} \\ \sigma^{22} \\ \sigma^{12} \end{pmatrix} = C_\sigma \begin{pmatrix} \epsilon_{11} \\ \epsilon_{22} \\ 2\epsilon_{12} \end{pmatrix} \tag{B.143}$$

where the 3×3 matrix is

$$C_\sigma = \begin{bmatrix} (\Lambda + 2\mu)A^{11}A^{11};\ \Lambda A^{11}A^{12} + 2\mu A^{12}A^{12};\ (\Lambda + 2\mu)A^{11}A^{12} \\ \Lambda A^{11}A^{22} + 2\mu A^{12}A^{12};\ (\Lambda + 2\mu)A^{22}A^{22};\ (\Lambda + 2\mu)A^{22}A^{12} \\ (\Lambda + 2\mu)A^{12}A^{11};\ (\Lambda + 2\mu)A^{12}A^{22};\ \Lambda A^{12}A^{12} + \mu(A^{11}A^{22} + A^{12}A^{21}) \end{bmatrix} \tag{B.144}$$

In finite element (a triangle in this case), we have calculated the restriction of the stress and strain tensors, as functions of shape functions and the local dof vector. The discrete variational equation is obtained by calculating the virtual internal work and external work per element and assembling over all the elements T_e of the mesh $\mathcal{T}_h$ as follows:

$$A(u, v) = \int_S \int_{h^-}^{h^+} \sigma^{\alpha\beta}(u)\epsilon_{\alpha\beta}(v) dS dz = \sum_{T_e \in \mathcal{T}_h} L(v)|_{T_e} \tag{B.145}$$

Hereafter $d\Omega \approx dSdz$.

Let us calculate the internal virtual work. We have per element:

$$\int_{T_e} \int_{h^-}^{h^+} \sigma^{\alpha\beta}(u)\epsilon_{\alpha\beta}(v) dS dz = \int_{T_e} \int_{h^-}^{h^+} \sigma^e(u) : \epsilon^e(v) dS dz \tag{B.146}$$

In an element we begin by calculating on Gauss points

(i) Geometric data

$(A_{\alpha\beta}),\ (A^{\alpha\beta}),\ A = \det(A_{\alpha\beta}),\ B_{\alpha\beta},\ B^\alpha_\beta,\ \nabla_\rho B^\alpha_\beta,\ \nabla_\rho B_{\alpha\beta},\ \Gamma^1_{\alpha\beta},\ \Gamma^2_{\alpha\beta}$

(ii) The strain tensor

$$\epsilon^e(U) = \begin{pmatrix} \epsilon_{11}(U) \\ \epsilon_{22}(U) \\ 2\epsilon_{12}(U) \end{pmatrix} = \left(C_e B^e_{mf} F^e(\eta) - z C_K B^e_{mf} F^e(\eta) + z^2 C_Q B^e_{mf} F^e(\eta)\right) U^e_m \tag{B.147}$$

$$= (\epsilon^e_e - z\epsilon^e_K + z^2\epsilon^e_Q) U^e_m$$

where

$$\epsilon^e_e = C_e B^e_{mf} F^e(\eta),\ \epsilon^e_K = C_K B^e_{mf} F^e(\eta,\ \epsilon^e_Q = C_Q B^e_{mf} F^e(\eta) \tag{B.148}$$

U^e_m is the nodal vector, m is the number of dof which depends on the choice of the finite element ($m = 9dof$ for the shift Lagrange element, $m = 12dof$ for the classical element and $m = 18dof$ for the strain deformation element).

$$\epsilon^e(V) = \begin{pmatrix} \epsilon_{11}(V) \\ \epsilon_{22}(V) \\ 2\epsilon_{12}(V) \end{pmatrix} = (C_e B^e_{mf} F^e(\eta) - zC_K B^e_{mf} F^e(\eta) + z^2 C_Q B^e_{mf} F^e(\eta))V^e_m \tag{B.149}$$

$$\sigma^e = \begin{pmatrix} \sigma^{11} \\ \sigma^{22} \\ \sigma^{12} \end{pmatrix} = C_\sigma \begin{pmatrix} \epsilon^e_{11}(U) \\ \epsilon^e_{22}(U) \\ 2\epsilon^e_{12}(U) \end{pmatrix} = C_\sigma(C_e B^e_{mf} F^e(\eta) - zC_K B^e_{mf} F^e(\eta) + z^2 C_Q B^e_{mf} F^e(\eta))U^e_m \tag{B.150}$$

We have

$$\int_{T_e}\int_{h^-}^{h^+} \sigma^e(u) : \epsilon^e(v) dz dS = \int_{T_e}\int_{h^-}^{h^+} C_\sigma(\epsilon^e_e - z\epsilon^e_K + z^2\epsilon^e_Q)U^e_m.(\epsilon^e_e - z\epsilon^e_K + z^2\epsilon^e_Q)V^e_m \tag{B.151}$$

We have

$$\begin{aligned}\int_{T_e}\int_{h^-}^{h^+} \sigma^e(u) : \epsilon^e(v) dz dS = \int_{T_e}\Big[\int_{h^-}^{h^+} \Big[&{}^t\epsilon^e_e C_\sigma \epsilon^e_e - z\,{}^t\epsilon^e_K C_\sigma \epsilon^e_e + z^2\,{}^t\epsilon^e_Q C_\sigma \epsilon^e_e - z\,{}^t\epsilon^e_e C_\sigma \epsilon^e_K \\ &+ z^2\,{}^t\epsilon^e_K C_\sigma \epsilon^e_K - z^3\,{}^t\epsilon^e_Q C_\sigma \epsilon^e_K + z^2\,{}^t\epsilon^e_e C_\sigma \epsilon^e_Q \\ &- z^3\,{}^t\epsilon^e_K C_\sigma \epsilon^e_Q + z^4\,{}^t\epsilon^e_Q C_\sigma \epsilon^e_Q\Big]dz\Big]U^e_m.V^e_m dS\end{aligned} \tag{B.152}$$

which reduces in the case of constant thickness ($h^- = -h/2, h^+ = h/2$) to $\mathbb{M}_e U^e_m.V^e_m$:

$$\int_{T_e}\int_{h/2}^{-h/2} \sigma^e(u)\epsilon^e(v) dz dS = \mathbb{M}_e U^e_m.V^e_m \tag{B.153}$$

where

$$\mathbb{M}_e = \int_{T_{ref}} \Big[h\,{}^t\epsilon^e_e C_\sigma \epsilon^e_e + \frac{h^3}{12}{}^t\epsilon^e_Q C_\sigma \epsilon^e_e + \frac{h^3}{12}{}^t\epsilon^e_K C_\sigma \epsilon^e_K + \frac{h^3}{12}{}^t\epsilon^e_e C_\sigma \epsilon^e_Q\Big] + \frac{h^5}{80}{}^t\epsilon^e_Q C_\sigma \epsilon^e_Q\Big] J\sqrt{A} d\eta \tag{B.154}$$

The element stiffness matrix $\mathbb{M}_e$ is a $m \times m$ matrix according to the choice of the finite elements, $m = 12$ in this case. Remind that the Gauss formula is used for all integrations. The elementary stiffness matrix $\mathbb{M}_e$ is the contribution of the element T_e to the global stiffness matrix $\mathbb{M}$. Let X denote the vector of all the dof. The global vector X can be rearranged in three parts: X_1 for unknown dof, X_2 for dof that interact with other structures such as soil-structure interaction on heavy duty (industrial foundations) and X_3 for imposed dof or dof that satisfies some constraints. After enumeration of each part, the global matrix $\mathbb{M}$ is filled with contributions from each element, i.e. the elementary stiffness matrix is immersed into $\mathbb{M}$ by expressing $\mathbb{M}_e$ in the global matrix $\mathbb{M}$. Let P_e denote the position vector matrix such that $U^e_m = P^e X$, $V^e_m = P^e X$. Then the contribution of $\mathbb{M}_e$ denoted $\mathbb{M}^g_e$ is ${}^t P^e \mathbb{M}_e P_e$ and

$$\mathbb{M} = \sum_{T_e \in \mathcal{T}_h} \mathbb{M}^g_e \tag{B.155}$$

Let N_d be the number of dof and N_e the number of elements, then P^e is a $m \times N_d$ matrix and N_e position matrices are needed to fill $\mathbb{M}$. In practice, the matrix P^e is not stored. Instead, each dof i in U^e_m or V^e_m is referenced in the total dof vector by a connector $Num(n_e, i)$ and $\mathbb{M}$ is filled as follows:

$$\begin{aligned} n_e &= 1, \ldots.., N_e \\ i, j &= 1, \ldots.m \\ I &= Num(n_e, i) \\ J &= Num(n_e, j) \\ \mathbb{M}(I, J) &= \mathbb{M}(I, J) + \mathbb{M}_e(i, j) \end{aligned} \tag{B.156}$$

The external virtual work per element reads

$$\int_{T_e} (P^i v_i dS + \int_{\gamma_1 \cap \partial T_e} q^i v_i d\gamma_1 + \int_{\gamma_1 \cap \partial T_e} (m^t \overline{\theta}_\nu - m^\nu \overline{\theta}_t) d\gamma \tag{B.157}$$

where $\overline{\theta}_\nu = -(\nabla_\alpha v_3 + B^\rho_\alpha v_\rho)\nu^\alpha$ and $\theta_t = \overline{\theta}.t$.

We have

$$\begin{aligned} v_1 &= v_{11}\eta^1 + v_{12}\eta^2 + v_{13}(1 - \eta^1 - \eta^2) \\ v_2 &= v_{21}\eta^1 + v_2\eta^2 + v_{23}(1 - \eta^1 - \eta^2) \\ v_3 &= v_{31}\eta^1 + v_{32}\eta^2 + v_{33}(1 - \eta^1 - \eta^2) + v_{34}(4\eta^1\eta^2) + v_{35}(4\eta^1\eta^3) + v_{36}(4\eta^2\eta^3) \end{aligned} \tag{B.158}$$

$$\begin{pmatrix} v_1 \\ v_2 \\ v_3 \end{pmatrix} = \begin{bmatrix} \eta^1 & \eta^2 & \eta^3 & 0 & 0 & 0 & 0 & 0 & 0 & 0 & 0 \\ 0 & 0 & 0 & \eta^1 & \eta^2 & \eta^3 & 0 & 0 & 0 & 0 & 0 & 0 \\ 0 & 0 & 0 & 0 & 0 & 0 & \eta^1 & \eta^2 & \eta^3 & 4\eta^1\eta^2 & 4\eta^1\eta^3 & 4\eta^2\eta^3 \end{bmatrix} V^e_m = L^e_m(\overline{\eta}) V^e_m \tag{B.159}$$

$$\int_{T_e} P^i v_i dS = \left(\int_{T_{ref}} {}^t L^e(\overline{\eta})[P^i]\sqrt{A} J d\overline{\eta} \right) V^e_m = F^e_S . V^e_m \tag{B.160}$$

where $J = 2|T_e|, {}^t[P^i] = [P^1, P^2, P^3]$.

Let us denote $x^1 = x$ and $x^2 = y$. The coordinates of the three summit-nodes of the triangle T_e are (x_1, y_1), (x_2, y_2) and (x_3, y_3). We still denote $x_i - x_j = x_{ij}$ and $y_i - y_j = y_{ij}$ and $J = 2|T_e|$, where $|T_e|$ is the area of T_e. The local coordinates or the barycentric coordinate systems is still denoted η^1, η^2, η^3 and $\eta^1 + \eta^2 + \eta^3 = 1$, $\bar{\eta} = (\eta^1, \eta^2)$, $x = x_{13}\eta^1 + x_{23}\eta^2 + x_3 = x(\bar{\eta})$, $y = y_{13}\eta^1 + y_{23}\eta^2 + y_3 = y(\bar{\eta})$. We also remind that if $v(x, y) = v(x(\bar{\eta}), y(\bar{\eta})) = v(\bar{\eta})$, then

$$\frac{\partial v}{\partial x} = \frac{y_{23}}{J}\frac{\partial \bar{v}}{\partial \eta^1} - \frac{y_{13}}{J}\frac{\partial \bar{v}}{\partial \eta^2}; \frac{\partial v}{\partial y} = -\frac{x_{23}}{J}\frac{\partial \bar{v}}{\partial \eta^1} + \frac{x_{13}}{J}\frac{\partial \bar{v}}{\partial \eta^2} \tag{B.161}$$

From the definition of $\theta_\alpha = v_{3,\alpha} + B^1_\alpha v_1 + B^2_\alpha v_2$, we obtain

$$\begin{aligned} \theta_1 &= v_{3,1} + B^1_1 v_1 + B^2_1 v_2, \\ \theta_2 &= v_{3,2} + B^1_2 v_1 + B^2_2 v_2 \end{aligned} \tag{B.162}$$

We deduce that

$$\begin{pmatrix} \theta_1 \\ \theta_2 \end{pmatrix} = L^e_\gamma . V^e_m \tag{B.163}$$

where the 2×12 matrix L^e_γ is defined as follows:

$$L^e_\gamma = \begin{bmatrix} B^1_1\eta^1 & B^1_1\eta^2 & B^1_1\eta^3 & B^2_1\eta^1 & B^2_1\eta^2 & B^2_1\eta^3 & \frac{y_{23}-y_{13}}{J} & \frac{y_{13}-y_{23}}{J} & l_1 & B^1_2\eta^1 \ldots \\ \ldots & B^1_2\eta^2 & B^1_2\eta^3 & & & & & & & \\ B^2_2\eta^1 & B^2_2\eta^2 & B^2_2\eta^3 & -\frac{x_{23}}{J} & \frac{x_{13}}{J} & \frac{x_{23}-x_{13}}{J} & l_2 & l_3 & \frac{4x_{23}\eta^2}{J} & \frac{x_{13}(4-8\eta^2)}{J} \ldots \\ \ldots & l_4 & l_5 & & & & & & & \end{bmatrix} \tag{B.164}$$

where $l_1 = \dfrac{4y_{23}\eta^2 - 4y_{13}\eta^1}{J}$, $l_2 = \dfrac{4x_{23}\eta^1 - 4x_{13}\eta^2}{J}$, $l_3 = -\dfrac{x_{23}}{J}(4 - 8\eta^2) - \dfrac{x_{13}\eta^1}{J}$, $l_4 = \dfrac{y_{23}}{J}(4 - 8\eta^1) + \dfrac{4y_{13}}{J}\eta^1$, $l_5 = -\dfrac{4y_{23}}{J}\eta^2 - \dfrac{y_{13}}{J}(4 - 8\eta^2)$.

Remind that the term on the border does not concern all the dof in V^e_m. On a border of a triangle only two coordinates are considered. So we either have $\eta^1 + \eta^2 = 1$, $\eta^1 + \eta^3 = 1$ or $\eta^2 + \eta^3 = 1$. The line element is $d\gamma = \sqrt{A^{\alpha\beta}\dfrac{dx^\alpha}{d\lambda}\dfrac{dx^\beta}{d\lambda}}$, where $x(\lambda)$ is the curve on the border. For example, consider a border between two nodes having (x_1, y_1) and (x_2, y_2) as coordinates. Then from $\eta^3 = 0$ and $\eta^1 + \eta^2 = 1$, we have

$$\begin{aligned} x &= x_1\eta^1 + x_2\eta^2 = x_{12}\eta^1 + x_2 \\ y &= y_1\eta^1 + y_2\eta^2 = y_{12}\eta^1 + y_2 \end{aligned} \tag{B.165}$$

Therefore, $\lambda = \eta^1$ and $\dfrac{dx}{d\eta^1} = x_{12}$, $\dfrac{dy}{d\eta^1} = y_{12}$, $d\gamma^2 = A_{11}x_{12}^2 + 2A_{12}x_{12}y_{12} + A_{22}y_{12}^2$.

If the border concerns these two nodes, then dof related to the third node is not to be considered. We shall define a diagonal Boolean 12 × 12 matrix, Δ with 0 on dof that should be ignored. We can now calculate border contribution to the external virtual work.

$$\int_{\partial T_e\cap\gamma_1}(m^t\overline{\theta}_\nu - m^\nu\overline{\theta}_t)d\gamma = \int_{\partial T_e\cap\gamma_1}(-m^t\theta_\nu + m^\nu\theta_t)d\gamma = \int_{\partial T_e\cap\gamma_1}(-m^t\theta.\vec{\nu} + m^\nu\theta.\vec{t})d\gamma \tag{B.166}$$

$$= \int_{\partial T_e\cap\gamma_1}(-m^t\nu^1.\theta_1 - m^t\nu^2.\theta_2 + m^\nu t_1\theta_1 + m^\nu t^2\theta_2)d\gamma$$

$$= \int_{\partial T_e\cap\gamma_1}(-m^t\nu^1 + m^\nu t^1, -m^t\nu^2 + m^\nu t^2)L^e_\gamma(\lambda)d\lambda\Delta V^e_m$$

$$= \int_{\partial T_e\cap\gamma_1}\begin{pmatrix}-\nu^1 & t^1\\ -\nu^2 & t^2\end{pmatrix}\begin{pmatrix}m^t\\ m^\nu\end{pmatrix}d\lambda L^e_\gamma(\lambda)\Delta V^e_m$$

$$= \underbrace{\int_{\partial T_e\cap\gamma_1}{}^t\Delta^t L^e_\gamma(\lambda)\begin{pmatrix}-\nu^1 & t^1\\ -\nu^2 & t^2\end{pmatrix}\begin{pmatrix}m^t\\ m^\nu\end{pmatrix}d\lambda}_{F^e_{\gamma m}}.V^e_m$$

$$= F^e_{\gamma m}.V^e_m$$

The term

$$\int_{\partial T_e\cap\gamma_1}q^i v_i d\gamma = \int_{\partial T_e\cap\gamma_1}\Delta^t L^e_s[q_i](\lambda)d\lambda.V^e_m = F_{\gamma q}.V^e_m \tag{B.167}$$

where L^e_s is the 3 × 12 matrix given by

$$L^e_s = \begin{bmatrix}\eta^1 & \eta^2 & \eta^3 & 0 & 0 & 0 & 0 & 0 & 0 & 0 & 0\\ 0 & 0 & 0 & \eta^1 & \eta^2 & \eta^3 & 0 & 0 & 0 & 0 & 0 & 0\\ 0 & 0 & 0 & 0 & 0 & 0 & \eta^1 & \eta^2 & \eta^3 & 4\eta^1\eta^2 & 4\eta^1\eta^3 & 4\eta^2\eta^3\end{bmatrix} \tag{B.168}$$

and $[q_i]$ reads

$$[q_i] = \begin{bmatrix}q_1\\ q_2\\ q_3\end{bmatrix}$$

Finally, the external force per element is

$$F^e = F^e_S + F^e_{\gamma q} + F^e_{\gamma m} \tag{B.169}$$

The global vector force is filled as follows:

$$\begin{aligned} n_e &= 1, \ldots.., N_e \\ i &= 1, \ldots., m \\ I &= Num(n_e, i) \\ F(I) &= F(I) + F^e(i) \end{aligned} \tag{B.170}$$

The final global force vector F_g is obtained by adding all the concentrated forces F_c:

$$F_g = F + F_c \tag{B.171}$$

The final discrete problem is now

$$\mathbb{M}X = F_g \tag{B.172}$$

This equation is rearranged before the resolution. The total dof vector is subdivided in (X_1, X_2, X_3) where X_3 denotes imposed dof and X_2 consists of dof that interacts with external structures such as soil-structure interaction. X_1 and X_2 are unknown dof. The equation now reads

$$\begin{aligned} \mathbb{M}_{11}X_1 + \mathbb{M}_{12}X_2 + \mathbb{M}_{13}X_3 &= F_1 \\ \mathbb{M}_{21}X_1 + \mathbb{M}_{22}X_2 + \mathbb{M}_{23}X_3 &= F_2 \\ \mathbb{M}_{31}X_1 + \mathbb{M}_{32}X_2 + \mathbb{M}_{33}X_3 &= F_3 \end{aligned} \tag{B.173}$$

F_3 is the reaction on imposed dof. Suppose F_2 is an elastic reaction due to soil for example and $F_2 = -K_s X_2$. The equation now reads

$$\begin{aligned} \mathbb{M}_{11}X_1 + \mathbb{M}_{12}X_2 &= F_1 - \mathbb{M}_{13}X_3 \\ \mathbb{M}_{21}X_1 + (\mathbb{M}_{22} + K_s)X_2 &= -\mathbb{M}_{23}X_3 \end{aligned} \tag{B.174}$$

$$X_2 = -(\mathbb{M}_{22} + K_s)^{-1}(\mathbb{M}_{21}X_1 - \mathbb{M}_{23}X_3) = -(\mathbb{M}_{22} + K_s)^{-1}\mathbb{M}_{21}X_1 - (\mathbb{M}_{22} + K_s)\mathbb{M}_{23}X_3 \tag{B.175}$$

which leads to

$$\mathbb{M}_{11}X_1 - \mathbb{M}_{12}(\mathbb{M}_{22} + K_s)^{-1}X_1 = F_1 - \mathbb{M}_{13}X_3 + (\mathbb{M}_{22} + K_s)^{-1}\mathbb{M}_{23}X_3$$

$$(\mathbb{M}_{11} - \mathbb{M}_{12}(\mathbb{M}_{22} + K_s)^{-1})X_1 = F_d \tag{B.176}$$

or

$$\overline{\mathbb{M}}_{11}X_1 = F_d \tag{B.177}$$

This final equation can be solved by various linear methods. The unknown X_2 is next deduced. From $X = (X_1, X_2, X_3)$ all the others $(e_{\alpha\beta}, K_{\alpha\beta}, Q_{\alpha\beta}, \epsilon_{\alpha\beta}, \sigma^{\alpha\beta})$ are calculated locally and the reactions $\sigma^{\alpha 3}$, σ^{33} are obtained by solving numerically the ordinary differential equations presented above.

Exercise 5.1

Calculate for an element the local strain and stress tensors of the N model by using the 15 dof finite element. Calculate the elementary stiffness matrix and nodal force.

Chapter 6: Other Models

B.39 Stiffened Shells

Stiffeners are extra thickness at intrados and extrados of the shell described by the functions $f(x, y)$ and $g(x, y)$, respectively.

The variational formulations of the models N-T or N consist in using the corresponding kinematics and to integrate with respect to z in the intervals $[h/2,\ f(x, y)]$, $[-h/2,\ h/2]$ and $[g(x, y), -h/2]$, i.e.

$$\int_S \int_{g(x,y)}^{\frac{-h}{2}} \sigma^{ij} \epsilon_{ij} \psi dz dS + \int_S \int_{\frac{-h}{2}}^{\frac{h}{2}} \sigma^{ij} \epsilon_{ij} \psi dz dS + \int_S \int_{\frac{h}{2}}^{f(x,y)} \sigma^{ij} \epsilon_{ij} \psi dz dS = L(v) \tag{B.178}$$

We can also use the N-T and N first-order models to establish variational equations of stiffened shells.

B.40 Thermoelastic Shell

In static analysis, thermal and mechanical strains and stress are decoupled, i.e. they act separately in linear elasticity. The total strain tensor reads $\epsilon = \varepsilon + \epsilon^{th}$ where ε is the mechanical strain tensor and $\epsilon^{th} = \kappa \Theta G$ the thermal strain tensor. The expansion coefficient is κ and $\Theta = T - T_0$; T_0 is the reference temperature of the material and T is the absolute temperature.

Considering the hypothesis $\sigma^{33} = 0$, we obtain the 2D thermoelastic constitutive relations

$$\begin{aligned}
\sigma^{\alpha\beta} &= \bar{\Lambda} \varepsilon^{\gamma}_{\gamma} G^{\alpha\beta} + 2\bar{\mu} \varepsilon^{\alpha\beta} = \bar{\Lambda} \epsilon^{\gamma}_{\gamma} G^{\alpha\beta} + 2\bar{\mu} \epsilon^{\alpha\beta} - 2(\bar{\Lambda} + \bar{\mu}) \kappa \Theta G^{\alpha\beta}, \\
&= \bar{\Lambda} \epsilon^{\gamma}_{\gamma} G^{\alpha\beta} + 2\bar{\mu} \epsilon^{\alpha\beta} - 3\eta E_v \kappa \Theta G^{\alpha\beta} \\
&= \frac{E\bar{\nu}}{1-\bar{\nu}^2} \epsilon^{\gamma}_{\gamma} G^{\alpha\beta} + \frac{E}{1+\bar{\nu}} \epsilon^{\alpha\beta} - \frac{E}{(1-\bar{\nu})} \kappa \Theta G^{\alpha\beta}
\end{aligned} \tag{B.179}$$

$$\bar{\Lambda} = \frac{2\bar{\mu}\bar{\lambda}}{\bar{\lambda} + 2\bar{\mu}} \ and \ \eta = \frac{1 - 2\bar{\nu}}{(1 - \bar{\nu})}$$

The variational equation comprises contributions of thermal origin. Indeed, in time-independent evolution, the equations are decoupled and the temperature field can be calculated separately. In this case, the N-T model variational equations read

$$A(u, v) = \int_S (N(u) : e(v) + M(u) : K(v) + M^*(u) : Q(v))dS = L(v) + L_\Theta(v);$$

$$L_\Theta(v) = \int_S \int_{\frac{-h}{2}}^{\frac{h}{2}} 3\eta E_\upsilon \kappa \Theta G^{\alpha\beta}(e_{\alpha\beta}(v) - zK_{\alpha\beta}(v) + z^2 Q_{\alpha\beta}(v))\psi dz dS \tag{B.180}$$

where $\Theta = T - T_0$. If $\Theta(x, z)$ is known either as the solution of separate equations on Θ or from hypotheses made through the thickness distribution of $\Theta(x, z)$, then $L_\Theta(v)$ can be integrated along the thickness.

B.41 Anisotropic Homogeneous Shell

Though the idealistic homogeneous isotropic material is widely used, some other materials such as orthotropic materials offer some specific properties preferable for the design of certain structures. Their constitutive law contains more than two constants; 9 for orthotropic materials and 21 for more general materials. The constitutive law reads

$$\sigma^{ij} = C^{ijkl}\epsilon_{kl}, \quad C^{ijkl} = C^{ijlk} = C^{jikl} = C^{klij} \tag{B.181}$$

and there exists a constant $a \rangle 0$ such that for any symmetric tensor τ, we have

$$C^{ijkl}\tau_{kl}\tau_{ij} \geqslant a\tau_{kl}\tau_{kl} \tag{B.182}$$

To establish the N-T model for such materials, the constitutive laws read

$$\sigma^{\alpha\beta} = A^{\alpha\beta\rho\gamma}\epsilon_{\rho\gamma} \tag{B.183}$$

calculated as follows:

$$\sigma^{\alpha 3} = C^{\alpha 3\gamma\delta}\epsilon_{\gamma\delta} + 2C^{\alpha 3\gamma 3}\epsilon_{\gamma 3} + C^{\alpha 333}\epsilon_{33} = 0 \tag{B.184}$$

$$\sigma^{33} = C^{33\gamma\delta}\epsilon_{\gamma\delta} + 2C^{33\gamma 3}\epsilon_{\gamma 3} + C^{3333}\epsilon_{33} = 0 \tag{B.185}$$

Let $X_\gamma = \epsilon_{\gamma 3}, X_3 = \epsilon_{33}, F^\alpha = -C^{\alpha 3\gamma\delta}\epsilon_{\gamma\delta}, F^3 = -C^{33\gamma\delta}\epsilon_{\gamma\delta}, M^{\alpha\gamma} = 2C^{\alpha 3\gamma 3}; M^{\alpha 3} = C^{\alpha 333}; M^{33} = C^{3333}, M^{-1} = Y$, then we have

$$M^{\alpha\gamma}X_{\gamma} + M^{\alpha 3}X_3 = F^{\alpha};\ \ M^{3\gamma}X_{\gamma} + M^{33}X_3 = F^3 \tag{B.186}$$

Then we solve the system

$$MX = F \tag{B.187}$$

We obtain

$$X = M^{-1}F \tag{B.188}$$

and

$$A^{\alpha\beta\gamma\delta} = C^{\alpha\beta 33}(Y_{3\nu}C^{\nu 3\gamma\delta} + Y_{33}C^{33\gamma\delta}) \tag{B.189}$$

The variational formulation is conducted as in the previous section.

B.42 Heterogeneous Shell

In real structures, for example composite structures, the tangent moduli depend on the position of the material point and may vary rapidly from one point to the other. Homogeneous equivalent moduli are preferable whenever it is possible to homogenize these rapidly varying quantities. In practice, some designers replace these quantities by a mean value (through averaging) in a Representative Elementary Volume of material. However, values proposed by the asymptotic analysis approach or the genuine two-scale convergence approach show that averaging leads to optimistic values in a periodic structure.

Periodic Shell

A periodic structure is described by a generic cell made of a finite number ($n \geq 2$) of different materials. The homogenized quantity is determined in a generic cell by solving a particular variational equation as follows:
(i) Solve the variational equation (6.39) for each χ^{mn}

(ii) The homogenized moduli are calculated by (6.42).

For a z-direction stratified periodic shell with a two-component generic cell Y Fig. 6.4, with (λ^{-}, μ^{-}) as moduli for material (1) occupying βY and (λ^{+}, μ^{+}) for material (2) which occupies $(1 - \beta)Y$, we first begin by solving the equation:

$$\chi^{kl}_{m,3}(x,y) = \delta^k_m G^{3l} - \frac{1}{2\bar{\mu}(y)} \frac{2\mu^- \mu^+}{(1-\beta^2)\mu^- + \beta^2 \mu^+} \delta^k_m G^{3l}$$
$$= \quad (1 - \frac{1}{\bar{\mu}(y)} \frac{\mu^- \mu^+}{(1-\beta^2)\mu^- + \beta^2 \mu^+}) \delta^k_m G^{3l} \tag{B.190}$$
$$= \quad \alpha^\star(y) \delta^k_m G^{3l}$$

Calculate the homogenized moduli as follows:

$$C_h^{ijkl}(x) = \int_0^1 (C^{ijkl}(x,y) - \alpha^\star(y) C^{ijm3}(x,y) \delta^k_m G^{3l}) dy$$
$$= \quad \int_0^1 (C^{ijkl}(x,y) - \alpha^\star(y) C^{ijk3}(x,y) G^{3l}) dy \tag{B.191}$$

which finally give

$$C_h^{ijk\alpha}(x) = C_h^{ij\alpha k}(x) = \int_0^1 C^{ijk\alpha}(x,y) dy, \tag{B.192}$$

$$C_h^{ijk3}(x) = C_h^{ij3k}(x) = \int_0^1 (1 - \alpha^\star(y)) C^{ijk3}(x,y)) dy, \tag{B.193}$$

$$C_h^{3333}(x) = \quad \int_0^1 (1 - \alpha^\star(y))(\bar{\lambda} + 2\bar{\mu}) dy$$
$$= \beta(\lambda^- + 2\mu^-) + (1-\beta)(\lambda^+ + 2\mu^+) \tag{B.194}$$
$$- \int_0^1 \alpha^\star(y)(\bar{\lambda} + 2\bar{\mu})(y) dy$$

These results confirm that averaging is optimistic.

B.43 Simplified Stratified N-T Model

$$\sigma^{\alpha\beta} = C^{\alpha\beta\gamma\delta}(x,0)\epsilon_{kl} \tag{B.195}$$

where $C^{\alpha\beta\gamma\delta} = A^{\alpha\beta\gamma\delta}$ calculated earlier under the plane stress and plane strain hypotheses. The variational equation is

$$
\begin{aligned}
&\int_S \sum_{m=1}^{n} \int_{z_{m-1}}^{z_m} C_m^{\alpha\beta\gamma\delta}(x,0)(e_{\gamma\delta}(u) - zK_{\gamma\delta}(u) + z^2 Q_{\gamma\delta}(u))e_{\alpha\beta}(v) \\
&\quad + C_m^{\alpha\beta\gamma\delta}(x)(-ze_{\gamma\delta}(u) + z^2 K_{\gamma\delta}(u) - z^3 Q_{\gamma\delta}(u))K_{\alpha\beta}(v) \\
&+ C_m^{\alpha\beta\gamma\delta}(x)(z^2 e_{\gamma\delta}(u) - z^3 K_{\gamma\delta}(u) + z^4 Q_{\gamma\delta}(u))Q_{\alpha\beta}(v))dzdS = L(v)
\end{aligned}
\tag{B.196}
$$

Let

$$
\int_S \sum_{m=1}^{n} N_m^{\alpha\beta}(u)e_{\alpha\beta}(v) + M_m^{\alpha\beta}(u)K_{\alpha\beta}(v) + M_m^{\star\alpha\beta}(u)Q_{\alpha\beta}(v))dS = L(v) \tag{B.197}
$$

$$
\begin{aligned}
N_m^{\alpha\beta}(u) = C_m^{\alpha\beta\gamma\delta}(x)[&(z_m - z_{m-1})e_{\gamma\delta}(u) - \frac{1}{2}(z_m^2 - z_{m-1}^2)K_{\gamma\delta}(u) \\
&+ \frac{1}{3}(z_m^3 - z_{m-1}^3)Q_{\gamma\delta}(u)]
\end{aligned}
$$

$$
\begin{aligned}
M_m^{\alpha\beta}(u) = C_m^{\alpha\beta\gamma\delta}(x)[&-\frac{1}{2}(z_m^2 - z_{m-1}^2)e_{\gamma\delta}(u) + \frac{1}{3}(z_m^3 - z_{m-1}^3)K_{\gamma\delta}(u) \\
&- \frac{1}{4}(z_m^4 - z_{m-1}^4)Q_{\gamma\delta}(u)]
\end{aligned}
$$

$$
\begin{aligned}
M_m^{\star\alpha\beta}(u) = C_m^{\alpha\beta\gamma\delta}(x)[&\frac{1}{3}(z_m^3 - z_{m-1}^3)e_{\gamma\delta}(u) - \frac{1}{4}(z_m^4 - z_{m-1}^4)K_{\gamma\delta}(u) \\
&+ \frac{1}{5}(z_m^5 - z_{m-1}^5)Q_{\gamma\delta}(u)]
\end{aligned}
$$

These resulting forces can be written in the form

$$
\begin{bmatrix} N_m(u) \\ M_m(u) \\ M_m^\star(u) \end{bmatrix} = C_m(x) \begin{bmatrix} (z_m - z_{m-1}) & -\frac{1}{2}(z_m^2 - z_{m-1}^2) & \frac{1}{3}(z_m^3 - z_{m-1}^3) \\ -\frac{1}{2}(z_m^2 - z_{m-1}^2) & \frac{1}{3}(z_m^3 - z_{m-1}^3) & -\frac{1}{4}(z_m^4 - z_{m-1}^4) \\ \frac{1}{3}(z_m^3 - z_{m-1}^3) & -\frac{1}{4}(z_m^4 - z_{m-1}^4) & \frac{1}{5}(z_m^5 - z_{m-1}^5) \end{bmatrix} \begin{pmatrix} e(u) \\ K(u) \\ Q(u) \end{pmatrix} \tag{B.198}
$$

We rewrite the variational equation as follows:

$$\int_s \sum_{m=1}^{n} (N_m(u) : e(v) + M_m(u) : K(v) + M_m^\star(u) : Q(v))dS = L(v) \tag{B.199}$$

B.44 Semi-analytical Models

All shell models depend on the different methods used in predicting the 3D behaviour of the shell by equations defined in its mid-surface. In other words, 2D equations are derived from 3D equation by integration along the thickness. Very many kinematics and material constitutive laws are derived under some hypotheses and lead to the existence of many models. Formal expansion in z of the displacement as a series of functions depending only on $x = (x^1, x^2)$ is another method which gives some satisfaction though boundary conditions are hardly satisfied. We hereafter give a non-exhaustive list.

$$\begin{cases} U_\alpha(x^1, x^2, z) = \left(1 + \frac{z}{R_\alpha}\right) u_\alpha(x^1, x^2) - \frac{z}{A_\alpha} \partial_\alpha u_3(x^1, x^2) + w(z)\gamma_\alpha(x^1, x^2) \\ U_3(x^1, x^2, z) = \quad u_3(x^1, x^2) \\ \gamma_\alpha(x^1, x^2) \quad = \quad \frac{1}{A_\alpha}\frac{\partial u_3(x^1,x^2)}{\partial x^\alpha} + f_\alpha(x^1, x^2) \end{cases} \tag{B.200}$$

$$U(x^1, x^2, z) = \sum_{n=0}^{n=N} z^n u^n(x^1, x^2) \tag{B.201}$$

$$U(x^1, x^2, z) = u^0(x^1, x^2) + \sum_{n=1}^{n=N} sin(\frac{n\pi z}{h}) u^n(x^1, x^2) \tag{B.202}$$

$$U(x^1, x^2, z) = u^0(x^1, x^2) + zu^1(x^1, x^2) + \sum_{n=1}^{n=N} cosin(\frac{n\pi z}{h}) u^{n+1}(x^1, x^2) \tag{B.203}$$

or z and the trigonometric functions $sine$, $cosine$

$$U(x^1, x^2, z) = u^0(x^1, x^2) + zu^1(x^1, x^2) + \sum_{n=1}^{n=N} (sin(\frac{n\pi z}{h}) u^{2n}(x^1, x^2) + cosin(\frac{n\pi z}{h}) u^{2n+1}(x^1, x^2)) \tag{B.204}$$

or z and an exponential expansion

$$U(x^1, x^2, z) = \sum_{n=0}^{n=N} e^{(nz/h)} u^n(x^1, x^2) \tag{B.205}$$

$$U(x^1, x^2, z) = u^0(x^1, x^2) + zu^1(x^1, x^2) + \sum_{n=1}^{n=N} e^{(nz/h)} u^{n+1}(x^1, x^2) \tag{B.206}$$

B.45 Cosserat Thick Shell Model

Some shells cannot be obtained by thickening the mid-surface in the normal direction. For conical shape shells for example, a direction D different from the normal direction is used. In other words, a material point in the thickness is

$$OM = Om + zD, \; m \in S, \; -\frac{h}{2} \leq z \leq \frac{h}{2}, \; D \in \mathbb{R}^3 \tag{B.207}$$

We denote by $\{A_1, A_2, A_3\}$, $\{A^1, A^2, A^3\}$, respectively, the covariant and contravariant bases of the mid-surface S and $\{a_1, a_2, a_3\}$, $\{a^1, a^2, a^3\}$ their counterparts on the deformed surface $\tilde{S}$. We shall use the same notations as in the previous chapters. Let $D(x) = D^i A_i = D_i A^i$, then we have

$$\begin{aligned} D_{,\alpha} &= (\nabla_\alpha D_\rho - D_3 B_{\alpha\rho})A^\rho + (\partial_\alpha D_3 + D_\rho B^\rho_\alpha)A^3 = D_{i\alpha} A^i \\ &= (\nabla_\alpha D^\rho - D^3 B^\rho_\alpha)A_\rho + (\partial_\alpha D^3 + D^\rho B_{\alpha\rho})A_3 = D^i_\alpha A_i \end{aligned} \tag{B.208}$$

Consider a vector $u(x) = u^i A_i = u_i A^i$, also we have

$$\begin{aligned} u_{,\alpha} &= (\nabla_\alpha u_\rho - u_3 B_{\alpha\rho})A^\rho + (\partial_\alpha u_3 + u_\rho B^\rho_\alpha)A^3 = u_{i\alpha} A^i \\ &= (\nabla_\alpha u^\rho - u^3 B^\rho_\alpha)A_\rho + (\partial_\alpha u^3 + u^\rho B_{\alpha\rho})A_3 = u^i_\alpha A_i \end{aligned} \tag{B.209}$$

The covariant base of the shell Ω is $\{G_1, G_2, G_3\}$ with $G_\alpha = OM_{,\alpha} = Om_{,\alpha} + zD_{,\alpha} = A_\alpha + zD_{,\alpha}$; $G_3 = D$. We assume that the parametrization of the shell is such that the covariant base and the contravariant base $\{G^1, G^2, G^3\}$ exist and that $\{A_1, A_2, D\}$ whose contravariant base is $\{A^1, A^2, D^*\}$ also constitute bases of the shell Ω. We also assume that $\{A_1, A_2, A_3\}$ and its dual $\{A^1, A^2, A^3\}$ constitute a base. As in the previous chapters, a vector can be expressed in any base.

We can write the following relations on any vector:

$$U(x, z) = U^i G_i = U_i G^i = \bar{U}^\rho A_\rho + \bar{U}^3 D = \bar{U}_\rho A^\rho + \bar{U}_3 D^* = u^i A_i = u_i A^i$$

We have

$$\bar{U}^\rho A_\rho + \bar{U}^3 D = (\bar{U}^\rho + \bar{U}^3 D^\rho)A_\rho + \bar{U}^3 D^3 A_3 = u^i A_i$$

We therefore deduce that

$$u^\rho = \bar{U}^\rho + \bar{U}^3 D^\rho, \quad u^3 = \bar{U}^3 D^3 = \bar{U}^3 D_3$$

$$\bar{U}^3 = \frac{u^3}{D_3}, \qquad \bar{U}^\rho = u^\rho - \frac{D^\rho}{D_3} u^3$$

$$\bar{U}_3 = \frac{u_3}{D_3}, \qquad \bar{U}_\rho = u_\rho - \frac{D_\rho}{D_3} u_3$$

From the above relations, we choose to express vectors in the bases $\{A_1, A_2, A_3\}$, $\{A^1, A^2, A^3\}$ in the reference configuration or in the bases $\{a_1, a_2, a_3\}$, $\{a^1, a^2, a^3\}$ in the deformed configuration. The directing vector is then $d(x) = d^i a_i = d_i a^i$.

The kinematic reads

$$u(x) + z\delta(x) = (u_i(x) + z\delta_i(x))A^i = (u^i(x) + z\delta^i(x))A_i \tag{B.210}$$

and where δ for some shell models reads

K-L ($\delta_\alpha = -\partial_\alpha u_3,\ \delta_3 = 0$); N-T ($\delta_\alpha = \partial_\alpha u_3 + B^\rho_\alpha u_\rho,\ \delta_3 = 0$); N ($\delta_\alpha = \partial_\alpha u_3 + B^\rho_\alpha u_\rho,\ \delta_3 = \bar{q}(x)\ with\ w(z) = z$); R-M ($\delta_\alpha = \theta_\alpha,\ \delta_3 = 0$)

The deformation of the shell depends on the displacement u of the mid-surface and δ, the variation of the direction D.

We also have as above on u:

$$\begin{aligned}\delta_{,\alpha} = (\nabla_\alpha \delta^\rho - \delta^3 B^\rho_\alpha)A_\rho + (\partial_\alpha \delta^3 + \delta^\rho B_{\alpha\rho})A_3 = \delta^{\alpha\rho} A^\rho + \delta^{\alpha 3} A^3 = \delta^{i\alpha}\\ (\nabla_\alpha \delta_\rho - \delta_3 B_{\rho\alpha})A^\rho + (\partial_\alpha \delta_3 + \delta_\rho B^\rho_\alpha)A_3 = \delta_{\alpha\rho} A^\rho + \delta_{\alpha 3} A^3 = \delta_{i\alpha}\end{aligned} \tag{B.211}$$

So

$$\begin{aligned}\epsilon_{\alpha\beta}(u, \delta) &= e_{\alpha\beta}(u) - zD_3 \Bbbk_{\alpha\beta}(u, \delta) + z^2 D_3^2 Q^c_{\alpha\beta}(u, \delta)\\ Q^c_{\alpha\beta}(u, \delta) &= (D^\rho_\alpha \delta_{\rho\beta} + D^\rho_\beta \delta_{\rho\alpha} + D_{3\alpha}\delta_{3\beta} + D_{3\beta}\delta_{3\alpha})/2D_3^2\\ \Bbbk_{\alpha\beta}(u, \delta) &= -(\delta_{\alpha\beta} + \delta_{\beta\alpha} + D^i_\beta u_{i\alpha} + D^i_\alpha u_{i\beta})/2D_3\\ e_{\alpha\beta}(u) &= \frac{1}{2}(\nabla_\alpha u_\beta + \nabla_\beta u_\alpha - 2u_3 B_{\alpha\beta})\end{aligned} \tag{B.212}$$

The transverse strains are rewritten as follows:

$$\epsilon_{\alpha 3}(u,\delta) = \frac{1}{2}(\vartheta_\alpha + z\varepsilon_\alpha)$$

$$\epsilon_{33}(u,\delta) = \varepsilon_3 \tag{B.213}$$

$$\vartheta_\alpha(u,\delta) = \delta_\alpha + D_i u^i_\alpha,\ \vartheta_3(u,\delta) = \delta_3 - D^\alpha u_{3\alpha}$$

$$\varepsilon_\alpha(u,\delta) = (D^i_\alpha \vartheta_i + \delta_{i\alpha} D^i),\ \varepsilon_3(u,\delta) = D^i \vartheta_i$$

We remark in the transverse strains $\epsilon_{\alpha 3}(u,\delta)$ and $\epsilon_{33}(u,\delta)$ the terms $\vartheta_\alpha(u,\delta)$ and $\vartheta_3(u,\delta)$ which are found in Cosseratats' theory, without the ε_i.

The following approximations will be adopted:

$$\sigma^{ij}(u,\delta) = (\bar{\lambda} G^{ij}(x,z)G^{kl}(x,z) + 2\bar{\mu} G^{ik}(x,z)G^{jl}(x,z))\epsilon_{kl}(u,\delta)$$

$$= A^{ijkl}(x,z)\epsilon_{kl}(u,\delta)$$

We use the simplification assumptions and we reformulate the stress-strain formulations:

$$G^{ij}(x,z) \simeq G^{ij}(x,0)$$

$$G^{\alpha\beta}(x,0) = A^{\alpha\beta},\quad G^{\alpha 3}(x,0) = -D^\alpha/D\cdot D,\quad G^{33}(x,0) = 1/D\cdot D$$

$$A^{ijkl}(x,z) = A^{ijkl}(x,0) = A^{ijkl}(x)$$

$$\sigma^{ij}(u,\delta) = (\bar{\lambda} G^{ij}(x,0)G^{kl}(x,0) + 2\bar{\mu} G^{ik}(x,0)G^{jl}(x,0))\epsilon_{kl}(u,\delta)$$

$$= A^{ijkl}(x)\epsilon_{kl}(u,\delta)$$

we also consider that

$d\Omega = (G_1, G_2, D)dxdz \simeq (A_1, A_2, D)dxdz = D_3(A_1, A_2, A_3)dxdz = D_3 dSdz$

Let $U_{ad} = H^1_{\bar{\gamma}_0}(S) \times H^1_{\bar{\gamma}_0}(S) \times H^2(S) \cap H^1_{\bar{\gamma}_0}(S)$.

The variational formulation of the model boundary value problem reads

$$\begin{cases} \text{find}(u,\delta) \in U_{ad} \times U_{ad} \text{ such that} \\ \int\limits_{\Omega} \sigma^{ij}(u,\delta)\epsilon_{ij}(v,w)d\Omega = \int\limits_{s}\int\limits_{-h/2}^{h/2} A^{ijkl}(x)\epsilon_{kl}(u,\delta)\epsilon_{ij}(v,w)D_3 dz dS \\ \qquad = L(v,w)\quad \forall (v,w) \in U_{ad}\times U_{ad} \end{cases} \tag{B.214}$$

The external virtual work $L(v, w)$ is expressed as follows:

$$L(v,w) = \int_S \left(\bar{p}^i v_i + \tilde{p}^i w_i\right) dS + \int_{\gamma_1} \left(\bar{q}^i v_i + \tilde{q}^3 w_3\right) d\gamma + \int_{\gamma_1} \tilde{m}^\alpha w_\alpha d\gamma \quad \text{(B.215)}$$

The internal virtual work is

$$\int_\Omega \sigma^{ij}(u,\delta)\epsilon_{ij}(v,w)d\Omega = \int_\Omega \sigma^{\alpha\beta}(u,\delta)\epsilon_{\alpha\beta}(v,w)d\Omega + 2\int_\Omega \sigma^{\alpha 3}(u,\delta)\epsilon_{\alpha 3}(v,w)d\Omega$$
$$+\int_\Omega \sigma^{33}(u,\delta)\epsilon_{33}(v,w)d\Omega$$

As in the previous chapters, we obtain the different tensors $N = (N^{\alpha\beta})$, $M = (M^{\alpha\beta})$, $M^* = (M^{*\alpha\beta})$, $T = (T^\alpha)$, $\tilde{T} = (\tilde{T}^\alpha)$, H and the variational equation now reads

$$\begin{cases} \text{Find}(u,\delta) \in U_{ad} \times U_{ad} \text{ such that} \\ E((u,\delta);(v,w)) = \int_S (N^{\alpha\beta}(u,\delta)e_{\alpha\beta}(v) + M^{\alpha\beta}(u,\delta)\Bbbk_{\alpha\beta}(v,w) \\ \quad +M^{*\alpha\beta}(u,\delta)Q^c_{\alpha\beta}(v,w))D_3 dS + \int_S (T^\alpha(u,\delta)\vartheta_\alpha(v,w) \\ \quad +\tilde{T}^\alpha(u,\delta)\varepsilon_\alpha(v,w))D_3 dS + \int_S H(u,\delta)\varepsilon_3(v,w)D_3 dS \\ \quad = \int_S \left(\bar{p}^i v_i + \tilde{p}^i w_i\right) dS + \int_{\gamma_1} \left(\bar{q}^i v_i + \tilde{q}^3 w_3\right) d\gamma + \int_{\gamma_1} \tilde{m}^\alpha w_\alpha d\gamma \\ \quad = L(v,w), \quad \forall (v,w) \in U_{ad} \times U_{ad} \end{cases} \quad \text{(B.216)}$$

The final variational problem obtained satisfies all the necessary conditions for the existence of a unique solution because

$$E((u,\delta);(v,w)) = \int_\Omega \sigma^{ij}(u,\delta)\epsilon_{ij}(v,w)d\Omega = \int_\Omega A^{ijkl}(x)\epsilon_{kl}(u,\delta)\epsilon_{ij}(v,w)d\Omega$$

and the tangent moduli tensor A^{ijkl} satisfies the usual material symmetric and ellipticity condition.

The terms $N = (N^{\alpha\beta})$, $M = (M^{\alpha\beta})$, $M^* = (M^{*\alpha\beta})$, $T = (T^{\alpha})$, $\tilde{T} = (\tilde{T}^{\alpha})$ and H depend on $e_{\alpha\beta}(u)$, $\Bbbk_{\alpha\beta}(u, \delta)$, $Q^c_{\alpha\beta}(u, \delta)$, $\vartheta_i(u, \delta)$, $\varepsilon_i(u, \delta)$.

Exercise 6.1

Calculate all the terms $\bar{p}^i$, $\tilde{p}^i$, $\bar{q}^i$, $\tilde{m}^{\alpha}$, found in $L(v, w)$ given above in (C.44).

Exercise 6.2

Calculate all the terms $N = (N^{\alpha\beta})$, $M = (M^{\alpha\beta})$, $M^* = (M^{*\alpha\beta})$, $T = (T^{\alpha})$, $\tilde{T} = (\tilde{T}^{\alpha})$ and H found in the variational problem (B.216).

Appendix C
Solution of Exercises

Solution 1.1

We consider the ball $B(0, R) = \Omega$. A point $M \in \Omega$ is defined by

$$\overrightarrow{OM} = (r \sin\theta cos\varphi, r \sin\theta \sin\varphi, r \cos\theta)\,, \quad x^1 = r, \;\; x^2 = \theta, \;\; x^3 = \varphi$$

The metric tensor is calculated as follows:
Let $G_i = OM_{,i}, \;\; i = r, \theta, \varphi$.

$$G_r = \begin{pmatrix} \sin\theta\cos\varphi \\ \sin\theta\sin\varphi \\ \cos\theta \end{pmatrix}, \; G_\theta = \begin{pmatrix} r\cos\theta\cos\varphi \\ r\cos\theta\sin\varphi \\ -r\sin\theta \end{pmatrix}, G_\varphi = \begin{pmatrix} -r\sin\theta\sin\varphi \\ r\sin\theta\cos\varphi \\ 0 \end{pmatrix} \tag{C.1}$$

The tensors (G_{ij}) and (G^{ij}) read

$$(G_{ij}) = \begin{pmatrix} 1 & 0 & 0 \\ 0 & r^2 & 0 \\ 0 & 0 & r^2\sin^2\theta \end{pmatrix}, \quad (G^{ij}) = \begin{pmatrix} 1 & 0 & 0 \\ 0 & 1/r^2 & 0 \\ 0 & 0 & 1/r^2\sin^2\theta \end{pmatrix}$$

The contravariant base vectors $G^r, \;\; G^\theta, \;\; G^\varphi$ are calculated as follows:

$$G^r = \begin{pmatrix} \sin\theta\cos\varphi \\ \sin\theta\sin\varphi \\ \cos\theta \end{pmatrix}, \; G^\theta = \frac{1}{r}\begin{pmatrix} \cos\theta\cos\varphi \\ \cos\theta\sin\varphi \\ -\sin\theta \end{pmatrix}, G^\varphi = \frac{1}{r\sin\theta}\begin{pmatrix} -\sin\varphi \\ \cos\varphi \\ 0 \end{pmatrix} \tag{C.2}$$

R. Nzengwa, *Plate and Shell Models*,
https://doi.org/10.1007/978-981-97-2780-3

We have also

$$G_{r,r} = 0, \quad G_{r,\theta} = \begin{pmatrix} \cos\theta\cos\varphi \\ \cos\theta\sin\varphi \\ -\sin\theta \end{pmatrix}, \quad G_{r,\varphi} = \begin{pmatrix} -\sin\theta\sin\varphi \\ \sin\theta\cos\varphi \\ 0 \end{pmatrix}$$

$$G_{\theta,r} = G_{r,\theta} \quad G_{\theta,\theta} = \begin{pmatrix} -r\sin\theta\cos\varphi \\ -r\sin\theta\sin\varphi \\ -r\cos\theta \end{pmatrix}, \quad G_{\theta,\varphi} = \begin{pmatrix} -r\cos\theta\sin\varphi \\ r\cos\theta\cos\varphi \\ 0 \end{pmatrix}$$

$$G_{\varphi,r} = G_{r,\varphi}, \quad G_{\varphi,\theta} = G_{\theta,\varphi} \quad G_{\varphi,\varphi} = \begin{pmatrix} -r\sin\theta\cos\varphi \\ -r\sin\theta\sin\varphi \\ 0 \end{pmatrix}$$

From $\Gamma_{ij}^k = G_{i,j}.G^k$, we deduce that

$$\Gamma_{ij}^r = \begin{pmatrix} 0 & 0 & 0 \\ 0 & -r & 0 \\ 0 & 0 & -r\sin\theta \end{pmatrix}, \quad \Gamma_{ij}^\theta = \begin{pmatrix} 0 & 0 & 0 \\ 0 & 0 & 0 \\ 0 & 0 & -\frac{1}{r}\sin\theta\cos\theta \end{pmatrix}, \quad \Gamma_{ij}^\varphi = \begin{pmatrix} 0 & 0 & \frac{1}{r} \\ 0 & 0 & \frac{1}{r}\coth\theta \\ \frac{1}{r} & \frac{1}{r}\coth\theta & 0 \end{pmatrix} \tag{C.3}$$

Euler's problem consists to solve

$$\begin{cases} \sigma^{ij}|_j + p^i = 0 \quad \text{in } \Omega \\ \sigma^{ij} = (\lambda G^{ij}G^{kl} + 2\mu G^{ik}G^{jl})\epsilon_{kl}(U), \\ \sigma^{ij}|_j = (\lambda G^{ij}G^{kl} + 2\mu G^{ik}G^{jl})\epsilon_{kl}(U)|_j \end{cases} \tag{C.4}$$

Here, $p^i = -\rho g\vec{k}.G^i, \quad i = r, \ \theta, \ \varphi$. The above equation becomes

$$\begin{cases} \sigma^{rr}|_r + \sigma^{r\theta}|_\theta + \sigma^{r\varphi}|_\varphi + p^r = 0 \\ \sigma^{\theta r}|_r + \sigma^{\theta\theta}|_\theta + \sigma^{\theta\varphi}|_\varphi + p^\theta = 0 \\ \sigma^{\varphi r}|_r + \sigma^{\varphi\theta}|_\theta + \sigma^{\varphi\varphi}|_\varphi + p^\varphi = 0 \end{cases} \tag{C.5}$$

where

$$\sigma^{ij}|_j = (\lambda G^{ij}G^{kl} + 2\mu G^{ik}G^{jl})\epsilon_{kl}(U)|_j,$$

$$\epsilon_{kl}|_j = \epsilon_{kl,j} - \Gamma_{kj}^m\epsilon_{ml} - \Gamma_{lj}^m\epsilon_{km},$$

$$\epsilon_{kl,j} = \frac{1}{2}\left(u_{k,l} + u_{l,k} - 2\Gamma_{kl}^m u_m\right)_{,j}$$

In Eq. (C.5), we have

$$\sigma^{rr}|_r = (\lambda + 2\mu)\epsilon_{rr}|_r + \frac{\lambda}{r^2}\epsilon_{\theta\theta}|_r + \frac{\lambda}{r^2 \sin^2\theta}\epsilon_{\varphi\varphi}|_r$$

$$\sigma^{r\theta}|_\theta = \frac{2\mu}{r^2}\epsilon_{r\theta}|_\theta$$

$$\sigma^{r\varphi}|_\varphi = \frac{2\mu}{r^2 \sin^2\theta}\epsilon_{r\varphi}|_\varphi$$

$$\sigma^{\theta r}|_r = \frac{2\mu}{r^2}\epsilon_{\theta r}|_r$$

$$\sigma^{\theta\theta}|_\theta = \frac{1}{r^4}(\lambda + 2\mu)\epsilon_{\theta\theta}|_\theta + \frac{\lambda}{r^4 \sin\theta}\epsilon_{\varphi\varphi}|_\theta + \frac{\lambda}{r^2}\epsilon_{rr}|_\theta$$

$$\sigma^{\theta\varphi}|_\varphi = \frac{2\mu}{r^4 \sin^2\theta}\epsilon_{\theta\varphi}|_\varphi$$

$$\sigma^{\varphi r}|_r = \frac{2\mu}{r^2 \sin^2\theta}\epsilon_{\varphi r}|_r$$

$$\sigma^{\varphi\theta}|_\theta = \frac{2\mu}{r^4 \sin^2\theta}\epsilon_{\varphi\theta}|_\theta +$$

$$\sigma^{\varphi\varphi}|_\varphi = \frac{1}{r^4 \sin^4\theta}(\lambda + 2\mu)\epsilon_{\varphi\varphi}|_\varphi + \frac{\lambda}{r^4 \sin^2\theta}\epsilon_{\theta\theta}|_\varphi + \frac{\lambda}{r^4 \sin\theta}\epsilon_{rr}|_\varphi$$

The above 3D equations (C.5) become

$$\begin{cases} (\lambda + 2\mu)\epsilon_{rr}|_r + \frac{1}{r^2}\epsilon_{\theta\theta}|_r + \frac{1}{r^2 \sin^2\theta}\epsilon_{\varphi\varphi}|_r + \frac{2\mu}{r^2}\epsilon_{r\theta}|_\theta + \frac{2\mu}{r^2 \sin^2\theta}\epsilon_{r\varphi}|_\varphi + p^r = 0, \\ \frac{2\mu}{r^2}\epsilon_{\theta r}|_r + \frac{1}{r^4}(\lambda + 2\mu)\epsilon_{\theta\theta}|_\theta + \frac{\lambda}{r^4 \sin\theta}\epsilon_{\varphi\varphi}|_\theta + \frac{\lambda}{r^2}\epsilon_{rr}|_\theta + \frac{2\mu}{r^4 \sin^2\theta}\epsilon_{\theta\varphi}|_\varphi + p^\theta = 0, \\ \frac{2\mu}{r^2 \sin^2\theta}\epsilon_{\varphi r}|_r + \frac{2\mu}{r^4 \sin^2\theta}\epsilon_{\varphi\theta}|_\theta + \frac{1}{r^4 \sin^4\theta}(\lambda + 2\mu)\epsilon_{\varphi\varphi}|_\varphi + \frac{\lambda}{r^4 \sin^2\theta}\epsilon_{\theta\theta}|_\varphi + \frac{\lambda}{r^4 \sin\theta}\epsilon_{rr}|_\varphi + p^\varphi = 0, \end{cases} \tag{C.6}$$

where the covariant derivatives of strain components read

$$\epsilon_{rr}|_r = \epsilon_{rr,r}, \quad \epsilon_{rr}|_\theta$$
$$= \epsilon_{rr,\theta}, \quad \epsilon_{rr}|_\varphi = \epsilon_{rr,\varphi} - \frac{2}{r}\epsilon_{r\varphi}$$

$$\epsilon_{r\theta}|_r = \epsilon_{r\theta,r}, \quad \epsilon_{r\theta}|_\theta = \epsilon_{r\theta,\theta} + r\epsilon_{rr}, \quad \epsilon_{r\theta}|_\varphi = \epsilon_{r\theta,\varphi} - \frac{2}{r}\epsilon_{r\varphi}$$

$$\epsilon_{r\varphi}|_r = \epsilon_{r\varphi,r} - \frac{1}{r}\epsilon_{\varphi\varphi}, \quad \epsilon_{r\varphi}|_\theta = \epsilon_{r\varphi,\theta} - \frac{1}{r}\coth\theta\epsilon_{\theta\varphi}, \quad \epsilon_{r\varphi}|_\varphi = \epsilon_{r\varphi,\varphi} - \frac{1}{r}\epsilon_{\varphi\varphi} + \frac{1}{r}\cos\theta\sin\theta\epsilon_{\varphi\theta} + r\sin\theta\epsilon_{\varphi 1}$$

$$\epsilon_{\theta\varphi}|_r = \epsilon_{\theta\varphi,r} - \frac{1}{r}\epsilon_{\varphi\varphi}, \quad \epsilon_{\theta\varphi}|_\theta = \epsilon_{\theta\varphi,\theta} + r\epsilon_{r\varphi} - \frac{1}{r}\coth\theta\epsilon_{\theta\varphi}, \quad \epsilon_{\theta\varphi}|_\varphi = \epsilon_{\theta\varphi,\varphi\varphi} - \frac{1}{r}\coth\theta\epsilon_{\varphi\varphi}$$

$$\epsilon_{\varphi\varphi}|_r = \epsilon_{\varphi\varphi,r} - \frac{1}{r}\epsilon_{\varphi\varphi}, \quad \epsilon_{\varphi\varphi}|_\theta = \epsilon_{\varphi\varphi,\theta} - \frac{2}{r}\coth\theta\epsilon_{\varphi\varphi}, \quad \epsilon_{\varphi\varphi}|_\varphi = \epsilon_{\varphi\varphi,\varphi} + r\sin\theta\epsilon_{r\varphi} + \frac{1}{r}\sin\theta\cos\theta\epsilon_{\theta\varphi}$$

with

$$\epsilon_{rr} = u_{r,r}, \quad \epsilon_{r\theta} = 1/2(u_{r,\theta} + u_{\theta,r}) \quad \epsilon_{r\varphi} = 1/2(u_{r,\varphi} + u_{\varphi,r}) - \frac{1}{r}u_\varphi$$

$$\epsilon_{\theta\theta} = u_{\theta,\theta} - ru_r, \quad \epsilon_{\theta\varphi} = 1/2(u_{\theta,\varphi} + u_{\varphi,\theta}) - \frac{1}{r}\coth\theta u_\varphi$$

$$\epsilon_{\varphi\varphi} = u_{\varphi,\varphi} + r\sin\theta u_r + \frac{1}{r}\sin\theta\cos\theta u_\theta$$

The covariant derivation of the displacements reads

$$u_r|_r = u_{r,r}, \quad u_r|_\theta = u_{r,\theta}, \quad u_r|_\varphi = u_{r,\varphi} - \frac{1}{r}u_\varphi$$

$$u_\theta|_r = u_{\theta,r}, \quad u_\theta|_\theta = u_{\theta,\theta} + ru_r, \quad u_\theta|_\varphi = u_{\theta,\varphi} - \frac{1}{r}\coth\theta u_\varphi$$

$$u_\varphi|_r = u_{\varphi,r} - \frac{1}{r}u_\varphi, \quad u_\varphi|_\theta = u_{\varphi,\theta} - \frac{1}{r}\coth\theta u_\varphi, \quad u_\varphi|_\varphi = u_{\varphi,\varphi} + r\sin\theta u_r + \frac{1}{r}\sin\theta\cos\theta u_\theta$$

The volume force components $p^i = -\rho g\vec{k}.G^i$ are given by

$$p^r = -\rho g\vec{k}.G^r = -\rho g \begin{pmatrix} 0 \\ 0 \\ 1 \end{pmatrix} . \begin{pmatrix} \sin\theta\cos\varphi \\ \sin\theta\sin\varphi \\ \cos\theta \end{pmatrix} = -\rho g\cos\theta, \tag{C.7}$$

$$p^\theta = -\rho g\vec{k}.G^\theta = -\rho g\frac{1}{r^2} \begin{pmatrix} 0 \\ 0 \\ 1 \end{pmatrix} . \begin{pmatrix} r\cos\theta\cos\varphi \\ r\cos\theta\sin\varphi \\ -r\sin\theta \end{pmatrix} = -\rho g r\sin\theta,$$

$$p^\varphi = -\rho g\vec{k}.G^\varphi = -\rho g\frac{1}{r^2\sin^2\theta} \begin{pmatrix} 0 \\ 0 \\ 1 \end{pmatrix} . \begin{pmatrix} -r\sin\theta\sin\varphi \\ r\sin\theta\cos\varphi \\ 0 \end{pmatrix} = 0$$

Using these above relations, we can derive by the different substitutions Euler's equation.

Solution 1.2

On a sphere of radius R,
(i) Calculation of the Riemann curvature tensor $R^\rho_{\alpha\beta\gamma}$. We have

$$\begin{aligned} A_{\alpha,\beta\gamma} &= (\Gamma^\rho_{\alpha\beta}A_\rho + B_{\alpha\beta}A_3)_{,\gamma} = \Gamma^\rho_{\alpha\beta,\gamma}A_\rho + \Gamma^\delta_{\alpha\beta}A_{\delta,\gamma} + B_{\alpha\beta,\gamma}A_3 + B_{\alpha\beta}A_{3,\gamma} \\ &= \Gamma^\rho_{\alpha\beta,\gamma}A_\rho + \Gamma^\delta_{\alpha\beta}\Gamma^\rho_{\delta\gamma}A_\rho + \Gamma^\delta_{\alpha\beta}B_{\delta\gamma}A_3 + B_{\alpha\beta,\gamma}A_{3,\gamma} - B^\rho_\gamma B_{\alpha\beta}A_\rho \\ &= \left(\Gamma^\rho_{\alpha\beta,\gamma} + \Gamma^\delta_{\alpha\beta}\Gamma^\rho_{\delta\gamma} + B^\rho_\gamma B_{\alpha\beta}\right)A_\rho + \left(\Gamma^\delta_{\alpha\beta}B_{\delta\gamma} + B_{\alpha\beta,\gamma}\right)A_3 \end{aligned} \tag{C.8}$$

$$\begin{aligned} A_{\alpha,\gamma\beta} &= (\Gamma^\rho_{\alpha\gamma}A_\rho + B_{\alpha\gamma}A_3)_{,\beta} \\ &= \left(\Gamma^\rho_{\alpha\gamma,\beta} + \Gamma^\delta_{\alpha\gamma}\Gamma^\rho_{\delta\beta} - B_{\alpha\gamma}B^\rho_\beta\right)A_\rho + \left(\Gamma^\rho_{\alpha\gamma}B_{\beta\rho} + B_{\alpha\gamma,\beta}\right)A_3 \end{aligned} \tag{C.9}$$

$$\begin{aligned} A_{\alpha,\beta\gamma} - A_{\alpha,\gamma\beta} &= \left(\Gamma^\rho_{\alpha\beta,\gamma} - \Gamma^\rho_{\alpha\gamma,\beta} + \Gamma^\delta_{\alpha\beta}\Gamma^\rho_{\delta\gamma} - \Gamma^\delta_{\alpha\gamma}\Gamma^\rho_{\delta\beta} + B_{\alpha\gamma}B^\rho_\beta - B_{\alpha\beta}B^\rho_\gamma\right)A_\rho \\ &\quad + \left(B_{\alpha\beta,\gamma} + \Gamma^\delta_{\alpha\beta}B_{\delta\gamma} - \left(B_{\alpha\gamma,\beta} + \Gamma^\rho_{\alpha\gamma}B_{\rho\beta}\right)\right)A_3 \\ &= \left(R^\rho_{\alpha\beta\gamma} + B_{\alpha\gamma}B^\rho_\beta - B_{\alpha\beta}B^\rho_\gamma\right)A_\rho + \left(B_{\alpha\beta,\gamma} + \Gamma^\delta_{\alpha\beta}B_{\delta\gamma} - \left(B_{\alpha\gamma,\beta} + \Gamma^\rho_{\alpha\gamma}B_{\rho\beta}\right)\right)A_3 \end{aligned} \tag{C.10}$$

where

$$R^\rho_{\alpha\beta\gamma} = \Gamma^\rho_{\alpha\beta,\gamma} - \Gamma^\rho_{\alpha\gamma,\beta} + \Gamma^\delta_{\alpha\beta}\Gamma^\rho_{\delta\gamma} - \Gamma^\delta_{\alpha\gamma}\Gamma^\rho_{\delta\beta}$$

If the mid-surface $S \in C^3$, then

$$R^\rho_{\alpha\beta\gamma} + B_{\alpha\gamma}B^\rho_\beta - B_{\alpha\beta}B^\rho_\gamma = 0$$

and

$$B_{\alpha\beta,\gamma} + \Gamma^{\rho}_{\alpha\beta} B_{\rho\gamma} - B_{\alpha\gamma,\beta} - \Gamma^{\rho}_{\alpha\gamma} B_{\rho\beta} - \Gamma^{\rho}_{\gamma\beta} B_{\alpha\rho} + \Gamma^{\rho}_{\gamma\beta} B_{\alpha\rho} = \nabla_{\alpha} B_{\alpha\beta} - \nabla_{\beta} B_{\alpha\gamma} = 0$$

from the Cirruti-Minardi formula. We deduce that

$$R^{\rho}_{\alpha\beta\gamma} = -B_{\alpha\gamma} B^{\rho}_{\beta} + B_{\alpha\beta} B^{\rho}_{\gamma}$$

(ii) Calculate the Ricci curvature tensor $R_{\alpha\gamma} = R^{\rho}_{\alpha\rho\gamma}$.
We have

$$R_{\alpha\gamma} = R^{1}_{\alpha 1\gamma} + R^{2}_{\alpha 2\gamma}, \quad (R_{\alpha\gamma}) = \begin{pmatrix} R_{11} & R_{12} \\ R_{21} & R_{22} \end{pmatrix}$$

That is

$$R_{11} = R^{1}_{111} + R^{2}_{121}, \quad R_{21} = R^{1}_{211} + R^{2}_{221}, \quad R_{12} = R^{1}_{112} + R^{2}_{122}, \quad R_{22} = R^{1}_{212} + R^{2}_{222}$$

We can calculate all these terms using the expression given above and the fact that

$$R^{\rho}_{\alpha\rho\gamma} = B_{\alpha\gamma} B^{\rho}_{\rho} - B_{\alpha\rho} B^{\rho}_{\gamma}, \ R^{\rho}_{\alpha\rho\gamma} = -R^{\rho_{\alpha\gamma\rho}} \quad \text{and} \quad R^{\rho}_{\alpha\beta\beta} = 0 \tag{C.11}$$

We obtain

$$R_{11} = R^{1}_{111} + R^{2}_{121} = R^{2}_{121} = 1, \quad R_{21} = R^{1}_{211} + R^{2}_{221} = 0$$

and

$$R_{12} = R^{1}_{112} + R^{2}_{122} = 0, \quad R_{22} = R^{1}_{212} + R^{2}_{222} = R^{1}_{212} = \sin^2\theta$$

$$(R_{\alpha\beta}) = \begin{pmatrix} 1 & 0 \\ 0 & \sin^2\theta \end{pmatrix}$$

(iii) Calculation of the trace or the scalar curvature of $R_{\alpha\beta}$ is given by

$$R^{\alpha}_{\alpha} = A^{\alpha\beta} R_{\alpha\beta}$$

We have

$$(A^{\alpha\beta}) = \begin{pmatrix} \dfrac{1}{R^2} & 0 \\ 0 & \dfrac{1}{R^2\sin^2\theta} \end{pmatrix}$$

$$R^{\alpha}_{\alpha} = A^{\alpha\beta} R_{\alpha\beta} = A^{11} R_{11} + A^{12} R_{12} + A^{21} R_{21} + A^{22} R_{22} = \frac{2}{R^2}$$

(iv) Calculation of the tensor $E_{\alpha\beta} = R_{\alpha\beta} - \dfrac{R^{\rho}_{\rho}}{2} A_{\alpha\beta}$.

$$E_{11} = R_{11} - \frac{R^{\rho}_{\rho}}{2} A_{11} = 1 - \frac{2}{2R^2} R^2 = 0$$

$$E_{12} = R_{12} - \frac{R^{\rho}_{\rho}}{2} A_{12} = 0$$

$$E_{21} = R_{21} - \frac{R^{\rho}_{\rho}}{2} A_{21} = 0$$

$$E_{22} = R_{22} - \frac{R^{\rho}_{\rho}}{2} A_{22} = \sin^2\theta - \frac{1}{R^2} R^2 \sin^2\theta$$

Therefore, we obtain

$$(E_{\alpha\beta}) = \begin{pmatrix} 0 & 0 \\ 0 & 0 \end{pmatrix}$$

It can be observed that the tensor E is different from zero if trace of R or the scalar curvature is divided by one instead of two. So also in a n-dimensional space if two is replaced by n over two and E by cE where c is a constant, then $1/cR$ and cE will verify similar equations.

Solution 1.3

Calculation of the total area of a sphere of radius R with contravariant x^{α} and covariant x_{α} parameters.

Calculation of the total area of a sphere with contravariant x^{α} parameter.

A position vector is

$$\overrightarrow{OM} = (R \sin\theta cos\varphi, R \sin\psi \sin\varphi, R \cos\psi), \quad x^1 = \theta, \quad x^2 = \varphi$$

and $(dx^1, dx^2) = (d\theta, d\varphi)$

We have

$$A_1 = \begin{pmatrix} R\cos\theta\cos\varphi \\ R\cos\theta\sin\varphi \\ -R\sin\theta \end{pmatrix}, \; A_2 = \begin{pmatrix} -R\sin\psi\sin\varphi \\ R\sin\psi\cos\varphi \\ 0 \end{pmatrix}, \; A_3 = \begin{pmatrix} \sin\theta\cos\varphi \\ \sin\theta\sin\varphi \\ \cos\theta \end{pmatrix}, \tag{C.12}$$

$$A_{\alpha\beta} = \begin{bmatrix} R^2 & 0 \\ 0 & R^2\sin^2\theta \end{bmatrix}$$

$$dS = |A_1 \times A_2| dx^1 dx^2$$

By a simple calculation, we find $|A_1 \times A_2| = R^2 \sin\theta$, and then the total surface reads

$$S = 2\pi R^2 \int_0^{\pi} \sin\theta dx^1 = 4\pi R^2 \tag{C.13}$$

Calculation of the Total Area of a Sphere with Covariant x_α

We calculate the metric tensor as follows:

$$A^1 = \frac{1}{R}\begin{pmatrix} \cos\theta\cos\varphi \\ \cos\theta\sin\varphi \\ -\sin\theta \end{pmatrix}, \; A^2 = \frac{1}{R\sin\theta}\begin{pmatrix} -\sin\varphi \\ \cos\varphi \\ 0 \end{pmatrix}, \; A^3 = \frac{1}{R^2}\begin{pmatrix} \sin\theta\cos\varphi \\ \sin\theta\sin\varphi \\ -\cos\theta \end{pmatrix} \tag{C.14}$$

The mid-surface metric tensor $A^{\alpha\beta} = \begin{bmatrix} 1/R^2 & 0 \\ 0 & 1/R^2\sin^2\psi \end{bmatrix}$

Using the variance relation we obtain $x_1 = A_{11}x^1 = R^2\theta$ and $x_2 = R^2\sin^2\theta\varphi$. The area element writes

$$dS = \sqrt{\det(A^{\alpha\beta})}dx_1dx_2 = \frac{1}{R^2\sin\theta}R^2 d\theta R^2 \sin\theta d\varphi$$

Therefore,

$$S = R^2(\int_0^{\pi} \sin\theta d\theta)(\int_0^{2\pi} d\varphi) = 4\pi R^2$$

Solution 2.1

We have $OM = \begin{pmatrix} x = R\cos\theta \\ y = y \\ z = R\sin\theta \end{pmatrix}$ and the coordinates are given by $x^1 = R\theta$ and $x^2 = y$.

The covariant base vector $\{A_1, A_2, A_3\}$ and metric tensor are

$$A_1 = OM_{,1} = \begin{pmatrix} -\sin\theta \\ 0 \\ \cos\theta \end{pmatrix}, A_2 = \begin{pmatrix} 0 \\ 1 \\ 0 \end{pmatrix}, \quad A_3 = \begin{pmatrix} -\cos\theta \\ 0 \\ -\sin\theta \end{pmatrix} \tag{C.15}$$

$$A^1 = A_1, \quad A^2 = A_2, \quad \Gamma^{\rho}_{\alpha\beta} = 0, \nabla_\alpha v_\beta = v_{\beta,\alpha}$$

$$f^1 = -\rho g k.A^1 = \rho g h \sin\theta, \quad f^2 = 0, \quad f^3 = -\rho g k.A^3 = -\rho g h \cos\theta$$

The kinematic is given by

$$u = u_1(x^1)A^1 + u_3(x^1)A^3, \quad u_2(x^1) = 0 \tag{C.16}$$

$$
\begin{aligned}
e_{11} &= \frac{1}{R}(u_{1,\theta} - u_3), \quad e_{12} = e_{22} = 0, \\
K_{11} &= \frac{1}{R^2}(u_{1,\theta} + u_{3,\theta\theta} - u_3), \quad K_{12} = K_{22} = 0, \\
Q_{11} &= \frac{1}{R^3}(u_{1,\theta} - u_{3,\theta\theta}), \quad Q_{12} = Q_{22} = 0.
\end{aligned}
$$

The best first-order N-T shell model gives

$$
N^{\alpha\beta} = \frac{Eh}{1-\nu^2}((1-\nu)e^{\alpha\beta} + \nu e^\rho_\rho A^{\alpha\beta}) + \frac{Eh^3}{12(1-\nu^2)}((1-\nu)Q^{\alpha\beta} + \nu Q^\rho_\rho A^{\alpha\beta}) = N_{\alpha\beta} \tag{C.17}
$$

Covariant and contravariant components are equal because the metric tensor is identity. So

$$
\begin{aligned}
N^{11} &= \frac{Eh}{1-\nu^2}((1-\nu)e_{11} + \nu e_{22} + \nu e_{11}) + \frac{Eh^3}{12(1-\nu^2)}Q_{11} = \frac{Eh}{1-\nu^2}e_{11} + \frac{Eh^3}{12(1-\nu^2)}Q_{11}, \\
N^{22} &= \frac{Eh}{1-\nu^2}\nu e_{11} + \frac{Eh^3}{12(1-\nu^2)}\nu Q_{11}, \\
M^{11} &= \frac{Eh^3}{12(1-\nu)}K_{11}, \quad M^{22} = \frac{Eh^3}{12(1-\nu^2)}\nu Q_{11} \\
M^{*11} &= \frac{Eh^5}{80(1-\nu^2)}Q_{11} + \frac{Eh^3}{12(1-\nu^2)}e_{11}, \\
M^{*22} &= \frac{Eh^5}{80(1-\nu^2)}\nu Q_{11} + \frac{Eh^3}{12(1-\nu^2)}\nu e_{11}
\end{aligned}
$$

The variational equation now reads

$$
\begin{aligned}
&\int_{\theta_0}^{\theta_1} (N^{11}(u,w)\bar{e}_{11} R d\theta + \int_{\theta_0}^{\theta_1} (M^{11}(u,w)\bar{K}_{11}) R d\theta \\
&\quad + \int_{\theta_0}^{\theta_1} (M^{*11}(u,w)\bar{Q}_{11}) R d\theta = \int_{\theta_0}^{\theta_1} (f^1 u + f^3 w) R d\theta
\end{aligned} \tag{C.18}
$$

We denote $u_1 = u$, $u_3 = w$ and the virtual displacement (tangent displacement) by δu and δw. So $\bar{e}_{11} = e_{11}(\delta u, \delta w)$, $\bar{K}_{11} = K_{11}(\delta u, \delta w)$, $\bar{Q}_{11} = Q_{11}(\delta u, \delta w)$

$$
\begin{aligned}
&\int_{\theta_0}^{\theta_1} \Big(N^{11}(u,w)e_{11}(\delta u, \delta w) \\
&\quad + M^{11}(u,w)K_{11}(\delta u, \delta w) + M^{*11}(u,w)Q_{11}(\delta u, \delta w)\Big) d\theta = \int_{\theta_0}^{\theta_1} (f^1 \delta u + f^3 \delta w) d\theta
\end{aligned} \tag{C.19}
$$

Now

$$
\begin{aligned}
N^{11}e_{11}(\delta u, \delta w) &= (\frac{Eh}{(1-\nu^2)}e_{11} + \frac{Eh^3}{12(1-\nu^2)}Q_{11})e_{11}(\delta u, \delta w) \\
&= \frac{1}{R^2}\frac{Eh}{1-\nu^2}(u_{,\theta} - w) + \frac{Eh^3}{12(1-\nu^2)R^4}(u_{,\theta} - w_{,\theta\theta})(\delta u_{,\theta} - \delta w),
\end{aligned} \tag{C.20}
$$

$$
M^{11}K_{11}(\delta u, \delta w) = \frac{1}{R^4}\frac{Eh^3}{12(1-\nu^2)}(u_{,\theta} - w_{,\theta\theta} - w)(\delta u_{,\theta} + \delta w_{,\theta\theta} - \delta u) \tag{C.21}
$$

$$
\begin{aligned}
M^{*11}Q_{11}(\delta u, \delta w) &= \frac{1}{R^4}\frac{Eh^3}{12(1-\nu^2)}(u_{,\theta} - w) \\
&+ \frac{Eh^5}{80(1-\nu^2)R^6}(u_{,\theta} - w_{,\theta\theta})(\delta u_{,\theta} + \delta w_{,\theta\theta})
\end{aligned} \tag{C.22}
$$

Therefore,

$$
\begin{aligned}
&N^{11}(u, w)e_{11}(\delta u, \delta w) + M^{11}(u, w)K_{11}(\delta u, \delta w) + M^{*11}(u, w)Q_{11}(\delta u, \delta w) \\
&= -D_e\delta w + D_e\delta u_{,\theta} - D_K\delta w + D_K\delta u_{,\theta} + D_K\delta w_{\theta\theta} + D_Q\delta u_{,\theta} - D_Q\delta w_{\theta\theta}
\end{aligned} \tag{C.23}
$$

where

$$
\begin{aligned}
D_e &= \frac{1}{R^2}\frac{Eh}{1-\nu^2}(u_{,\theta} - w) + \frac{Eh^3}{12(1-\nu^2)R^4}(u_{,\theta} - w_{,\theta\theta}) \\
D_K &= \frac{1}{R^4}\frac{Eh^3}{12(1-\nu^2)}(u_{,\theta} - w_{,\theta\theta} - w) \\
D_Q &= \frac{1}{R^4}\frac{Eh^3}{12(1-\nu^2)}(u_{,\theta} - w) + \frac{1}{R^6}\frac{Eh^5}{80(1-\nu^2)}(u_{,\theta} - w_{,\theta\theta})
\end{aligned} \tag{C.24}
$$

The variational formulation

$$
\begin{aligned}
&\int_{\theta_0}^{\theta_1} \big(-D_e\delta w + (D_e + D_K + D_Q)(u, w)\delta u_{,\theta} \\
&+ (D_K - D_Q)(u, v)\delta w_{,\theta\theta}\big)\, d\theta = \int_{\theta_0}^{\theta_1} (f^1\delta u + f^3\delta w)d\theta
\end{aligned} \tag{C.25}
$$

is equivalent to

$$\int_{\theta_0}^{\theta_1} \big((-D_e + (D_K - D_Q)_{,\theta\theta}\big)\,\delta w$$

$$+(D_e + D_K + D_Q)(u, w)_{,\theta}\delta u\big)\,d\theta = \int_{\theta_0}^{\theta_1} (f^1\delta u + f^3\delta w)d\theta \tag{C.26}$$

We deduce Euler's equation

$$\begin{cases} -[D_e + D_K + D_Q]_{,\theta} = f^1 = \rho g h \sin\theta \\ -D_e + (D_K - D_Q)_{,\theta\theta} = f^3 = -\rho g h \cos\theta \\ u(\theta_0) = u(\theta_1) = 0 \\ w(\theta_0) = w_{,\theta}(\theta_0) = w(\theta_1) = w_{,\theta}(\theta_1) = 0 \end{cases} \tag{C.27}$$

Recall that u and w are functions of $x^1 = R\theta$. So

$$\frac{\partial u}{\partial \theta} = \frac{\partial u}{\partial x^1}\frac{\partial x^1}{\partial \theta} = R\frac{\partial u}{\partial x^1}, \quad \text{and} \quad \frac{\partial w}{\partial \theta} = R\frac{\partial w}{\partial x^1}, \quad \frac{\partial^2 w}{\partial \theta^2} = R^2\frac{\partial^2 w}{(\partial x^1)^2} \tag{C.28}$$

Solution 3.1

The position of a point on the mid-surface is given by
$\overrightarrow{Om} = (R\cos\theta, R\sin\theta, z)$ and the coordinates are $(z, \theta) = (x^1,\ x^2) = x$.

$$A_1 = \begin{pmatrix} 0 \\ 0 \\ 1 \end{pmatrix}, \quad A_2 = \begin{pmatrix} -R\sin\theta \\ R\cos\theta \\ 0 \end{pmatrix} \quad A_3 = \begin{pmatrix} -\cos\theta \\ -\sin\theta \\ 0 \end{pmatrix}$$

The metric tensors are

$$(A_{\alpha\beta}) = \begin{pmatrix} 1 & 0 \\ 0 & R^2 \end{pmatrix}, \quad (A^{\alpha\beta}) = \begin{pmatrix} 1 & 0 \\ 0 & 1/R^2 \end{pmatrix}$$

The curvature tensors are

$$(B_{\alpha\beta}) = \begin{pmatrix} 0 & 0 \\ 0 & R \end{pmatrix}, \quad (B^{\alpha}_{\beta}) = \begin{pmatrix} 0 & 0 \\ 0 & 1/R \end{pmatrix}$$

The stress-strain relation reads

$$\sigma^{\alpha\beta} = (\overline{\Lambda} A^{\alpha\beta}A^{\gamma\delta} + 2\mu A^{\alpha\gamma}A^{\beta\delta})\epsilon_{\gamma\delta}(u)$$

where

$$\begin{aligned}\epsilon_{\alpha\beta} &= e_{\alpha\beta} - zK_{\alpha\beta} + z^2 Q_{\alpha\beta},\\ \epsilon_{i3} &= 0\end{aligned} \tag{C.29}$$

The variational equation is

$$B\left(\frac{d^2}{dt^2}u, v\right) + A\left(u\left(x,t\right), v\left(x\right)\right) = L\left(v\right) \tag{C.30}$$

with initial conditions $u(x,0)$; $\dot{u}(x,0)$ and the boundary condition $u(x,t)=0$, at $x^1=0 \quad \partial_z u_3(x,t) = 0\text{at} \quad x^1 = 0$

$$\begin{cases} B\left(\frac{d^2}{dt^2}u, v\right) = \int\limits_{\Omega} \rho \frac{\partial^2 U}{\partial t^2} V d\Omega = \frac{d^2}{dt^2} B\left(u, v\right); \\ A\left(u\left(x,t\right), v\left(x\right)\right) = \int\limits_{S} [N(u(x,t)) : e\left(v\right) + M(u(x,t)) : K\left(v\right) \\ \qquad + M^*(u(x,t)) : Q\left(v\right)] dS \\ \frac{d^2}{dt^2} B\left(u, v\right) = \int\limits_{s} \left(I^\beta\left(\ddot{u}\right) v_\beta + I^3\left(\ddot{u}\right) v_3\right) dS \end{cases} \tag{C.31}$$

$$L(v) = \int\limits_{S} - p(x_1) v_3(x) dS \tag{C.32}$$

We shall calculate the initial terms I^β and I^3.

$$\begin{aligned} I^\beta\left(\ddot{u}\right) = \rho h A^{\alpha\beta} \ddot{u}_\alpha + \rho\{\frac{h^3}{12}\{A^{\alpha\tau} B^\beta_\nu B^\nu_\tau \ddot{u}_\alpha \\ + 2A^{\alpha\tau} B^\beta_\tau \left(\partial_\alpha \ddot{u}_3 + 2B^\nu_\alpha \ddot{u}_\nu\right) + A^{\alpha\beta}(B^\nu_\tau B^\tau_\alpha \ddot{u}_\nu \\ + B^\tau_\alpha \partial_\tau \ddot{u}_3)\} + \frac{h^5}{80} A^{\delta\alpha} B^\beta_\gamma B^\gamma_\delta (B^\nu_\tau B^\tau_\alpha \ddot{u}_\nu + B^\tau_\alpha \partial_\tau \ddot{u}_3)\} \end{aligned}$$

$$\begin{aligned} I^3\left(\ddot{u}\right) = \rho h \ddot{u}_3 - \rho \frac{h^3}{12} \partial_\beta \left(A^{\alpha\tau} B^\beta_\tau \ddot{u}_\alpha + A^{\alpha\beta}\left(\partial_\alpha \ddot{u}_3 + 2B^\tau_\alpha \ddot{u}_\tau\right)\right) \\ -\rho \frac{h^5}{80} \partial_\beta \left(A^{\alpha\tau} B^\beta_\tau \left(B^\nu_\delta B^\delta_\alpha \ddot{u}_\nu + B^\nu_\alpha \partial_\nu \ddot{u}_3\right)\right) \end{aligned}$$

For the cylindrical shell, we have

$$
\begin{aligned}
I^1 &= \rho h \ddot{u}_1, \\
I^2 &= \rho h \frac{1}{R^2}\ddot{u}_2 + \rho\frac{h^3}{12}(\frac{1}{R^4}\ddot{u}_2 - \frac{2}{R^3}(\partial_\theta \ddot{u}_3 - \frac{2}{R}\ddot{u}_2) \\
&\quad + \frac{1}{R^2}(\frac{1}{R^2}\ddot{u}_2 + \frac{2}{R}\partial_\theta \ddot{u}_3) - \frac{h^5}{80}\frac{1}{R^4}(\frac{1}{R^2}\ddot{u}_2 + \frac{1}{R}\partial_\theta \ddot{u}_3) \\
I^3 &= \rho h \ddot{u}_3 - \rho\frac{h^3}{12}\partial_\theta\left(\frac{1}{R^3}\ddot{u}_2 + \frac{2}{R^2}(\frac{2}{R}\ddot{u}_2 + \partial_\theta \ddot{u}_3)\right) + \frac{h^5}{80}\partial_\theta(\frac{1}{R^3}(\frac{2}{R^2}\ddot{u}_2 + \frac{1}{R}\partial_\theta \ddot{u}_3))
\end{aligned}
$$

We remind that metric tensors are dimensionless and b^α_β and $A^{\alpha\beta}$ have same unit $(1/m)$, $b^\rho_\beta v_\rho = b^\beta_\beta v_\beta$ without summation.

At the first order

$$
\frac{d^2}{dt^2}B_1\left(u(x,t), v\right) + A_1(u(x,t), v) = L(v) \tag{C.33}
$$

with

$$
A_1(u(x,t), v) = \int_S \left[\mathcal{N} : e\left(v\right) + M : K\left(v\right) + \mathcal{M} : Q\left(v\right)\right] dS
$$

$$
\mathcal{N} = N + \overline{N},\ \mathcal{M} = \overline{M} + \overline{\overline{M}}
$$

$$
\bar{M}^{\alpha\beta} = \frac{Eh^3}{12\left(1-\bar{\nu}^2\right)}\left[(1-\bar{\nu})\, e^{\alpha\beta}(u(x,t)) + \bar{\nu} e^\rho_\rho(u(x,t))A^{\alpha\beta}\right]
$$

$$
\overline{\overline{M}}^{\alpha\beta} = \frac{Eh^5}{80\left(1-\bar{\nu}^2\right)}\left[(1-\bar{\nu})\, Q^{\alpha\beta}\left(u(x,t)\right) + \bar{\nu} Q^\rho_\rho\left(u(x,t)\right) A^{\alpha\beta}\right]
$$

$$
\bar{N}^{\alpha\beta} = \frac{Eh^3}{12\left(1-\bar{\nu}^2\right)}\left[(1-\bar{\nu})\, Q^{\alpha\beta}(u(x,t)) + \bar{\nu} Q^\rho_\rho(u(x,t))A^{\alpha\beta}\right]
$$

$$
M^{\alpha\beta} = \frac{Eh^3}{12\left(1-\bar{\nu}^2\right)}\left[(1-\bar{\nu})\, K^{\alpha\beta}\left(u(x,t)\right) + \bar{\nu} K^\rho_\rho\left(u(x,t)\right) A^{\alpha\beta}\right]
$$

$$
N^{\alpha\beta} = \frac{Eh}{1-\bar{\nu}^2}\left[(1-\bar{\nu})\, e^{\alpha\beta}\left(u(x,t)\right) + \bar{\nu} e^\rho_\rho\left(u(x,t)\right) A^{\alpha\beta}\right]
$$

For a cylindrical shell, we have

$$(e_{\alpha\beta}) = \frac{1}{2}\begin{bmatrix} 2u_{,x} & v_{,x} + \frac{1}{R}u_{,\theta} \\ * & \frac{2}{R}(v_{,\theta} + w) \end{bmatrix}, \tag{C.34}$$

$$(K_{\alpha\beta}) = \begin{bmatrix} w_{,xx} & \frac{1}{R}(w_{,x\theta} - v_{,x}) \\ * & -\frac{1}{R^2}(2v_{,\theta} + w - w_{,\theta\theta}) \end{bmatrix}, \tag{C.35}$$

$$(Q_{\alpha\beta}) = \frac{1}{2}\begin{bmatrix} 0 & \frac{1}{R^2}(v_{,x} - w_{,x\theta}) \\ * & -\frac{2}{R^3}(v_{,\theta} - w_{,\theta\theta}) \end{bmatrix} \tag{C.36}$$

Solution 4.1

The desk of length a and width b is subject to self-weight $P(x, y)$ and pre-stress $p(x)$ at $x = 0$ and $x = a$. The Euler equation is

$$D_\alpha D_\beta M^{\alpha\beta} - N^{\alpha\beta} b_{\alpha\beta} = P$$

Let the displacement be $u, v, b_{\alpha\beta} = B_{\alpha\beta} + K_{\alpha\beta} = w_{,\alpha\beta}$. Because of the pre-stress $N^{xx}(x, y) = p(a)$.

$$M^{\alpha\beta} = \frac{Eh^3}{12(1-\nu^2)}[(1-\nu)K^{\alpha\beta} + \nu K^{\rho\rho}\delta^{\alpha\beta}].$$

$$D_\alpha D_\beta M = D\Delta\Delta w, \quad D = \frac{Eh^3}{12(1-\nu^2)}$$

The equation now reads

$$D\Delta\Delta w - p(a)w_{,xx} = P$$

Note that $N^{xy} = N^{yy} = 0$

Suppose

$$P = P_{mn}w_{mn}(x, y), w(x, y) = a_{mn}w_{mn}(x, y), \quad w_{mn}(x, y) = \sin\frac{m\pi x}{a}\sin\frac{n\pi y}{b}$$

$D\Delta\Delta w - p(a)w_{,xx} = P$ is equivalent to

$$D((\frac{m\pi}{a})^2 + (\frac{n\pi}{b})^2)^2 + p(\frac{m\pi}{a})^2)a_{mn} = P_{mn}$$

This algebraic equation can be solved provided

$$D((\frac{m\pi}{a})^2 + (\frac{n\pi}{b})^2)^2 + p(\frac{m\pi}{a})^2 \neq 0$$

For example for $m = 1, n = 1$, we have

$$D(\frac{\pi^2}{a^2} + \frac{\pi^2}{b^2})^2 + p(\frac{\pi}{a})^2 = 0 \text{ if } p = -\frac{D(\frac{\pi^2}{a^2} + \frac{\pi^2}{b^2})^2}{(\frac{\pi}{a})^2}$$

Solution 4.2

The vertical suspended tank is of radius R, thickness h and height H. The liquid's density is ρ.
(i) The lateral part undergoes a membrane deformation. Let z be the vertical coordinate on the generic line. A generic point m of the mid-surface is referenced by the angular parameter $x^1 = R\theta$ and $x^2 = z$. The pressure is hydrostatic. The coordinate $z = 0$ at the top edge and $z = H$ at the bottom edge. Therefore at z, $p(z) = \rho g z \overrightarrow{n}$, where $\overrightarrow{n}$ is the unit outer normal vector of the tank. The radius is therefore negative and the curvature tensor is
$\begin{pmatrix} -\frac{1}{R} & 0 \\ 0 & 0 \end{pmatrix}$
From third Euler's equation on membrane deformation, we obtain

$$N^{\alpha\beta} B_{\alpha\beta} = -p(z)$$

and

$$-\frac{N^{11}}{R} = -p(z) = -\rho g z, \text{ and } N^{11} = p(z)R = \rho g z R \text{ and } \sigma_l^{11} = \frac{\rho g H R}{h}$$

The equilibrium of a vertical portion leads to the equation

$$2\pi R N^{22}(z) = g\rho\pi R^2 H, \ \ N^{22} = \frac{\rho g R H}{2}, \ \ \sigma_l^{22} = \frac{\rho g H R}{2h}$$

(ii) On the bottom section the pressure $p = \rho g H$ is constant and the displacement is $w(r)$. At the junction between the disc and the lateral part $w(r) = w'(r) = 0$
(iii) The maximum moment at the junction is given by

$$M^{\nu\nu} = \frac{\rho g H R^2}{8} = M^{22}.$$

which leads to

$$\sigma_d^{22} = \pm\frac{\rho g H R^2}{4}\frac{3}{h^2}$$

In order to choose the thickness h of the material, we should consider the maximum between
$\frac{\rho g H R^2}{4}\frac{3}{h^2}$ and $\frac{\rho g H R}{2h}$.

In order to determine the thickness of tank, the following condition should be satisfied:

$$\max\{\sigma_d^{22}, \sigma_l^{22}\} < \frac{f_e}{\gamma_s}$$

where f_e is the bearing capacity of the material and γ_s the design security factor.

It should be remarked that at the junction there is discontinuity on σ^{22} due to the fact that the calculation model is too simplified.

Solution 5.1

Here, we calculate for an element the local strain and stress tensors of the N model by using the 15 dof finite element. Next we calculate the elementary stiffness matrix and nodal force.

The Local Strain and Stress Tensors of the N Model by Using the 15 Dof Finite Element

The N shell model reads

$$\varepsilon_{\alpha\beta}(u) = e_{\alpha\beta}(\overline{u}) - zk_{\alpha\beta}(\overline{u}) + z^2 Q_{\alpha\beta}(\overline{u}) + q(x^\alpha, z)\Upsilon_{\alpha\beta}(x^\alpha, z),$$

$$\varepsilon_{\alpha 3}(u) = \partial_\alpha q/2,$$

$$\varepsilon_{33}(u) = \partial_z q,$$

$$e_{\alpha\beta}(\overline{u}) = \left(\nabla_\alpha \xi_\beta + \nabla_\beta \xi_\alpha - 2b_{\alpha\beta}\xi_3\right)/2,$$

$$k_{\alpha\beta}(\overline{u}) = \nabla_\alpha b_\beta^\upsilon \xi_\upsilon + b_\alpha^\upsilon \nabla_\beta \xi_\upsilon + b_\beta^\upsilon \nabla_\alpha \xi_\upsilon + \nabla_\alpha \nabla_\beta \xi_3 - b_\alpha^\tau b_{\tau\beta}\xi_3,$$

$$Q_{\alpha\beta}(\overline{u}) = \left(b_\alpha^\upsilon \nabla_\beta b_\upsilon^\tau \xi_\tau + b_\alpha^\upsilon b_\upsilon^\tau \nabla_\beta \xi_\tau + b_\alpha^\upsilon \nabla_\beta \nabla_\upsilon \xi_3 + b_\beta^\upsilon \nabla_\alpha b_\upsilon^\tau \xi_\tau + b_\beta^\upsilon b_\upsilon^\tau \nabla_\alpha \xi_\tau + b_\beta^\upsilon \nabla_\alpha \nabla_\upsilon \xi_3\right)/2,$$

$$\Upsilon_{\alpha\beta}(x^\alpha, z) = -\left(\mu_\alpha^\nu b_{\nu\beta} + \mu_\beta^\nu b_{\nu\alpha}\right)/2$$

Let

$$q(X^\alpha, z)\gamma_{\alpha\beta}(X^\alpha, z) = w(z)\overline{Q}_{\alpha\beta}(\overline{q})$$

We denote the covariant derivation by $\frac{\partial}{\partial x_i}$ and the classical derivation by $\frac{\partial'}{\partial x_i}$.

The deformation tensor reads

$$\begin{pmatrix} e_{11} \\ e_{22} \\ 2e_{12} \end{pmatrix} = \begin{bmatrix} \frac{\partial'}{\partial x_1} - \Gamma^1_{11} & -\Gamma^2_{11} & -b_{11} & 0 \\ -\Gamma^1_{22} & \frac{\partial'}{\partial x_2} - \Gamma^2_{22} & -b_{22} & 0 \\ \frac{\partial'}{\partial x_2} - \Gamma^1_{12} - \Gamma^1_{21} & \frac{\partial'}{\partial x_1} - \Gamma^2_{12} - \Gamma^2_{21} & -b_{12} & 0 \end{bmatrix} \begin{pmatrix} u \\ v \\ w \\ q \end{pmatrix} = \left[\partial'\right]_e \begin{Bmatrix} u \\ v \\ w \\ q \end{Bmatrix}$$

$$\begin{pmatrix} K_{11} \\ K_{22} \\ 2K_{12} \end{pmatrix} = \begin{bmatrix} k_1 & k_2 & k_3 & 0 \\ k_4 & k_5 & k_6 & 0 \\ k_7 & k_8 & k_9 & 0 \end{bmatrix} \begin{pmatrix} u \\ v \\ w \\ q \end{pmatrix}$$

$$k_1 = \frac{\partial'}{\partial x_1} b^1_1 + \Gamma^1_{12} b^2_1 - \Gamma^2_{11} b^1_2 + 2b^1_1 \left(\frac{\partial'}{\partial x_1} - \Gamma^1_{11}\right) - 2b^1_1 \Gamma^1_{12}$$

$$k_2 = \frac{\partial'}{\partial x_1} b^2_1 + \Gamma^2_{12} b^2_1 - \Gamma^2_{11}\left(-b^1_1 + b^2_2\right) - \Gamma^1_{11} b^2_1 + 2b^2_1 \left(\frac{\partial'}{\partial x_1} - \Gamma^2_{12}\right) - 2b^1_1 \Gamma^2_{11}$$

$$k_3 = \frac{\partial'^2}{\partial {x_1}^2} - \Gamma^1_{11} \frac{\partial'}{\partial x_1} - \Gamma^2_{11} \frac{\partial'}{\partial x_2} - b^1_1 b_{11} - b^2_1 b_{21}$$

$$k_4 = \frac{\partial'}{\partial x_2} b^1_2 + \Gamma^1_{21} b^1_2 - \Gamma^1_{22} b^2_1 + 2b^1_2 \left(\frac{\partial'}{\partial x_2} - \Gamma^1_{21}\right) - 2b^2_2 \Gamma^1_{22}$$

$$k_5 = \frac{\partial'}{\partial x_2} b^2_2 + \Gamma^2_{21} b^1_2 - \Gamma^1_{22} b^2_1 + 2b^2_2 \left(\frac{\partial'}{\partial x_2} - \Gamma^2_{22}\right) - 2b^1_2 \Gamma^2_{21}$$

$$k_6 = \frac{\partial'^2}{\partial {x_2}^2} - \Gamma^1_{22} \frac{\partial'}{\partial x_1} - \Gamma^2_{22} \frac{\partial'}{\partial x_2} - b^1_2 b_{12} - b^2_2 b_{22}$$

$$k_7 = 2\left(\frac{\partial'}{\partial x_1} b^1_2 + \left(\Gamma^1_{11} - \Gamma^2_{12}\right) b^1_2 - \Gamma^1_{12} b^2_1 + b^1_1 \left(\frac{\partial'}{\partial x_2} - \Gamma^1_{21}\right) + b^1_2 \frac{\partial'}{\partial x_1}\right)$$

$$k_8 = 2\left(\frac{\partial'}{\partial x_2} b^2_2 + \Gamma^2_{11} b^1_2 - \Gamma^1_{12} b^2_1 + b^1_1 \left(\frac{\partial'}{\partial x_2} - \Gamma^2_{22}\right) + b^2_2 \frac{\partial'}{\partial x_1} - \Gamma^2_{21} b^1_1\right)$$

$$k_9 = 2\left(\frac{\partial'^2}{\partial x_1 \partial x_2} - \Gamma^1_{12} \frac{\partial'}{\partial x_1} - \Gamma^2_{12} \frac{\partial'}{\partial x_2} - b^1_2 b_{12} - b^2_1 b_{22}\right)$$

$$\begin{pmatrix} K_{11} \\ K_{22} \\ 2K_{12} \end{pmatrix} = [\partial']_K \begin{pmatrix} u \\ v \\ w \\ q \end{pmatrix}$$

We have also about the Gauss tensor:

$$\begin{pmatrix} Q_{11} \\ Q_{22} \\ 2Q_{12} \end{pmatrix} = \begin{bmatrix} q_1 & q_2 & q_3 & 0 \\ q_4 & q_5 & q_6 & 0 \\ q_7 & q_8 & q_9 & 0 \end{bmatrix} \begin{pmatrix} u \\ v \\ w \\ q \end{pmatrix}$$

$$q_1 = b_1^1 \frac{\partial'}{\partial x_1} b_1^1 - \Gamma_{11}^2 b_2^1 + b_1^2 \left(\frac{\partial'}{\partial x_1} b_2^1 + \Gamma_{11}^2 b_2^1 + \Gamma_{12}^2 b_2^2 - \Gamma_{12}^2 b_1^1 - \Gamma_{12}^2 b_2^1 \right)$$

$$q_2 = b_1^1 \left(\frac{\partial'}{\partial x_1} b_1^2 + \Gamma_{11}^2 b_1^1 + \Gamma_{11}^2 b_1^2 - \Gamma_{11}^1 b_2^2 \right) + b_1^2 \left(\frac{\partial'}{\partial x_1} b_1^1 + \Gamma_{11}^2 b_2^1 - \Gamma_{12}^1 b_1^1 \right)$$

$$q_3 = b_1^1 \left(\frac{\partial'^2}{\partial x_1^2} - \Gamma_{11}^1 \frac{\partial'}{\partial x_1} \right) + b_1^2 \left(\frac{\partial'^2}{\partial x_1 \partial x_2} - \Gamma_{12}^1 \frac{\partial'}{\partial x_1} - \Gamma_{12}^2 \frac{\partial'}{\partial x_2} \right)$$

$$q_4 = b_2^1 \left(\frac{\partial'}{\partial x_1} b_1^1 + \Gamma_{21}^1 b_1^1 + \Gamma_{22}^1 b_2^2 - \Gamma_{21}^1 b_1^1 - \Gamma_{21}^2 b_2^1 \right) + b_2^2 \left(\frac{\partial'}{\partial x_1} b_2^1 + \Gamma_{21}^2 b_2^1 - \Gamma_{12}^1 b_1^2 \right)$$

$$q_5 = (b_2^1 + b_2^2) \left(\frac{\partial'}{\partial x_2} b_2^2 + \Gamma_{21}^2 b_2^1 \right)$$

$$q_6 = b_2^1 \left(\frac{\partial'^2}{\partial x_1 \partial x_2} - \Gamma_{12}^1 \frac{\partial'}{\partial x_1} - \Gamma_{12}^2 \frac{\partial'}{\partial x_2} \right) + b_2^2 \left(\frac{\partial'^2}{\partial x_2^2} - \Gamma_{22}^1 \frac{\partial'}{\partial x_1} - \Gamma_{12}^2 \frac{\partial'}{\partial x_2} \right)$$

$$\begin{aligned} q_7 = {} & b_1^1 \left(\frac{\partial'}{\partial x_2} b_1^1 + \Gamma_{21}^1 b_1^1 + \Gamma_{22}^1 b_2^2 - \Gamma_{21}^1 b_1^1 - \Gamma_{21}^2 b_2^1 \right) \\ & + (b_1^2) \left(\frac{\partial'}{\partial x_2} b_2^2 + \Gamma_{21}^2 b_2^1 \right) \\ & + b_2^1 \left(\frac{\partial'}{\partial x_1} b_1^1 + \Gamma_{12}^1 b_1^2 - \Gamma_{11}^2 b_2^1 \right) + \end{aligned}$$

$$b_2^2 \left(\frac{\partial'}{\partial x_1} b_2^1 + \Gamma_{11}^1 b_2^1 + \Gamma_{12}^1 b_2^2 - \Gamma_{12}^1 b_2^1 \right)$$

$$q_8 = b_1^2\left(\frac{\partial'}{\partial x_2}b_1^2 + \Gamma_{21}^2 b_1^1 + \Gamma_{22}^2 b_1^2 - \Gamma_{21}^1 b_1^2 - \Gamma_{21}^2 b_2^2\right)$$
$$+ (b_2^1)\left(\frac{\partial'}{\partial x_2}b_2^2 + \Gamma_{21}^2 b_2^1\right)$$
$$+ b_2^1\left(\frac{\partial'}{\partial x_1}b_1^2 + \Gamma_{11}^2 b_1^1 + \Gamma_{11}^2 b_1^2 - \Gamma_{11}^1 b_1^2 - \Gamma_{11}^2 b_2^2\right) +$$

$$b_2^2\left(\frac{\partial'}{\partial x_1}b_2^2 + \Gamma_{11}^2 b_2^1 + \Gamma_{22}^2 b_2^2 - \Gamma_{12}^1 b_1^2 - \Gamma_{12}^2 b_2^2\right)$$

$$q_9 = b_1^2\left(\frac{\partial'^2}{\partial x_2^2} - \Gamma_{22}^1\frac{\partial'}{\partial x_1} - \Gamma_{22}^2\frac{\partial'}{\partial x_2}\right)$$
$$+ b_2^1\left(\frac{\partial'^2}{\partial x_1^2} - \Gamma_{11}^1\frac{\partial'}{\partial x_1}\right) + (b_1^1 + b_2^2)$$
$$\left(\frac{\partial'^2}{\partial x_1 \partial x_2} - \Gamma_{12}^1\frac{\partial'}{\partial x_1} - \Gamma_{12}^2\frac{\partial'}{\partial x_2}\right)$$

Therefore

$$\begin{pmatrix} Q_{11} \\ Q_{22} \\ 2Q_{12} \end{pmatrix} = [\partial']_Q \begin{pmatrix} u \\ v \\ w \\ q \end{pmatrix}$$

$$\begin{pmatrix} \overline{Q}_{11} \\ \overline{Q}_{22} \\ 2\overline{Q}_{12} \end{pmatrix} = \begin{bmatrix} 0\,0\,0 & -\left(\mu_1^1 b_{11} + \mu_1^2 b_{21}\right) \\ 0\,0\,0 & -\left(\mu_2^1 b_{12} + \mu_2^2 b_{12}\right) \\ 0\,0\,0 & -\left(\mu_1^1 b_{12} + \mu_1^2 b_{22} + \mu_2^1 b_{11} + \mu_2^2 b_{22}\right) \end{bmatrix} \begin{pmatrix} u \\ v \\ w \\ q \end{pmatrix}$$

$$P_1(\xi,\eta) = -(1-\xi-\eta).(1-2(1-\xi-\eta));$$
$$P_2(\xi,\eta) = -\xi(1-2\xi);$$
$$P_3(\xi,\eta) = -\eta(1-2\eta);$$
$$P_4(\xi,\eta) = 4.\xi(1-\xi-\eta);$$
$$P_5(\xi,\eta) = 4.\xi\eta;$$
$$P_6(\xi,\eta) = 4.\eta(1-\xi-\eta).$$

$$N_{T6-m} = \begin{pmatrix} a\ \xi\ \eta\ 0\ 0\ 0\ 0 & 0 & 0 & 0 & 0 & 0\ 0\ 0\ 0 \\ 0\ 0\ 0\ a\ \xi\ \eta\ 0 & 0 & 0 & 0 & 0 & 0\ 0\ 0\ 0 \\ 0\ 0\ 0\ 0\ 0\ 0\ P_1 & P_2 & P_3 & P_4 & P_5 & P_6\ 0\ 0\ 0 \\ 0\ 0\ 0\ 0\ 0\ 0\ 0 & 0 & 0 & 0 & 0 & 0\ a\ \xi\ \eta \end{pmatrix}$$

where $a = 1 - \xi - \eta$.

$$q(\xi,\eta)=\sum_{i=1}^{3}N_i q_i \quad U(\xi,\eta)=\begin{pmatrix} u(\xi,\eta)=\sum_{i=1}^{3}N_i u_i \\ v(\xi,\eta)=\sum_{i=1}^{3}N_i v_i \\ w(\xi,\eta)=\sum_{i=1}^{6}P_i w_i \end{pmatrix}$$

$${}^t\hat{U}_e=\left[\hat{u}_1\ \hat{u}_2\ \hat{u}_3\ \hat{v}_1\ \hat{v}_2\ \hat{v}_3\ \hat{w}_1\ \hat{w}_2\ \hat{w}_3\ \hat{w}_4\ \hat{w}_5\ \hat{w}_6\quad \hat{q}_1\ \hat{q}_2\ \hat{q}_3\right]$$

we have

$$\begin{pmatrix} e_{11} \\ e_{22} \\ 2e_{12} \end{pmatrix}=\mathcal{P}_e\begin{pmatrix} u_1\\u_2\\u_3\\v_1\\v_2\\v_3\\w_1\\w_2\\w_3\\w_4\\w_5\\w_6\\q_1\\q_2\\q_3 \end{pmatrix},\quad \begin{pmatrix} K_{11} \\ K_{22} \\ 2K_{12} \end{pmatrix}=\mathcal{P}_K\begin{pmatrix} u_1\\u_2\\u_3\\v_1\\v_2\\v_3\\w_1\\w_2\\w_3\\w_4\\w_5\\w_6\\q_1\\q_2\\q_3 \end{pmatrix},\quad \begin{pmatrix} Q_{11} \\ Q_{22} \\ 2Q_{12} \end{pmatrix}=\mathcal{P}_Q\begin{pmatrix} u_1\\u_2\\u_3\\v_1\\v_2\\v_3\\w_1\\w_2\\w_3\\w_4\\w_5\\w_6\\q_1\\q_2\\q_3 \end{pmatrix}$$

$$P_i=\left[\partial'\right]_l N_{T6-m},\ l=e,K,Q$$

$$\begin{pmatrix} \frac{\partial' N_i}{\partial\xi}=\frac{\partial' N_i}{\partial x_1}\frac{\partial' x_1}{\partial\xi}+\frac{\partial' N_i}{\partial x_2}\frac{\partial' x_2}{\partial\xi} \\ \frac{\partial' N_i}{\partial\eta}=\frac{\partial' N_i}{\partial x_1}\frac{\partial' x_1}{\partial\eta}+\frac{\partial' N_i}{\partial x_2}\frac{\partial' x_2}{\partial\eta} \end{pmatrix}$$

$$x_1=x_1^1N_1+x_1^2N_2+x_1^3N_3$$

$$x_2=x_2^1N_1+x_2^2N_2+x_2^3N_3$$

Thus

$$\begin{pmatrix} \frac{\partial' N_i}{\partial x_1} \\ \frac{\partial' N_i}{\partial x_2} \end{pmatrix}=\begin{bmatrix} -x_1^1+x_1^2 & -x_2^1+x_2^2 \\ -x_1^1+x_1^3 & -x_2^1+x_2^3 \end{bmatrix}^{-1}\begin{pmatrix} \frac{\partial' N_i}{\partial\xi} \\ \frac{\partial' N_i}{\partial\eta} \end{pmatrix}$$

$$\begin{pmatrix} \frac{\partial' P_i}{\partial x_1} \\ \frac{\partial' P_i}{\partial x_2} \end{pmatrix} = \begin{bmatrix} -x_1^1 + x_1^2 & -x_2^1 + x_2^2 \\ -x_1^1 + x_1^3 & -x_2^1 + x_2^3 \end{bmatrix}^{-1} \begin{pmatrix} \frac{\partial' P_i}{\partial \xi} \\ \frac{\partial' P_i}{\partial \eta} \end{pmatrix}$$

We have

$$\left[\begin{array}{ccc} [N] & \frac{\partial' N_i}{\partial \xi} & \frac{\partial' N_i}{\partial \eta} \\ \hline 1-\xi-\eta & -1 & -1 \\ \xi & 1 & 0 \\ \eta & 0 & 1 \\ \hline [P] & \frac{\partial' P_i}{\partial \xi} & \frac{\partial' P_i}{\partial \eta} \\ \hline P_1 & 1-4(1-\xi-\eta) & 1-4(1-\xi-\eta) \\ P_2 & -1+4\xi & 0 \\ P_3 & 0 & -1+4\eta \\ P_4 & 4(1-2\xi-\eta) & -4\xi \\ P_5 & 4\eta & 4\xi \\ P_6 & 4\eta & 4(1-\xi-2\eta) \\ \hline \frac{\partial'^2 P}{\partial \xi^2} & \frac{\partial'^2 P}{\partial \xi \partial \eta} & \frac{\partial'^2 P}{\partial \eta^2} \\ \hline 4 & 4 & 4 \\ 4 & 0 & 0 \\ 0 & 0 & 4 \\ -8 & -4 & 0 \\ 0 & 4 & 0 \\ 0 & -4 & -8 \end{array}\right]$$

$$\begin{pmatrix} \overline{Q}_{11} \\ \overline{Q}_{22} \\ 2\overline{Q}_{12} \end{pmatrix} = \begin{bmatrix} 0\ 0\ 0 & -\left(\mu_1^1 b_{11} + \mu_1^2 b_{21}\right) \\ 0\ 0\ 0 & -\left(\mu_2^1 b_{12} + \mu_2^2 b_{12}\right) \\ 0\ 0\ 0 & -\left(\mu_1^1 b_{12} + \mu_1^2 b_{22} + \mu_2^1 b_{11} + \mu_2^2 b_{22}\right) \end{bmatrix} N_{T6-m} \begin{pmatrix} u_1 \\ u_2 \\ u_3 \\ v_1 \\ v_2 \\ v_3 \\ w_1 \\ w_2 \\ w_3 \\ w_4 \\ w_5 \\ w_6 \\ q_1 \\ q_2 \\ q_3 \end{pmatrix}$$

$$\varepsilon_{\alpha\beta} = \begin{pmatrix} \varepsilon_{11} \\ \varepsilon_{22} \\ 2\varepsilon_{12} \end{pmatrix}$$

$$= \left[[\partial']_e N_{T6-m} - z[\partial']_K N_{T6-m} + z^2[\partial']_Q N_{T6-m} + w(z) Y N_{T6-m} \right] \begin{bmatrix} u_1 \\ u_2 \\ u_3 \\ v_1 \\ v_2 \\ v_3 \\ w_1 \\ w_2 \\ w_3 \\ w_4 \\ w_5 \\ w_6 \\ q_1 \\ q_2 \\ q_3 \end{bmatrix} \quad \text{(C.37)}$$

$$Y = \begin{bmatrix} 0\,0\,0 & -\left(\mu_1^1 b_{11} + \mu_1^2 b_{21}\right) \\ 0\,0\,0 & -\left(\mu_2^1 b_{12} + \mu_2^2 b_{12}\right) \\ 0\,0\,0 & -\left(\mu_1^1 b_{12} + \mu_1^2 b_{22} + \mu_2^1 b_{11} + \mu_2^2 b_{22}\right) \end{bmatrix}$$

$$\varepsilon_{\alpha\beta} = [B_0]\,[U], \quad [\varepsilon] = \begin{pmatrix} \varepsilon_{\alpha\beta} \\ \varepsilon_{33} \\ \varepsilon_{\alpha 3} \end{pmatrix} = [B]\,[U]$$

$$K_e = \int\limits_{T_e} \int\limits_{-h/2}^{h/2} [B]^t H\,[B]\,ds dz = \int\limits_{-h/2}^{h/2} \int\limits_{T_e} [B]^t H\,[B]\,\sqrt{A} dx_1 dx_2 dz$$

$$= |J| \int\limits_0^1 \int\limits_{-h/2}^{h/2} \int\limits_0^{1-\xi} [B]^t H\,[B]\,\sqrt{A} d\xi d\eta dz$$

Calculation of the Local Nodal Force

$$L^w(v, \overline{y}) = \int\limits_{T_e} \left(\overline{P}^i v_i + \overline{P}^4 \overline{y} \right) dS + \int\limits_{\partial T_e \gamma_1} (\overline{q}^i v_i + q^4 \overline{y}) d\gamma + \int\limits_{\partial T_e \gamma_1} m^\alpha \overline{\theta}_\alpha(v) d\gamma$$

$$\overline{P}^4 = \int_{-h/2}^{h/2} f^3 w\varphi dz + w(h/2)P_+^3 - w(-h/2)P_-^3$$

$$q^4 = \int_{-h/2}^{h/2} P^3 w dz$$

$$L^w(v, \overline{y})|T_e = \left(\int_{T_e} |J| \begin{bmatrix} p^1 \\ p^2 \\ p^3 \\ p^4 \end{bmatrix} N_{T_6-m} d\eta d\xi + \int_{\partial T_e \cap \gamma_1} |l| \begin{bmatrix} q^1 \\ q^2 \\ q^3 \\ q^4 \end{bmatrix} N_{T_6-m} d\xi + \int_{\partial T_e \cap \gamma_1} |l| \begin{bmatrix} m^1\bar{\theta}_1 \\ m^2\bar{\theta}_2 \\ 0 \\ 0 \end{bmatrix} N_{T_6-m} d\xi \right) [V_e]^t \tag{C.38}$$

where

$$V_e = \left[\hat{u}'_1, \hat{u}'_2, \hat{u}'_3, \hat{v}'_1, \hat{v}'_2, \hat{v}'_3, \hat{w}'_1, \hat{w}'_2, \hat{w}'_3, \hat{w}'_4, \hat{w}'_5, \hat{w}'_6, \hat{q}'_1, \hat{q}'_2, \hat{q}'_3 \right]$$

is the virtual field displacement.

The dynamic variational equation reads

$$\begin{cases} \text{find } (u(x,t), \bar{q}(x,t)) \in U_{ad}, \forall t \in]0, T] \\ B^w\left((\ddot{u}, \ddot{\bar{q}}); (v, \bar{y})\right) + A^w\left((u, \bar{q}); (v, \bar{y})\right) = L^w(v, \bar{y}) \end{cases} \tag{C.39}$$

where

$$B^w\left((\ddot{u}, \ddot{\bar{q}}); (v, \bar{y})\right) = \int_s \left(J^\alpha(\ddot{u})v_\alpha + J^3(\ddot{u}, \ddot{\bar{q}})v_3 + J^4(\ddot{u}, \ddot{\bar{q}})\overline{y}\right) dS \tag{C.40}$$

with

$$J^3(\ddot{u}, \ddot{\bar{q}}) = \quad J^3(\ddot{u}) + \ddot{\bar{q}} \int_{-\frac{h}{2}}^{\frac{h}{2}} w\psi dz$$

$$J^4(\ddot{u}, \ddot{\bar{q}}) = \ddot{\bar{q}} \int_{\frac{-}{h}2}^{\frac{h}{2}} w^2\psi dz + \ddot{u}_3 \int_{\frac{-}{h}2}^{\frac{h}{2}} w\psi dz$$

We have

$$J^{\beta}(\ddot{u}) = \rho h A^{\alpha\beta}\ddot{u}_{\alpha} + \rho\{\tfrac{h^3}{12}\{A^{\alpha\tau}B^{\beta}_{\nu}B^{\nu}_{\tau}\ddot{u}_{\alpha}$$

$$+2A^{\alpha\tau}B^{\beta}_{\tau}\left(\partial_{\alpha}\ddot{u}_3 + 2B^{\nu}_{\alpha}\ddot{u}_{\nu}\right) + A^{\alpha\beta}(B^{\nu}_{\tau}B^{\tau}_{\alpha}\ddot{u}_{\nu}$$

$$+B^{\tau}_{\alpha}\partial_{\tau}\ddot{u}_3)\} + \tfrac{h^5}{80}A^{\delta\alpha}B^{\beta}_{\gamma}B^{\gamma}_{\delta}(B^{\nu}_{\tau}B^{\tau}_{\alpha}\ddot{u}_{\nu} + B^{\tau}_{\alpha}\partial_{\tau}\ddot{u}_3)\}$$

$$J^3(\ddot{u}) = \rho h\ddot{u}_3 - \rho\frac{h^3}{12}\partial_{\beta}\left(A^{\alpha\tau}B^{\beta}_{\tau}\ddot{u}_{\alpha} + A^{\alpha\beta}\left(\partial_{\alpha}\ddot{u}_3 + 2B^{\tau}_{\alpha}\ddot{u}_{\tau}\right)\right)$$

$$-\rho\frac{h^5}{80}\partial_{\beta}\left(A^{\alpha\tau}B^{\beta}_{\tau}\left(B^{\nu}_{\delta}B^{\delta}_{\alpha}\ddot{u}_{\nu} + B^{\nu}_{\alpha}\partial_{\nu}\ddot{u}_3\right)\right)$$

$$B_1 = \begin{pmatrix} a_1 & a_2 & a_3 & 0 \\ a_4 & a_5 & a_6 & 0 \\ a_7 & a_8 & a_9 & \int_{-\frac{h}{2}}^{\frac{h}{2}} w\varphi dz \\ 0 & 0 & \int_{-\frac{h}{2}}^{\frac{h}{2}} w\varphi dz & \int_{-\frac{h}{2}}^{\frac{h}{2}} w^2\varphi dz \end{pmatrix}$$

where

$$a_1 = \rho h A^{11} + -\rho\tfrac{h^3}{3}A^{\alpha\tau}B^{1}_{\alpha}B^{1}_{\tau} + -\rho\tfrac{h^3}{12}A^{\alpha 1}B^{1}_{\tau}B^{\tau}_{\alpha} + \rho\tfrac{h^5}{80}A^{\delta\alpha}B^{1}_{\gamma}B^{\gamma}_{\delta}B^{1}_{\tau}B^{\tau}_{\alpha},$$

$$a_2 = \rho h A^{21} + \rho\tfrac{h^3}{3}A^{\alpha\tau}B^{1}_{\alpha}B^{2}_{\tau} - \rho\tfrac{h^3}{12}A^{\alpha 1}B^{2}_{\tau}B^{\tau}_{\alpha} + \rho\tfrac{h^5}{80}A^{\delta\alpha}B^{1}_{\gamma}B^{\gamma}_{\delta}B^{2}_{\tau}B^{\tau}_{\alpha},$$

$$a_3 = -\rho\frac{h^3}{6}A^{\alpha 1}B^{\tau}_{\alpha}\partial_{\tau} + \rho\frac{h^5}{80}A^{\delta\alpha}B^{1}_{\gamma}B^{\gamma}_{\delta}B^{\tau}_{\alpha}\partial_{\tau},$$

$$a_4 = \rho h A^{12} + \rho\tfrac{h^3}{3}A^{\alpha\tau}B^{2}_{\alpha}B^{1}_{\tau} + \rho\tfrac{h^3}{12}A^{\alpha 2}B^{1}_{\tau}B^{\tau}_{\alpha} + \rho\tfrac{h^5}{80}A^{\delta\alpha}B^{2}_{\gamma}B^{\gamma}_{\delta}B^{1}_{\tau}B^{\tau}_{\alpha},$$

$$a_5 = \rho h A^{12} + -\rho\tfrac{h^3}{3}A^{\alpha\tau}B^{2}_{\alpha}B^{1}_{\tau} - \rho\tfrac{h^3}{12}A^{\alpha 2}B^{1}_{\tau}B^{\tau}_{\alpha} + \rho\tfrac{h^5}{80}A^{\delta\alpha}B^{2}_{\gamma}B^{\gamma}_{\delta}B^{1}_{\tau}B^{\tau}_{\alpha},$$

$$a_6 = -\rho\tfrac{h^3}{6}A^{\alpha\tau}B^{2}_{\alpha}\partial_{\tau} - \rho\tfrac{h^3}{6}A^{\alpha 2}B^{\tau}_{\alpha}\partial_{\tau} + \rho\tfrac{h^5}{80}A^{\delta\alpha}B^{2}_{\gamma}B^{\gamma}_{\delta}B^{\tau}_{\alpha}\partial_{\tau},$$

$$a_7 = -\rho\tfrac{h^3}{12}\partial_{\beta}\left(A^{1\tau}B^{\beta}_{\tau} + 2A^{\alpha\beta}B^{1}_{\alpha}\right) - \rho\tfrac{h^5}{80}\partial_{\beta}\left(A^{\alpha\tau}B^{\beta}_{\tau}B^{1}_{\delta}B^{\delta}_{\alpha}\right),$$

$$a_8 = -\rho\tfrac{h^3}{12}\partial_{\beta}\left(A^{2\tau}B^{\beta}_{\tau} + 2A^{\alpha\beta}B^{2}_{\alpha}\right) - \rho\tfrac{h^5}{80}\partial_{\beta}\left(A^{\alpha\tau}B^{\beta}_{\tau}B^{2}_{\delta}B^{\delta}_{\alpha}\right),$$

$$a_9 = \rho h - \rho\tfrac{h^3}{12}\partial_{\beta}\left(A^{\alpha\beta}\partial_{\alpha}\right) - \rho\tfrac{h^5}{80}\partial_{\beta}\left(A^{\alpha\tau}B^{\beta}_{\tau}B^{\nu}_{\alpha}\partial_{\nu}\right).$$

$$B^{w}\left((\ddot{u}, \ddot{\bar{q}}); (v, \bar{y})\right) = \int_{T_e}[J^1\ J^2\ J^3\ J^4]^t[V_e]^t N_{T_6-m}d\xi d\eta dz \tag{C.41}$$

$$\begin{bmatrix} J^1 \\ J^2 \\ J^3 \\ J^4 \end{bmatrix} = B_1 \begin{pmatrix} \ddot{u}_1 \\ \ddot{u}_2 \\ \ddot{u}_3 \\ \ddot{\bar{q}} \end{pmatrix} = B_1 N_{T_6-m} \begin{pmatrix} \hat{\ddot{u}}_1 \\ \hat{\ddot{u}}_2 \\ \hat{\ddot{u}}_3 \\ \hat{\ddot{v}}_1 \\ \hat{\ddot{v}}_2 \\ \hat{\ddot{v}}_3 \\ \hat{\ddot{w}}_1 \\ \hat{\ddot{w}}_2 \\ \hat{\ddot{w}}_3 \\ \hat{\ddot{w}}_4 \\ \hat{\ddot{w}}_5 \\ \hat{\ddot{w}}_6 \\ \hat{\ddot{q}}_1 \\ \hat{\ddot{q}}_2 \\ \hat{\ddot{q}}_3 \end{pmatrix}$$

$$B^w\left((\ddot{u}, \ddot{\bar{q}}); (v, \bar{y})\right) = \int_{T_e} \mathcal{B} \begin{pmatrix} \hat{\ddot{u}}_1 \\ \hat{\ddot{u}}_2 \\ \hat{\ddot{u}}_3 \\ \hat{\ddot{v}}_1 \\ \hat{\ddot{v}}_2 \\ \hat{\ddot{v}}_3 \\ \hat{\ddot{w}}_1 \\ \hat{\ddot{w}}_2 \\ \hat{\ddot{w}}_3 \\ \hat{\ddot{w}}_4 \\ \hat{\ddot{w}}_5 \\ \hat{\ddot{w}}_6 \\ \hat{\ddot{q}}_1 \\ \hat{\ddot{q}}_2 \\ \hat{\ddot{q}}_3 \end{pmatrix} [V_e]^t \sqrt{A} dx^1 dx^2 = |J| \int_0^1 \int_0^{1-\xi} \mathcal{B} \begin{pmatrix} \hat{\ddot{u}}_1 \\ \hat{\ddot{u}}_2 \\ \hat{\ddot{u}}_3 \\ \hat{\ddot{v}}_1 \\ \hat{\ddot{v}}_2 \\ \hat{\ddot{v}}_3 \\ \hat{\ddot{w}}_1 \\ \hat{\ddot{w}}_2 \\ \hat{\ddot{w}}_3 \\ \hat{\ddot{w}}_4 \\ \hat{\ddot{w}}_5 \\ \hat{\ddot{w}}_6 \\ \hat{\ddot{q}}_1 \\ \hat{\ddot{q}}_2 \\ \hat{\ddot{q}}_3 \end{pmatrix} [V_e]^t$$

where $\mathcal{B} = {N_{T_6-m}}^t B_1 N_{T_6-m}$

Solution 6.1

Calculation of $\bar{p}^i$, $\tilde{p}^i$, $\tilde{m}^\alpha$ from the variational formulation is given by

$$\begin{cases} \text{find}(u, \delta) \in U_{ad} \times U_{ad} \text{such that} \\ \int\limits_{\Omega} \sigma^{ij}(u, \delta)\epsilon_{ij}(v, w) d\Omega = \int\limits_{S} \int\limits_{-h/2}^{h/2} A^{ijkl}(x)\epsilon_{kl}(u, \delta)\epsilon_{ij}(v, w) D_3 dz dS \\ \qquad = \quad L(v, w) \quad \forall (v, w) \in U_{ad} \times U_{ad} \end{cases} \tag{C.42}$$

The external power forces read

$$P^{ext} = \int_\Omega \left(f^\alpha V_\alpha + f^3 V_3\right) d\Omega + \int_{\Gamma_-} \sigma \vec{n}\, V dS + \int_{\Gamma_+} \sigma \vec{n}\, V dS + \int_{\Gamma_1} \sigma \vec{n}\, V dS$$
$$= \int_\Omega \left(f^\alpha V_\alpha + f^3 V_3\right) d\Omega + \int_{\Gamma_-} p.V dS + \int_{\Gamma_+} p.V dS + \int_{\Gamma_1} p.V dS \tag{C.43}$$

with

$$V_\alpha = v_\alpha + z w_\alpha, \quad V_3 = v_3 + z w_3$$

and with

$$\Gamma_- = S \times \left\{-\frac{h}{2}\right\}, \quad \Gamma_+ = S \times \left\{\frac{h}{2}\right\}, \quad \Gamma_1 = \bar{\gamma}_0 \times [-\frac{h}{2}, \frac{h}{2}]$$

We have

$$\int_{-\frac{h}{2}}^{\frac{h}{2}} \left(f^\alpha V_\alpha + f^3 V_3\right) D_3 dz = \int_{-\frac{h}{2}}^{\frac{h}{2}} f^\alpha v_\alpha D_3 dz + \int_{-\frac{h}{2}}^{\frac{h}{2}} z(f^\alpha w_\alpha + f^3 w_3) D_3 dz + \int_{-\frac{h}{2}}^{\frac{h}{2}} f^3 v_3 D_3 dz$$

$$\int_{\Gamma_-} p.V dS + \int_{\Gamma_+} p.V dS + \int_{\Gamma_1} p.V dS$$
$$= \int_{\Gamma_-} \left(p_-^\alpha (v_\alpha + z w_\alpha) + p_-^3 (v_3 + z w_3)\right) d\Gamma_-$$
$$+ \int_{\Gamma_+} \left(p_+^\alpha (v_\alpha + z w_\alpha) + p_+^3 (v_3 + z w_3)\right) d\Gamma_+ + \int_{\Gamma_1} \left(p^\alpha (v_\alpha + z w_\alpha) + p^3 (v_3 + z w_3)\right) d\Gamma_1$$

the external force rewritten as

$$L(v, w) = \int_S \left(\bar{p}^i v_i + \tilde{p}^i w_i\right) dS + \int_{\gamma_1} \left(\bar{q}^i v_i + \tilde{q}^3 w_3\right) d\gamma + \int_{\gamma_1} \tilde{m}^\alpha w_\alpha d\gamma \tag{C.44}$$

where

$$\bar{p}^i = \int_{-\frac{h}{2}}^{\frac{h}{2}} f^i D_3 dz + p_-^i(-\frac{h}{2}) + p_+^i(\frac{h}{2})$$

$$\tilde{p}^i = \int_{-\frac{h}{2}}^{\frac{h}{2}} z f^i D_3 dz - \frac{h}{2} p_-^i + \frac{h}{2} p_+^i$$

$$\bar{q}^i = \int_{-\frac{h}{2}}^{\frac{h}{2}} p^i D_3 dz, \quad \tilde{q}^3 = \int_{-\frac{h}{2}}^{\frac{h}{2}} zp^3 D_3 dz$$

$$\tilde{m}^\alpha = \int_{-\frac{h}{2}}^{\frac{h}{2}} zp^\alpha D_3 dz$$

Solution 6.2

The internal virtual work is

$$\int_\Omega \sigma^{ij}(u,\delta)\epsilon_{ij}(v,w)d\Omega = \int_\Omega \sigma^{\alpha\beta}(u,\delta)\epsilon_{\alpha\beta}(v,w)d\Omega + 2\int_\Omega \sigma^{\alpha 3}(u,\delta)\epsilon_{\alpha 3}(v,w)d\Omega$$
$$+\int_\Omega \sigma^{33}(u,\delta)\epsilon_{33}(v,w)d\Omega$$

i.e.

$$\int_\Omega \sigma^{\alpha\beta}(u,\delta)\epsilon_{\alpha\beta}(v,w)d\Omega = \int_S \int_{-h/2}^{h/2} (A^{\alpha\beta\rho\gamma}(x)\epsilon_{\rho\gamma}(u,\delta)$$
$$+ 2A^{\alpha\beta\rho 3}(x)\epsilon_{\rho 3}(u,\delta) + A^{\alpha\beta 33}(x)\epsilon_{33}(u,\delta))\epsilon_{\alpha\beta}(v,w)D_3 dz dS$$
$$= \int_S (N^{\alpha\beta}e_{\alpha\beta}(v) + M^{\alpha\beta}\mathbb{k}_{\alpha\beta}(v,w) + M^{*\alpha\beta}Q^c_{\alpha\beta}(v,w))D_3 dS$$

We also have

$$2\int_\Omega \sigma^{\alpha 3}(u,\delta)\epsilon_{\alpha 3}(v,w)d\Omega = \int_S 2\int_{-h/2}^{h/2} ((A^{\alpha 3\rho\gamma}(x)\epsilon_{\rho\gamma}(u,\delta)$$
$$+ 2A^{\alpha 3\rho 3}(x)\epsilon_{\rho 3}(u,\delta) + A^{\alpha 333}(x)\epsilon_{33}(u,\delta))\epsilon_{\alpha 3}(v,w)D_3 dz dS$$
$$= \int_S (T^\alpha\vartheta_\alpha(v,w) + \tilde{T}^\alpha\varepsilon_\alpha(v,w))D_3 dS \tag{C.45}$$

Using the following relations:

$$
\begin{aligned}
\epsilon_{\alpha 3}(u,\delta) &= \frac{1}{2}(\vartheta_\alpha + z\varepsilon_\alpha),\\
\epsilon_{33}(u,\delta) &= \varepsilon_3,
\end{aligned}
\tag{C.46}
$$

$$
\vartheta_\alpha(u,\delta) = \delta_\alpha + D_i u^i_{\alpha},\ \vartheta_3(u,\delta) = \delta_3 - D^\alpha u_{3\alpha},
$$

$$
\varepsilon_\alpha(u,\delta) = (D^i_{\alpha}\vartheta_i + \delta_{i\alpha}D^i),\ \varepsilon_3(u,\delta) = D^i\vartheta_i.
$$

We find by simple manipulation the following results:

$$
N^{\alpha\beta} = \int_{\frac{-h}{2}}^{\frac{h}{2}} \left(A^{\alpha\beta\rho\gamma}e_{\rho\gamma} + z\varepsilon_\alpha) + A^{\alpha\beta 33}\varepsilon_3\right) dz \tag{C.47}
$$

$$
M^{\alpha\beta} = \int_{\frac{-h}{2}}^{\frac{h}{2}} \left(-zA^{\alpha\beta\rho\gamma}K_{\rho\gamma} - zA^{\alpha\beta 33}\varepsilon_3\right) D_3 dz \tag{C.48}
$$

$$
M^{*\alpha\beta} = \int_{\frac{-h}{2}}^{\frac{h}{2}} \left(z^2 D_3^2 A^{\alpha\beta\rho\gamma}Q_{\rho\gamma} + z^2 A^{\alpha\beta 33}\varepsilon_3\right) D_3^2 dz \tag{C.49}
$$

$$
T^{\alpha} = \int_{\frac{-h}{2}}^{\frac{h}{2}} \left(A^{\alpha 3\rho 3}(\vartheta_\rho + z\varepsilon_\rho) + A^{\alpha 333}\varepsilon_3\right) dz, \tag{C.50}
$$

$$
\tilde{T}^{\alpha} = \int_{\frac{-h}{2}}^{\frac{h}{2}} \left(-zA^{\alpha 3\rho 3}(\vartheta_\rho + z\varepsilon_\rho) + zA^{\alpha 333}\varepsilon_3\right) dz, \tag{C.51}
$$

and

$$
H = \int_{\frac{-h}{2}}^{\frac{h}{2}} \left(A^{33\rho\gamma}e_{\rho\gamma} - zA^{33\rho\gamma}K_{\rho\gamma} + z^2 A^{33\rho\gamma}Q_{\rho\gamma} + 2A^{33\rho 3}(\vartheta_\rho + z\varepsilon_\rho) + A^{3333}\varepsilon_3\right) dz \tag{C.52}
$$

References

1. FEUMO Achille Germain , NZENGWA Robert, and NKONGHO ANYI Joseph , 'Finite Element Model for Linear Elastic Thick Shells Using Gradient Recovery Method, Mathematical Problems in Engineering , Article ID 5903503,14 pages (2017).
2. ADAMS R. A.: Sobolev spaces, Academic Press, New York 1975.
3. SADOWSKI A.J. and ROTTER J.M., " Solid or Shell finite Elements to model thick cylindrical tubes and shells under global bendind". Inter J of Mechanical Science 74,143-154 (2013)
4. ALLAIRE, G. Homogenization and two-scale convergence, SIAM J. Math. Anal., Vol 23 (1) pp 482-518 (1992)
5. AMBARTSUMIAN SA. On theory of bending plates. Isz Otd Tech Nauk AN SSSR 1958;5:69-77.
6. Analyse Multi échelle et systèmes physiques couplés : Actes du symposium du 28-29 aot 1997, Paris, Presses de l'Ecole Nationale des Ponts et Chaussées.
7. NAGA A. and ZHANG Z., The polynomial preserving recovery for higher order finite element methods in 2D and 3D, Discrete and continuous dynamical systems series B, 5(2005),769-708.
8. ANNIN B.D.; KALAMKAROV A.L.; KOLPAKOV A. G.; (1990); Analysis of local stresses in high-modulus fibre composites. Pro.int. conf. on localized Damage Computer aided Assesment and control, Vol. 2 Comput. Mechanics Publ. Southampton, pp. 231- 244.
9. ARON H., (1874) "Das Gleichgewicht und die Bewegung einer unendlich dénnen, beliebig gekrémmten" J.Maths 78-136
10. BAKHKALOV, N.S.; PANASENKO, G.P. 1989: Homogenization: Averaging Processes in Periodic Media, Mathematics and its Applications 36 (Dordrecht:Kluwer).
11. BAKHKALOV, N.S.; PANASENKO, G.P. 1984: Homogenization in periodic Media, Mathematical problems of mechanics of composite Materials, Nanka, Mo.
12. BENSOUSSAN, A.; LIONS, J.L.; PAPANICOLAOU, G. 1979 Asymptotic Analysis for periodic structures, North-Holland, Amsterdam.
13. BOURQUIN, F.; CIARLET, P. G 1989: Modeling and justification of Eigenvalue Problems for Junctions between Elastic Structures. J. Funct. Anal. 87, 392-427.
14. BERNADOU, M. ; CIARLET P.G. : Sur l'ellipticité du modèle linéaire des coques de W.T. Koiter, Computing Methods in Sciences and Engineering, Eds R. Glowinsky, J. L. Lions, Lecture notes in Economics and systems, Vol 34 (1976), Springer-Verlag, Berlin, p 89-136.

R. Nzengwa, *Plate and Shell Models*,
https://doi.org/10.1007/978-981-97-2780-3

15. BERNADOU M, EIROA PM, TROUVE P. ON THE APPROXIMATION OF GENERAL LINEAR THIN SHELL PROBLEMS BYD. KT METHODS. Computation and Applied Mathematics. 1991;10:103.
16. BERNADOU M, CIARLET P.G., Miara B. Existence theorems for two-dimensional linear shell theories. Journal of Elasticity. 1994;34:111-38.
17. BISCH P. : Cours de Plaques et Coques, Tome I et II, polycopiés de l'Ecole Nationale des Ponts et Chaussées, Paris 1998.
18. BLOUZA A., LE DRET H., Existence and uniqueness for the linear Koiter model for shells with little regularity, Comptes Rendus de l'Académie des Sciences (CRAS) Paris, 1995, Séries I, 317, p 327-329.
19. BREZIS H. :Analyse fonctionnelle, Théorie et Application, Masson Paris 1983.
20. BREZZI F, FORTIN M. Mixed and hybrid finite element methods: Springer Science & Business Media; 2012.
21. BUSSE S., CIARLET P.G., MIARA B., Justification d'un modèle linéaire bi-dimensionnel de coques <<faiblement courbées>> en coordonnées curvilignes, RAIRO, M2AN, vol 31, no3 (1997) p 409-434.
22. CARRERA E, GIUNTA G, NALI P, PETROLO M., Refined beam elements with arbitrary cross-section geometries. Computers & Structures. 2010;88:283-93.
23. CARRERA E, BRISCHETTO S, CINEFRA M, SOAVE M., Effects of thickness stretching in functionally graded plates and shells. Composites Part B: Engineering. 2011;42:123-33.
24. CHAPELLE D, BATHE K-J, Fundamental considerations for the finite element analysis of shell structures. Computers & Structures. 1998;66:19-36.
25. CHAPELLE D, BATHE K-J. The finite element analysis of shells-Fundamentals: Springer Science & Business Media; 2010.
26. CIARLET P.G., Plates and junctions in elastic multi-structures, an asymptotique analysis, Masson, Paris Springer-Verlag, Heilderberg 1990.
27. CIARLET , P.G., LE DRET H., NZENGWA R. : Junctions between three-dimensional and two-dimensional linearly elastic structures, J. Math Pures et Appl. 68, (1989), 261-295.
28. CIARLET P.G , MIARA B. :Justification of the two dimensional equations of a linearly elastic shallow shell,Comm. Pure and Appl. Math, Vol XLV (1992) p 327-360, John Wiley and Sons Inc.
29. CIARLET P.G. (2001). Mathematical Elasticity. Volume III, Theory of shells. Amsterdam, Elseviezer.
30. CIARLET P.G., LODS V., Asymptotic analysis of linearly elastic shells I, Justification of membrane shell equations, Arch. Rational Mech. Anal 136 (1996) p 119-161.
31. , Asymptotic analysis of linearly elastic shells III, Justifications of Koiter's shell equations, Arch. Rational Mech. Anal 136 (1996) p 191-200.
32. CIARLET P.G., LODS V. ,MIARA B.: Asymptotic analysis of linearly elastic shells II, Justification of Flexural shell equations, Arch. Rational Mech. Anal 136 (1996) p 163-190.
33. CIARLET P.G., (1978) The Finite Element Method for Elliptic Problems, North Holland, Amsterdam.
34. CIARLET P.G, LE DRET H. , NZENGWA R., Modélisation de la jonction entre un corps élastique tridimensionnel et une plaque, Comptes Rendus de l'Académie des Sciences, (C.R.A.S) Paris, série I, 305, 55-58 (1987).
35. CIARLET P.G, LE DRET H. , NZENGWA R., Junction between three-dimensional and two dimensional linearly elastic structures, J. Math Pures Appl, 68, 1989, P. 261- 295.
36. COMBESCURE A., Les Travaux d'Habilitations à Diriger les Recherches "HDR No 83901, Laboratoire de Mécanique et Technologie "LMT, Ecole Normale Supérieure de Cachan
37. COSSERAT F. and COSSERAT E., "Théorie des Corps Déformables" Edition Hermann Archives (2009)
38. DELLA CROCE L, VENINI P., Finite elements for functionally graded Reissner-Mindlin plates. Computer Methods in Applied Mechanics and Engineering. 2004;193:705-25.
39. DESTUYNDER P., Une théorie asymptotique des plaques minces en élasticité linéaire, Masson Paris 1986.

40. DO CARMO, M 1976 Differential geometry of curves and surfaces, Prentice Hall
41. DOYLE JF. Nonlinear analysis of thin-walled structures: statics, dynamics, and stability: Springer Science & Business Media; 2013.
42. CARRERA E.,VALVANO S.,"Shell elements with through-the-thicknes variable kinematic for the analysis of laminated composite and sandwich structures" Elsevier BV Vol 111, 15 February 2017 P294-314
43. ECHTER R, OESTERLE B, BISCHOFF M. A hierarchic family of isogeometric shell finite elements. Computer Methods in Applied Mechanics and Engineering. 2013;254:170-80.
44. FANG Y-g, PAN J-h, CHEN W-X. Theory of thick-walled shells and its application in cylindrical shell. Applied Mathematics and Mechanics. 1992;13:1055-65.
45. GERMAIN, P.; MULLER, : Introduction Ã la mécanique des milieux continus, Masson, Paris.(1980)
46. MUNGLANI G.,WITTEL F.K.,VETTER R.,BIANCHI F. and HERMANN H.J., "Collapse of orthotropic Spherical Shells". Physical Review Letter 123,058002 (2019).
47. GILBERT R.P., HSIAO G.C., SCHNEIDER M.: The two dimensional linear orthotropic plate; Applicable Analysis, Vol. 15 (1983) p 147-169.
48. GOLDENVEIZER A.L. : The principles of reducing three-dimensional problems of the theory of plates and shells, Proceedings of the 11th International Congress of theoretical and applied mechanics (H.*Görtler*, Editor) Springer-Verlag Berlin (1964) p 306-311.
49. Gol'DENSVEIZER, A.L. (1961). Theory of Elastic thin shells. New York Pergamon Press
50. MARCO AMABILI, Non linear Mechanics of Shells and Plates Composite soft and biological materials, Cambridge University Press, (2018)
51. HOLM ALTENBACH, JACEK CHROSCIELEWISKY, VICTOR A. EREMEYER, KRZYSZTOF WISSNIEWSKI, Recent Developments in the Theory of Shells, Springer Nature SWITZERLAND AG, (2019).
52. ANSEL C. UGURAL, Plates and Shells Theory and Analysis, CRC Press Taylor and Francis Group, (2018).
53. VADIM A KRYSKO, MAXIM V.ZHIGALOV, VALERY F KIRICHENKO, ANTON V. KRYSKO, Mathematical models of higher orders shells in temperature, Springer Nature SWITZERLAND , (2018).
54. HILLER J-F, Bathe K-J. Measuring convergence of mixed finite element discretizations: an application to shell structures. Computers & Structures. 2003;81:639-54.
55. JAWAD M. Theory and design of plate and shell structures: Springer Science & Business Media; 2012.
56. NKONGHO ANYI J. , NZENGWA R., AMBA J. C. , and NGAYIHI ABBE C. V. , "Approximation of linear elastic shells by curved triangular finite elements based on elastic thick shells theory,"Mathematical Problems in Engineering, vol. 2016, Article ID8936075, 12 pages, 2016.
57. NKONGHO ANYI J., Amba J.C., Essola D., Ngayihi Abbe C.V., Bodol Moonha M. and NZENGWA R., "Generalised assumed strain curved shell finite elements(CSFE-sh) with shifted-Lagrange and application on N-T's shells theory" DE GRUYTER, Vol 2020, Research Article curved and layer.struct,2020;7:125_138
58. HOEFAKKER J. Theory review for cylindrical shells and parametric study of chimneys and tanks: Eburon Uitgeverij BV; 2010.
59. KARAMA M, AFAQ KS, MISTOU S. Mechanical behavior of laminated composite beam by the new multilayered laminated composite structures model with transverse shear stress continuity. Int J Solid Struct 2003;40(6): 1525-46.
60. KIM D-N, BATHE K-J. A triangular six-node shell element. Computers & Structures. 2009;87:1451-60.
61. KIRCHHOFF, G.: *Über* das Gleichgewicht und die Bewegung einer elastischen Scheibe, J. Reine angew. Math. 40 (1850) 51-58.
62. KOITER, W. T. 1970 On the foundation of the linear theory of thin elastic shells, I&II, roc. Kon. Ned. Akad. Wetensch. B73, 169-195

63. KOITER WT, SIMMONDS JG. Foundations of shell theory. In: Becker E, Mikhailov G, editors. Theoretical and Applied Mechanics: Springer Berlin Heidelberg; 1973. p. 150-76.
64. LE DRET, H. 1991 Problèmes variationnels dans les multi-domaines: modélisation des jonctions et applications, Masson Paris.
65. LEVINSON M. An accurate simple theory of the statics and dynamics of elastic plates. Mech Res Commun 1980;7:343-50.
66. LIBAI, A. & SIMMONDS,J.G. (1998). The Nonlinear Theory of Elastic Shells. Cambridge: Cambridge University Press.
67. LIONS, . L., MAGENES, E. : Problème aux limites non homogènes et applications, Vol. 1 Dunod Paris 1968.
68. LOVE, A.E.H. : A treatise on the mathematical theory of elasticity, 4th edition, Cambridge University Press, Cambridge 1934.
69. MANTARI J.L, OKTEM AS, GUEDES SOARES C. A new higher order shear deformation theory for sandwich and composite laminated plates. Composites: Part B 2011. http://dx.doi.org/10.1016/j.compositesb.2011.07.017
70. MINDLIN, R. D. 1951: Influence of rotatory inertia and shear on flexural motions of isotropic elastic plates. J. Appl.Mech. 18, 31-38.
71. MOHAROS I, OLDAL I, SZEKRÉNYES A, . Finite element methods: Typotex Publishing House; 2012.
72. MOROZOV N.F (1967) “Nonlinear vibration of thin plates with allowance for rotational inertia”, Dokl.Akad.Nauk 176(Soviet Math, 8,(1967), 1136-1141)
73. MURTHY M.V., An improved transverse shear deformation theory for laminated anisotropic plates. NASA Technical Paper 1903; 1981
74. NAGHDI, P.M.: The theory of shells and plates, in Hanbuch der Physik, Vol. VI a/2, S. Flugge and C. Truesdell, Editors,(1970) Berlin p 425-640.
75. NAGHDI P.M. Foundations of elastic shell theory. In ‘Progress in Solid Mechanics’ Volume 4. North Holland Publishing Company. 1963
76. BOURBAKI N. : Espaces vectoriels topologiques,Hermann, Paris, 1965.
77. NEJAD M.Z., JABBARI M, GHANNAD M. Elastic analysis of axially functionally graded rotating thick cylinder with variable thickness under non-uniform arbitrarily pressure loading. International Journal of Engineering Science. 2015;89:86-99.
78. NGATCHA NDENGNA Arno Roland , Renaud Ngouanom, Edmon Mbangue and Pandong Achille. Two dimensional static mechanic analysis of laminated composite tube using ABCDE matriix with no correction factor, International Journal of Mechanic 2021, 1(15), 107-120. https://doi.org/10.46300/9104.2021.15.12
79. NGATCHA NDENGNA. A. R., Ngouanom, R. and Pandong, A. A Two-Dimensional Model to Analyze the Static and Dynamic Mechanical Behavior of Multilayered shell Structures.," Composite Structures, 2022, 295(4), 115754. https://doi.org/10.1016/j.compstruct.2022.115754
80. NGATCHA NDENGNA Arno Roland, Ngouanom Gnidakouong, R., Pandong, A. New three-dimensional Mathematical Modellings to Analyse of Mechanical behavior of Laminated Anisotropic Elastic Shell (LCES) according to First order Shear Deformation Theory(FSDT). hal-03229256ff, 2021.
81. NGATCHA NDENGNA Arno Roland, Mbangue Ekmon, Joel Renaud Ngouanom Gnidakouong, Joseph Nkongho Anyi and Robert Nzengwa. A new approach to study the mechanical behavior of 2D/3D anisotropic shell structures: case of uniform cylinder composite tube To appear.
82. NGUETSENG G 1989 A general convergence result for a functional related to the theory of homogenization, SIAM J. Math. Anal., Vol 20(3), pp 608-623
83. NGUETSENG G 1990 Asymptotic analysis for a stiff variational problem arising in mechanics, SIAM J. Math. Anal., Vol 21(6), pp 1394-1414
84. NGUYEN-THOI T, PHUNG-VAN P, THAI-HOANG C, NGUYEN-XUAN H. A cell-based smoothed discrete shear gap method (CS-DSG3) using triangular elements for static and free vibration analyses of shell structures. International Journal of Mechanical Sciences. 2013;74:32-45.

85. NGUYEN-XUAN H, TRAN L.V., THAI C.H, NGUYEN-THOI T. Analysis of functionally graded plates by an efficient finite element method with node-based strain smoothing. Thin-Walled Structures. 2012;54:1-18.
86. NZENGWA R., Incremental methods in nonlinear, three-dimensional incompressible elasticity, RAIRO MAN, vol. 22, no2, 1988, P. 311-342.
87. NZENGWA R., TAGNE SIMO B.H., A two-dimensional model for linear Elastic thick shells, SYMPOSIUM ST VENANT, Paris, 28-29 1997, P. 425-431.
88. R. NZENGWA, Asymptotic 2D-modelling for dynamics of linear elastic thick shells. In W. Pietraszkiewicz & C. Syzmczak (Eds), Proc. Shell Structures Theory and Application 8th International Conference, Gdansk, Poland 12-14 october 2005, Balkema, P. 157-161.
89. NZENGWA R., Eigenvalue problems for Linear elastic thick shells. In W. Pietraszkiewicz & C. Syzmczak (Eds), Proc. Shell Structures Theory and Application 8th International Conference, Gdansk, Poland 12-14 october 2005, Balkema, P. 403-407.
90. NZENGWA R., A 2d model for dynamics of linear elastic thick shells with Transversal strains variation, In J. Awrejcewicz (Ed). Proc 8th Conference on Dynamical Systems Theory and Applications, Lodz, Poland, 12-15 december, 2005 P. 769-776.
91. VETTER R., STOOP N. and al: Subdivision Shell Elements with anisotropic growth. arXIV:1208.4434V2 [cs.NA] 27 march 2013
92. REDDY J. N, LIU C.F., A higher-order shear deformation theory of laminated elastic shells. International Journal for Numerical Methods in Engineering, Vol 23,319 -30 (1985)
93. REISSNER, E. 1945: The effect of transverse shear deformations on the bending of elastic plates. J. Appl. Mech. 12, A69-A77.
94. REISSNER, E. 1944: On the theory of bending of elastic plates, J. Math. and Phys. 23, 184-191.
95. SÃVIK S. Theoretical and experimental studies of stresses in flexible pipes. Computers & Structures. 2011;89:2273-91.
96. SALENCON,J.; 1988; Mécanique des Milieux Continus,Tome II: Elasticité Milieu Curvilignes, Ellipses, Paris.
97. SCHWARTZ, L. 1966 Théorie des distributions, Hermann,Paris
98. SOLDATOS K. P. A transverse shear deformation theory for homogeneous monoclinic plates. Acta Mech 1992;94:195-220.
99. SPIVAK, M.: Differential geometry, vols 3,4,5,Publish of Perish 1975.
100. STOLARSKI H, BELYTSCHKO T. Membrane Locking and Reduced Integration for Curved Elements. Journal of Applied Mechanics 1982;49:172-6.
101. TIMOSHENKO, S.; WOINOWSKY-KREIGER, W. 1958: Theory of plates and shells, Mc Graw Hill, New-York.
102. TORNABENE F, LIVERANI A, CALIGIANA G. General anisotropic doubly-curved shell theory: A differential quadrature solution for free vibrations of shells and panels of revolution with a free-form meridian. Journal of Sound and Vibration. 2012;331:4848-69.
103. TOURATIER M. An efficient standard plate theory. Int J Eng Sci 1991;29(8): 901-16.
104. TUTEK, Z. 1986 A Homogenized model of rod in linear elasticity in Proceedings of the International Conference, Ecole Normale Supérieure, Paris November 24-28.
105. VINSON JR. The behavior of thin walled structures: beams, plates, and shells: Springer Science & Business Media; 2012.
106. ZEIGHAMPOUR H, TADI Beni Y. Cylindrical thin-shell model based on modified strain gradient theory. International Journal of Engineering Science. 2014;78:27-47.
107. P.G. Ciarlet, Mathematical Elasticity Volume I ;Three-dimensional Elasticity, North Holland, Amsterdam, 1988.

Printed in the USA
CPSIA information can be obtained
at www.ICGtesting.com
CBHW061016090924
14265CB00003B/21

9 789819 727797